decision mathematics

Sue de Pomerai
Seaford Head Community College
Sussex

John Berry
Centre for Teaching Mathematics
University of Plymouth

contributory author
Mike Herring

project contributors
Steve Dobbs, Roger Fentem, Bob Francis, Ted Graham, Howard Hampson, Penny Howe, Rob Lincoln, Claire Rowland, Stuart Rowlands, Stewart Townend, John White

Collins

Published by HarperCollins*Publishers* Limited
77–85 Fulham Palace Road
Hammersmith
London W6 8JB

www.**Collins**Education.com
On-line Support for Schools and Colleges

First published 2001
ISBN 000 322527 5

British Library Cataloguing in Publication Data:
A catalogue record for this book is available from the British Library.

Production: Kathryn Botterill
Cover Design: Terry Bambrook
Internal Design: Ann Miller
Project Editor: Joan Miller
Illustrations by Ken Vail Graphic Design, Ann Miller and Moondisks, Cambridge
Commissioned by Mark Jordan
Indexing Specialist: Susan Leech
Printed and bound by Scotprint, Haddington

About the authors

Sue de Pomerai is currently Head of Mathematics at Seaford Head Community College in Sussex. She has many years' experience of teaching Decision Maths at A-level and is an examiner and reviser for OCR Decision Maths. She is a member of MEI Decision and Discrete Development Group.

John Berry is Professor of Mathematics Education and Director of the Centre for Teaching Mathematics at the University of Plymouth. John Berry worked with QCA on the development of the new A-level syllabuses, and the applications of new technology on the teaching and assessment of A-level mathematics.

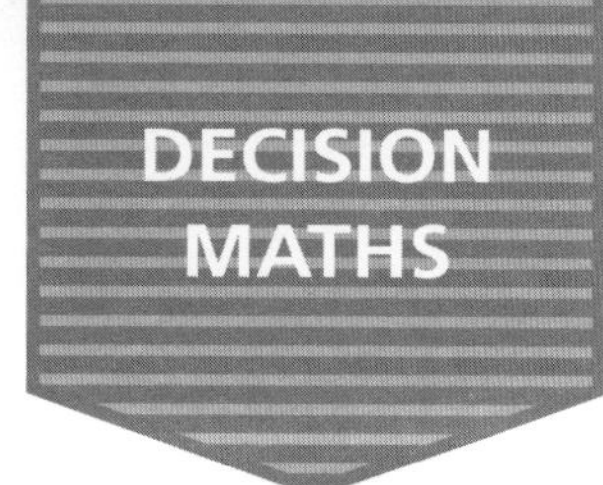

Contents

Acknowledgements

We are grateful to the following Examination Boards for permission to reproduce questions from their past examination papers and from specimen papers. Full details are given with each question. The Examination groups accept no responsibility whatsoever for the accuracy or method of working in the answers given, which are solely the responsibility of the author and publisher.

Associated Examining Board (*AEB*)
Edexcel foundation
Northern Examinations and Assessment Board (*NEAB*); also School Mathematics project, 16–19 (*SMP 16–19*)
Oxford and Cambridge Schools Examination Board (*OCSEB*)
Oxford, Cambridge and RSA examinations (*OCR*)
Scottish Examination Board (*SEB*)
University of Cambridge Local Examinations Syndicate (*UCLES*)
University of London Examinations and Assessment Council (*ULEAC*)
University of Oxford Delegacy of Local Examinations (*UODLE*); also Nuffield Advanced Mathematics (*Nuffield*)
Welsh Joint Examinations Council (*WJEC*)

We are also grateful to the following for permission to reproduce copyright photographs:

Mary Evans Picture Library 60 (left), 201; by permission of the President and Council of the Royal Society 2, 60 (right), 155.

Every effort has been made to contact the holders of copyright material, but if any have been inadvertently overlooked the publishers will be pleased to make the necessary arrangements at the first opportunity.

Preface

discovering advanced mathematics

Mathematics is not just an important subject in its own right, but also a tool for solving problems. Mathematics A-level is changing to reflect this; during the A-level course, you must study at least one area of the *application* of mathematics. This is what we mean by 'mathematical modelling'. Of course, mathematicians have been applying mathematics to problems in mechanics and statistics for years. But now, the process has been formally included throughout A-level maths.

Decision mathematics is a fairly new area of applied mathematics in the A-level curriculum. Until recently most students following a course in advanced mathematics would have studied mechanics and/or statistics. There are many decision-making problems in business and commerce for which a different problem-solving approach is required; one based on an algorithmic strategy instead of the traditional skills of algebra and calculus. The development of cheap and powerful computers has been an incentive for mathematicians to develop efficient algorithms for solving these decision-making problems.

This book in the *discovering advanced mathematics* series is designed as an introduction to Decision Mathematics.

We have revised *discovering advanced mathematics* to meet the needs of the new A and AS specifications of Curriculum 2000. The books in the series provide opportunities to study advanced mathematics while learning about modelling and problem solving.

In every chapter in this book, you will find:

- an introduction that explains a new idea or technique in a helpful context;
- plenty of worked examples to show you how the techniques are used;
- exercises in two sets: classwork problems for work in class and homework problems that mirror the classwork problems so that you can practise the same work in self study sessions;
- consolidation exercises that test you in the same way as real exam questions;
- questions from the awarding Boards.

And then, once you have finished the chapter:

- modelling and problem-solving exercises to help you pull together all the ideas in the chapter.

We hope that you will enjoy Decision Mathematics by working through this book.

We thank the many people involved in developing *discovering advanced mathematics*. In particular, we thank Mike Herring for many of the homework exercises and modelling activities and Karen Eccles for typing the manuscript.

John Berry, Sue de Pomerai *July 2001*

D & D Syllabus mapping

The table below shows the topics covered by each specification and the relevant chapters of this book. Where a topic covered by the book occurs in a D1 specification, it is shown as a 1 in the table and where a topic is in a D2 specification it is shown as a 2 in the table. The book covers all topics in each of the D1 specifications, and many of the topics for D2 as well.

Topic	Algorithm	Edexcel	AQA A	AQA B	OCR A	MEI	Chapters
Algorithms	Sorting	1	1		1	1	1
	Searching	1			1	1	1
	Packing	1			1	1	1
Graph		1	1	1	1	1	2, 3, 4, 5, 7
Networks	Prim	1	1	1	1	1	3
	Kruskal	1	1	1	1	1	3
	Dijkstra	1	1	1	1	1	2
	TSP	2	1	2	1	2	4
	CPP	1	1	2	1	2	5
	Flows	1	2	2	2		11
CPA		1	2	1	2	1	6
Optimisation	Matchings	1	1		2		7
	LP graphical	1	1	1	1	1	8
	LP Simplex	1	2	2	1	2	9
Simulation						1	10
Boolean Algebra				1		2	12

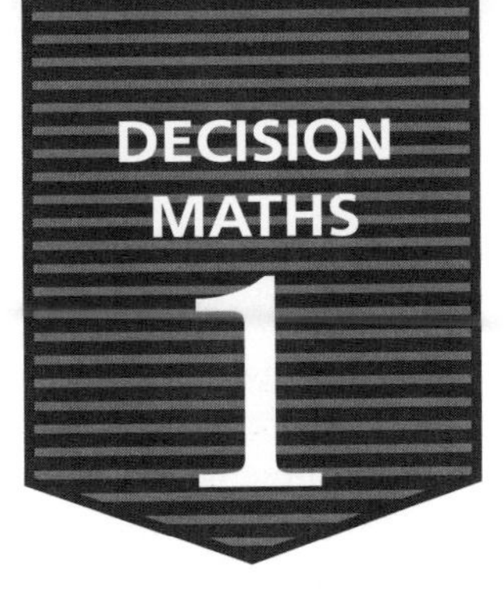

Algorithms

By the end of this chapter you should be able to:

- *understand what is meant by an algorithm*
- *understand an algorithm presented in a variety of forms*
- *know how to sort a list of numbers, using a variety of algorithms*
- *use techniques to assess the speed and efficiency of an algorithm*
- *use algorithms for searching*
- *use algorithms for packing.*

The best way to start a course in decision mathematics is to look at the sorts of problems that you will be able to solve by the time you reach the end of this book.

- How do you find the shortest route between two places?
- How would you design the fire exit routes in a building?
- How would you plan a job, to make the most efficient use of your time?
- How can you best match the skills of workers to jobs that need to be done?
- How can a manufacturer maximise profits?

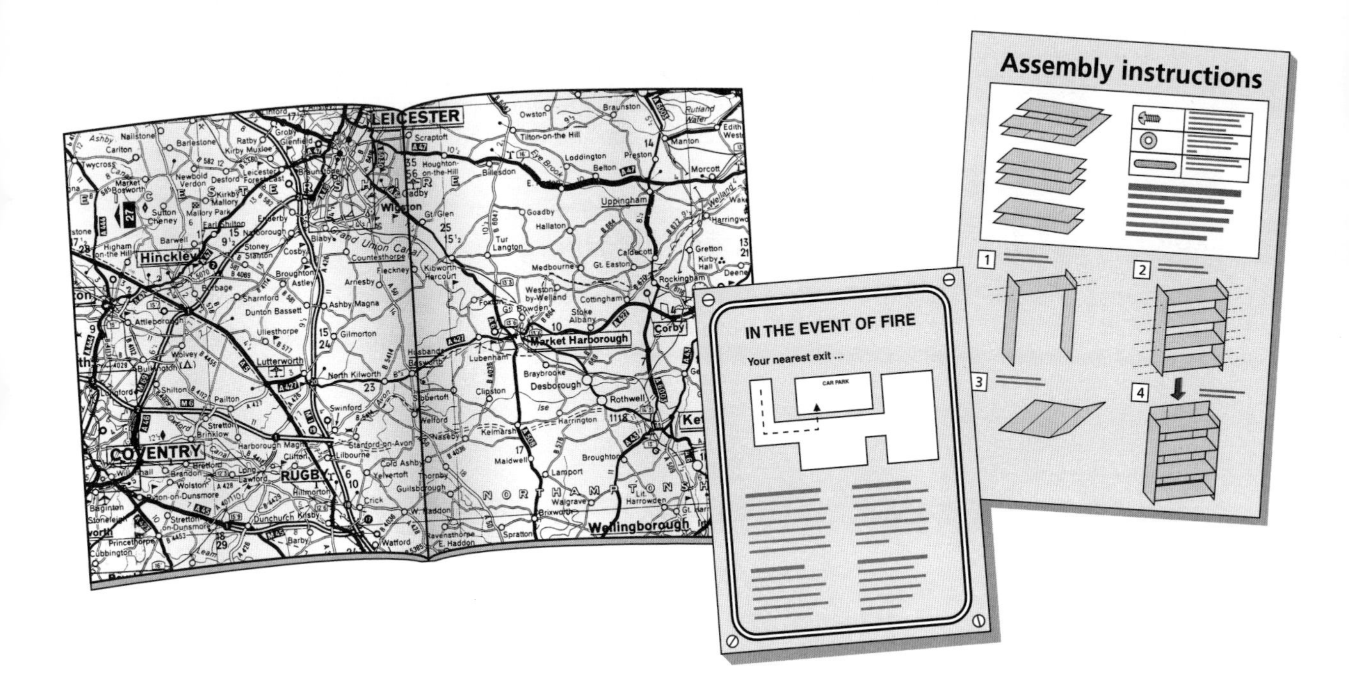

WHAT IS DECISION MATHS?

In the past 40 years, the development of methods for solving complex problems like those listed above has been accompanied by advances in computer technology that have allowed us to generate real solutions. Although recent years have seen rapid progress in the development of techniques for solving decision-making problems, this is not a new branch of mathematics. In 1666, the German mathematician Gottfried Leibniz published *The Art of Combinatorics* – what he called the study of placing, ordering and choosing objects.

We consider all decision-making problems under four headings.

Existence: Does a solution to the problem exist?

Construction: If a solution does exist, how can you construct a method to find the solution?

Enumeration: How many solutions are there? Can you list them all?

Optimisation: If there are several solutions, which is the best one? How do you know that this is the best solution?

Gottfried Liebniz (1646–1716)

DOES A SOLUTION TO THE PROBLEM EXIST?

Most of the problems that you will have met in your study of mathematics have an answer, so you may not always think to ask if a solution exists. But the following problem shows that this is not always an easy question to answer. It is one of the most famous examples of a decision-making problem.

Exploration 1.1

The bridges of Konigsberg

The town of Konigsberg in East Prussia was built on the banks of the River Pregel, with islands that were linked to each other and the river banks by seven bridges.

Konigsberg was renamed Kaliningrad and, due to bombing in World War II, only four of the bridges now remain. We shall consider this problem again in Chapter 4, Networks 3: Route inspection problems.

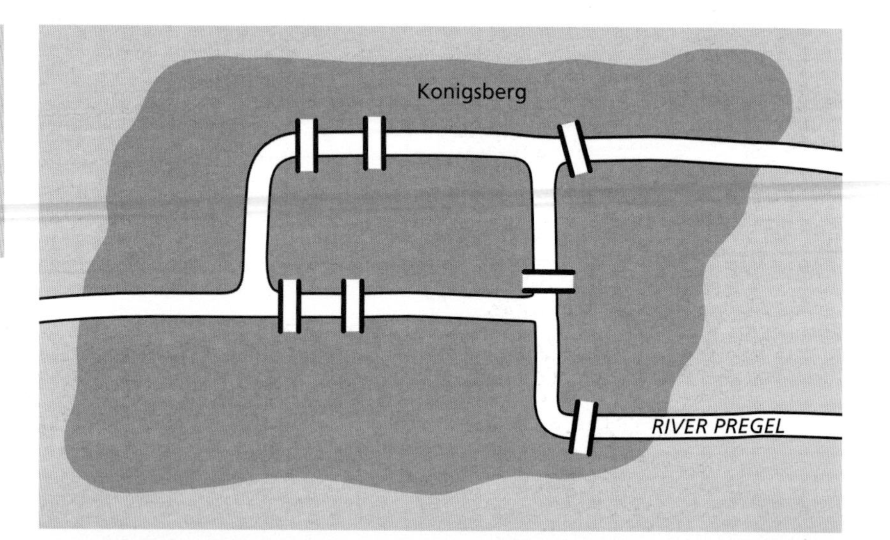

For many years, the citizens of the town tried to find a route for a walk that would cross each bridge only once and allow them to end their walk where they had started from.

Can you find a suitable route?

Hint: Don't spend too long trying to find a solution!

Exploration 1.2

League positions

We often need to sort things into a particular order, either alphabetically or numerically. This table gives the end of season results for twelve school football teams listed *alphabetically*.

What is the final position of each team in the league?

Team	P	W	D	L	Points
Ashleigh	11	0	5	6	5
Beechwood	11	5	3	3	18
Enford	11	3	4	4	13
Hamworth	11	8	2	1	26
Lyndhurst	11	0	3	8	3
Manor	11	3	5	3	14
Morristone	11	5	1	5	16
Raneham	11	6	3	2	21
St John's	11	4	3	4	15
Springfield	11	3	3	5	12
Tallington	11	9	1	1	28
Yatley	11	2	3	6	9

How is it done?

To find the final positions, we need to sort the list by points. How would you do this?

It is clear that a solution to this problem does exist. The next step is to develop a method by which to *find* the solution. It should become clear that you need to be systematic when sorting the information. Once you have decided on a method, it is important to keep to the same method throughout the problem (although it may be a different method to someone else's).

The name for a systematic process that will produce the required solution is an **algorithm**. An algorithm consists of a set of input data and a list of instructions; to solve a problem, you take the initial data and apply the instructions one at a time until you reach a solution.

The word algorithm comes from the Persian mathematician Mohammed al-Khowarizmi who lived in the ninth century. His most famous work Al-jabr wal-muqabalah *(the science of equations) gives us the words algebra and algorithm.*

Al-Khowarizmi (c. 825 AD)

All algorithms must have the following properties.

1. Each step must be defined precisely.
2. The algorithm must work for any set of inputs.
3. The answer must depend only on the inputs for a particular problem.
4. It must produce an output and stop after a finite number of steps.

COMMUNICATING AN ALGORITHM

We use algorithms in many areas of mathematics. You may already be familiar with the one we use for finding the real roots of a quadratic equation, known as **completing the square**.

Example 1.1

List the steps for completing the square as a method for solving a quadratic equation and apply the steps to solve $3x^2 + 10x + 8 = 0$.

Solution

Step 1	*Divide the equation by the coefficient of x^2.*	$x^2 + \frac{10}{3}x + \frac{8}{3} = 0$
Step 2	*Subtract the constant term from both sides.*	$x^2 + \frac{10}{3}x = -\frac{8}{3}$
Step 3	*Halve the coefficient of x, square it and add it to both sides.*	$x^2 + \frac{10}{3}x + \left(\frac{5}{3}\right)^2 = -\frac{8}{3} + \left(\frac{5}{3}\right)^2$
Step 4	*Rewrite the left-hand side of the equation as a perfect square.*	$\left(x + \frac{5}{3}\right)^2 = \frac{1}{9}$
Step 5	*Take the square root of both sides of the equation.*	$\left(x + \frac{5}{3}\right) = \pm\frac{1}{3}$
Step 6	*Solve for x.*	$x = -\frac{5}{3} \pm \frac{1}{3}$

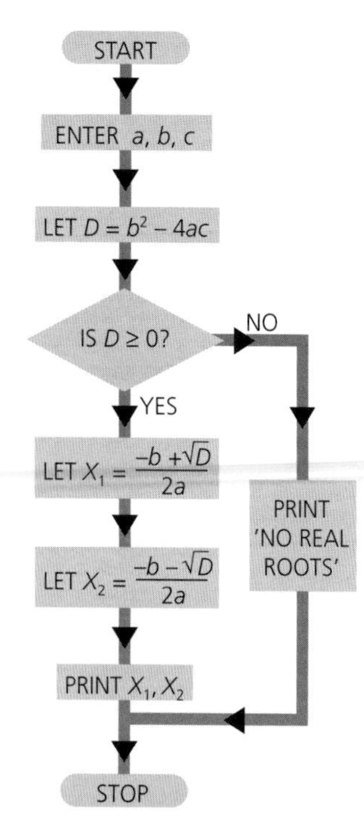

Listing the steps in this way is one way of communicating an algorithm. Another common technique is to use a flowchart. For example, to find the real roots of a quadratic equation, $ax^2 + bx + c = 0$, we can use the algorithm expressed by the formula:

$$x = \frac{\left(-b \pm \sqrt{b^2 - 4ac}\right)}{2a}$$

Each time you complete one loop, you have performed an **iteration** of the algorithm. A simple algorithm such as this is easy and straightforward to apply. Many complex, real-life problems are less easy and we often need to write the algorithm as a computer program.

We can write this algorithm as a set of instructions for a programmable calculator; for example for the TI–80 range of calculators the program is as follows.

```
:clrHome
:Disp "Y = AX² + BX + C"
:Disp "VALUE A"
:Input A
:Disp "VALUE B"
:Input B
:Disp "VALUE C"
:Input C
:(-B+ √(B²-4AC))/2A STO> D
:(-B- √(B²-4AC))/2A STO> E
:Disp"X ="
:Disp D
:Disp E
```

EXERCISES

1.1 CLASSWORK

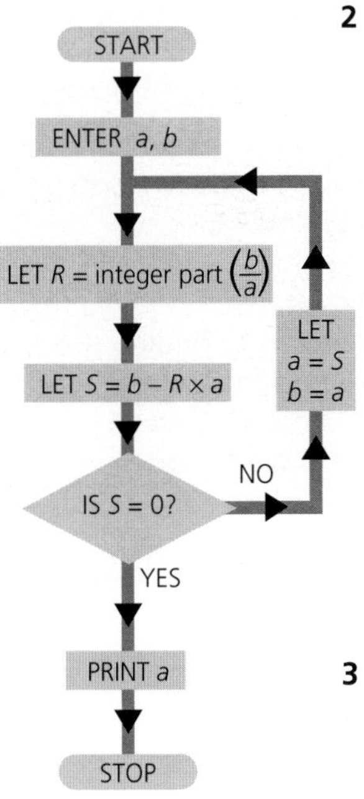

1 Use the above algorithm for completing the square to find the roots of the equation $2x^2 - 5x + 1 = 0$.

2 The algorithm in this flowchart finds the highest common factor of two positive integers.

a) Use it to work out the highest common factor of a = 105 and b = 180. Fill in the table to show the stages in your working.

a	105			
b	180			
R				
S				

b) How many iterations does it take to find the HCF of 105 and 180?
c) What happens if you enter a = 180 and b = 105?
d) Write a program for your calculator based on the flowchart. Check the program with different values of a and b.

3 The diagram below shows a maze. Find a path from the centre (C) to the exit (E). Think of how you might devise an algorithm for escaping from a maze if you had no map.

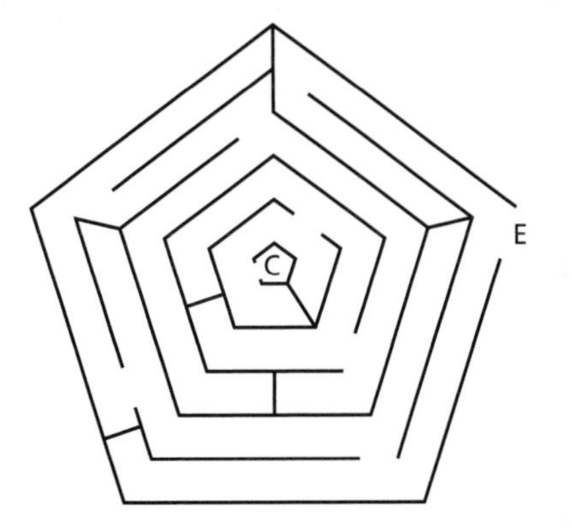

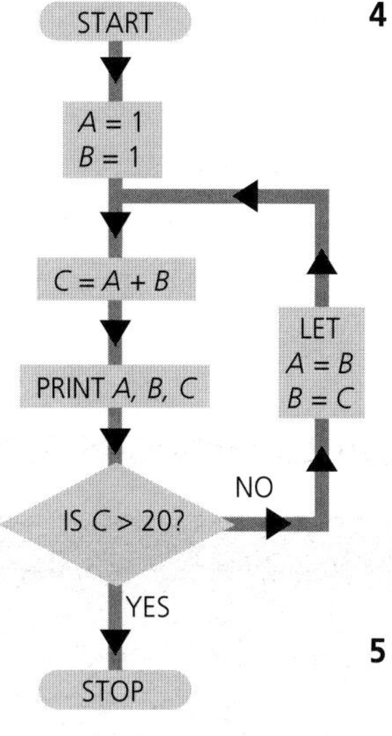

4 Consider the flowchart on the left.

a) Complete this table to show the output.

A	B	C
1	1	

b) What does the algorithm produce?

5 Draw a flowchart or write a short program to set out an algorithm of your choice (for example: multiplication of two 2-digit numbers, the sine rule or cosine rule).

EXERCISES

1.1 HOMEWORK

1 Use the above algorithm for completing the square to find the roots of equation $5x^2 + 2x - 12 = 0$.

2 **a)** Use the algorithm from Exercises 1.1 CLASSWORK, question 2, to work out the highest common factor of 135 and 220.

b) How many iterations does it take?

3 The 'interval bisection method' is a numerical means of finding the root of a non-linear equation $f(x) = 0$. The idea is to find two values of x, $x = a$ and $x = b$ say, such that $f(a) < 0$ and $f(b) > 0$. We then choose c to be the mid-point between a and b i.e. we define $c = \frac{1}{2}(a + b)$.

If $f(c) < 0$ then the root lies between c and b; if $f(c) > 0$ then the root lies between a and c.

The process is repeated until the interval $[a, b]$ containing the root is narrow enough to give the desired accuracy.

a) Apply the interval bisection method to find a root of the cubic equation $x^3 - 7x + 3 = 0$.
b) Write the interval bisection method as an algorithm.
c) Write a program for your calculator based on your algorithm. Check the program by solving $x^3 - 7x + 3 = 0$.

4 A unique code for referencing, called the International Standard Book Number (ISBN), is used to identify published books. For example, the ISBN of this book is 000 322372 8. It has ten digits, the last being a check digit which is introduced to reduce the likelihood of copying down the number incorrectly.

The check digit is found like this.

- Multiply the first digit by 1, multiply the second digit by 2, the third digit by 3, … , the ninth digit by 9.
- Add all the results of the multiplications.

- Divide the total sum by 11 and note the remainder.
- If the remainder is 0 to 9 then this is the check digit, if it is 10 then the check digit is X.

Write a program for your calculator that evaluates the check digit. Use your program to check the ISBNs of several more of your books.

5 Write an algorithm to find the smallest prime number greater than a given positive integer N.

ALGORITHMS FOR SORTING

We can use the results for the school football teams, on page 3, to investigate some algorithms for sorting. Since we are only interested in their total number of points we can simplify the original information.

Team	Points
Ashleigh	5
Beechwood	18
Enford	13
Hamworth	26
Lyndhurst	3
Manor	14
Morristone	16
Raneham	21
St John's	15
Springfield	12
Tallington	28
Yatley	9

Bubble sort

Bubble sorts work by placing one number in its correct position in the list each time we run through the algorithm. (Remember, this is called an iteration or **pass**). Numbers that are below their correct position rise up to their proper place like bubbles in a fizzy drink, hence the name.

For our example we want to use a **descending bubble sort** as we want the team with the greatest number of points at the top. The algorithm starts by comparing the first two numbers in the list; if the second is larger than the first, the numbers are exchanged. It then compares the second and third numbers and exchanges them if necessary, and so on through the list. After one iteration we obtain the following table of results.

5	18	18	18	18	18	18	18	18	18	18	18
18	5	13	13	13	13	13	13	13	13	13	13
13	13	5	26	26	26	26	26	26	26	26	26
26	26	26	5	5	5	5	5	5	5	5	5
3	3	3	3	3	14	14	14	14	14	14	14
14	14	14	14	14	3	16	16	16	16	16	16
16	16	16	16	16	16	3	21	21	21	21	21
21	21	21	21	21	21	21	3	15	15	15	15
15	15	15	15	15	15	15	15	3	12	12	12
12	12	12	12	12	12	12	12	12	3	28	28
28	28	28	28	28	28	28	28	28	28	3	9
9	9	9	9	9	9	9	9	9	9	9	3

First pass

From the list of twelve numbers, eleven comparisons and ten exchanges have been made. By the end of this first pass the smallest number, 3, is in position at the bottom of the list.

The process is repeated in the second pass, but there is no need to consider the last number, since it is in the correct position. This time there are only ten comparisons to be made, leaving the last two numbers in their correct positions. On the third pass nine comparisons are made, and so on until all the numbers are in their correct places.

2nd	3rd	4th	5th	6th	7th	8th	9th	10th	11th
18	26	26	26	26	26	26	26	28	28
26	18	18	18	21	21	21	28	26	26
13	14	16	21	18	18	28	21	21	21
14	16	21	16	16	28	18	18	18	18
16	21	15	15	28	16	16	16	16	16
21	15	14	28	15	15	15	15	15	15
15	13	28	14	14	14	14	14	14	14
12	28	13	13	13	13	13	13	13	13
28	12	12	12	12	12	12	12	12	12
9	9	9	9	9	9	9	9	9	9
5	5	5	5	5	5	5	5	5	5
3	3	3	3	3	3	3	3	3	3

pass number

The final solution to the problem looks like this.

Team	Points
Tallington	28
Hamworth	26
Raneham	21
Beechwood	18
Morristone	16
St John's	15
Manor	14
Enford	13
Springfield	12
Yatley	9
Ashleigh	5
Lyndhurst	3

The algorithm can be expressed like this.

Step 1: Compare the first two numbers.
Step 2: If the second number is larger than the first, exchange the numbers.
Step 3: Repeat steps 1 and 2 for all pairs of numbers until you reach the end of the list.
Step 4: Repeat steps 1 to 3 until no more exchanges are made.

In all we made eleven passes, although the last pass did not involve any exchanges. It simply confirmed that all the entries were in the correct position. We made a total of 66 comparisons and 36 exchanges. Work through the sort again and check this.

Quicksort

Quicksort is another algorithm for sorting. It takes the first number in the list as a **pivot**, then separates the rest of the numbers into two subsets:

- those numbers that are smaller than the pivot
- those that are larger than the pivot.

Don't re-order these subsets.

To illustrate the Quicksort algorithm, consider the football teams once again: the first pass of the algorithm will look like this.

pivot →	5	18	18	18	18	18	18	18	18	18	18
	18	5	13	13	13	13	13	13	13	13	13
	13	13	5	26	26	26	26	26	26	26	26
	26	26	26	5	14	14	14	14	14	14	14
	3	3	3	3	5	16	16	16	16	16	16
	14	14	14	14	3	5	21	21	21	21	21
	16	16	16	16	16	3	5	15	15	15	15
	21	21	21	21	21	21	3	5	12	12	12
	15	15	15	15	15	15	15	3	5	28	28
	12	12	12	12	12	12	12	12	3	5	9
	28	28	28	28	28	28	28	28	28	3	5
	9	9	9	9	9	9	9	9	9	9	3

At the end of this pass the algorithm has separated the numbers into two subsets:

- numbers greater than 5
- numbers less than 5.

We have made eleven comparisons and ten exchanges.

We now repeat the process on each of the two subsets and then on all subsequent subsets created, until we have the list in the desired order.

The darker shading shows the numbers which were fixed after each pass and the lighter shading shows the pivots for the next pass. Work through the sort again and find the total number of comparisons and exchanges made.

2nd	3rd	4th	5th	6th
18	26	28	28	28
13	21	26	26	26
26	28	21	21	21
14	18	18	18	18
16	13	14	16	16
21	14	16	15	15
15	16	15	14	14
12	15	13	13	13
28	12	12	12	12
9	9	9	9	9
5	5	5	5	5
3	3	3	3	3

Other versions of Quicksort use different numbers as the pivot point. One way of doing this is to select the middle number in the list. If we consider the example, we could select either 14 or 16 as the pivot.

Exploration 1.3

Changing the pivot

Repeat the Quicksort algorithm using 14 as the initial pivot. Consider the number of comparisons and exchanges you make, and say whether you think this method may have some advantage over the method above, which pivoted about the top number in the list.

The following figures show the first pass in full and the effects of each subsequent pass.

Quicksort – alternative method

5	28	28	28	28
18	18	18	18	18
13	13	15	15	15
26	26	26	26	26
3	3	3	21	21
14	14	14	14	16
16	16	16	16	14
21	21	21	3	3
15	15	13	13	13
12	12	12	12	12
28	5	5	5	5
9	9	9	9	9

First pass

28	28	28	28	28
18	18	18	26	26
15	16	21	21	21
26	26	26	18	18
21	21	16	16	16
16	15	15	15	15
14	14	14	14	14
3	13	13	13	13
13	3	12	12	12
12	12	3	9	9
5	5	5	5	5
9	9	9	3	3
2nd	3rd	4th	5th	6th

Pass number | Final result

Speed and efficiency of an algorithm

When we have the choice of more than one algorithm for a task, we want to choose the best, most **efficient** one for the job. We may define the efficiency in terms of how easy the calculations are, or how long it takes to find the solution. If we are running an algorithm on a computer, we may also consider the memory needed to run the program.

We can compare two of the sorting algorithms: the Bubblesort and the Quicksort, to see which performed the sort more efficiently, using the total number of comparisons and exchanges made in each case, for the same set of numbers. In this table, c is the number of comparisons and e is the number of exchanges.

Bubble sort	$c = 66$	$e = 36$
Quicksort	$c = 31$	$e = 19$

Quicksort seems to be more efficient in this case.

In general terms, if the bubble sort takes the maximum number of passes, then the total number of comparisons needed for a list of 12 numbers is:

$$11 + 10 + 9 + 8 + 7 + 6 + 5 + 4 + 3 + 2 + 1 = 66$$

and the maximum number of exchanges will also be 66.

For a list of N items there will be a maximum of $N - 1$ comparisons on the first pass, $N - 2$ comparisons on the second pass and so on until we get to zero. Thus the total number of comparisons (c) can be expressed by: $c \leq 0 + 1 + 2 + 3 + \ldots + (N - 2) + (N - 1)$

This is an arithmetic progression, with $a = 0$ and $d = 1$.

The sum of an arithmetic progression is given by:

$$S_n = \frac{n}{2}\left[2a + (n-1)d\right]$$

which gives us:

$$c \leq \frac{N}{2}\left[0 + (N-1)1\right] \Rightarrow c \leq \frac{N}{2}(N-1)$$

and the maximum number of exchanges (e) will be:

$$e \leq \frac{N}{2}(N-1)$$

Thus for $N = 12$, $c \leq 6 \times 11 = 66$ and $e \leq 66$.

It turns out that for Quicksort the largest number of comparisons and exchanges is the same as for Bubble sort, but this only occurs when the initial list is in reverse order from what is required. A list that is randomly arranged will be more efficient on average, especially if the initial pivot chosen is close to the median value.

Interchange sort

In an Interchange sort, the largest number (or smallest number if you are sorting in ascending order) is interchanged with the first number in the list. The largest number remaining is then interchanged with the second number and so on until the list is sorted.

Example 1.2

Sort this list in descending order 6, 19, 12, 21, 2, 15, 9

Original list	1st pass	2nd pass	3rd pass	4th pass	5th pass	6th pass
6	21	21	21	21	21	21
19	19	19	19	19	19	19
12	12	12	15	15	15	15
21	6	6	6	12	12	12
2	2	2	2	2	9	9
15	15	15	12	6	6	6
9	9	9	9	9	2	2

This list contained seven numbers and six passes were needed to complete the sort. Although no interchanges took place on the 2nd and 6th passes, it is still necessary to complete these passes to ensure that the list is correctly sorted. The number of comparisons on the first pass was 6, on the second pass 5 and so on giving a total of
$6 + 5 + 4 + 3 + 2 + 1 = 21$ comparisons.
The maximum number of exchanges will be one each pass, in this case 6.
In general, for a list of length n, the Interchange sort takes a maximum of $n–1$ passes to complete the sort, so $P < n - 1$.
The maximum number of comparisons is $(n - 1) + (n - 2) + \ldots + 2 + 1$ and the maximum number of exchanges will be $n - 1$.

Shuttle sort

The Shuttle sort works by comparing pairs of numbers and exchanging them if necessary. It can be stated in this way:

Step 1: Compare the first two numbers and exchange if necessary.
Step 2: Compare the second and third numbers and exchange if necessary, then compare the second and first numbers and exchange if necessary.
Step 3: Compare the third and fourth numbers and exchange if necessary, compare the second and third numbers and exchange if necessary, compare the second and first numbers and exchange if necessary.
Step 4: For a list of length n, continue until n passes have been performed.

The Shuttle sort is illustrated in this example, using the list from the previous example.

	6	19	19	21	21	21	21
	19	6	12	19	19	19	19
	12	12	6	12	12	15	15
	21	21	21	6	6	12	12
	2	2	2	2	2	6	9
	15	15	15	15	15	2	6
	9	9	9	9	9	9	2
Comparisons	1	2	3	1	4	3	total 14
Exchanges	1	1	3	0	3	2	total 10

For a list of length n, the Shuttle sort will always need $n - 1$ passes. The maximum number of comparisons will be $1 + 2 + 3 + \ldots + (n - 1)$. The maximum number of exchanges will also be $1 + 2 + 3 + \ldots + (n - 1)$. In what situation would you need the maximum number of comparisons and exchanges?

SEARCHING

Exploration 1.4

Guess the number

Working in pairs, you choose a number between one and a hundred; the other person's task is to find your number with the least number of guesses. Each time your partner guesses you may only answer 'bigger' or 'smaller'. Swap over and try it again.

- How many guesses did you each take?
- What strategies did you use to minimise the number of guesses you took?

We often have to find a piece of information that is contained within a large set for example, a telephone number in a directory or a word in a dictionary. From the exploration, you will have realised that the task is much easier if you have a strategy for performing the search so in this section we will consider some of the methods for performing a search efficiently.

Linear search

A linear search is an exhaustive method, checking each piece of data in turn. It is the simplest to perform but is very inefficient.

Sequential search

The data is first ordered and subdivided into lists. An extra list, the **index**, is then created giving the first and last entry in each subdivision. For a given entry, first search the index for the correct subdivision, then carry out a linear search on the data in that subdivision to find the correct item. An example of this would be the telephone directory, which is in alphabetical order and gives the first and last entries at the top of each page.

Binary search

This is the method you probably used in the exploration to locate a number between one and a hundred. The data is first ordered then you look at the middle item (or one next to the middle if you have an even number of items). If this is not the correct item, you select which half of the list the item is in and choose the middle item again. You continue in this way, halving the search area until the correct item is found.

The binary search is most efficient when the number of items is $2^n - 1$. Can you suggest why this is the case?

EXERCISES

1.2 CLASSWORK

1 Sort these sets of data in ascending order using:

a) Bubble sort **b)** Quicksort
c) Interchange sort **d)** Shuttle sort.

set 1 8, 7, 6, 5, 4, 3, 2, 1
set 2 1, 3, 5, 7, 2, 4, 6, 8
set 3 5, 1, 4, 7, 3, 2, 8, 6

In each case compare the efficiency of the sorts.

2 Sort this list in *ascending* order, using both Bubble sort and Quicksort.

50, 48, 76, 30, 12, 4, 28, 56, 63, 77, 21, 57, 29, 41

3 Sort this list of marks using: **a)** Interchange sort
b) Shuttle sort.

14, 25, 12, 31, 18, 13, 20, 28

4 For a list of n numbers {L(1), L(2), L(3), ... L(n)}, the Interchange sort can be stated like this.

Step 1: Look for the smallest number in L(i).
Step 2: Exchange with number in L(1).
Step 3: Look for smallest L(i) remaining.
Step 4: Exchange with next L(i).
Step 5: Repeat steps 3 and 4 until the list is sorted.

a) Use the Interchange sort to complete the sort on this list (the first pass is done for you).

28	20	13	18	31	12	25	14	L ↓
12	20	13	18	31	28	25	14	first pass

Complete the sort.

b) How many comparisons and exchanges were made?

c) Comment on the efficiency of the exchange sort.

5 Work with a partner.

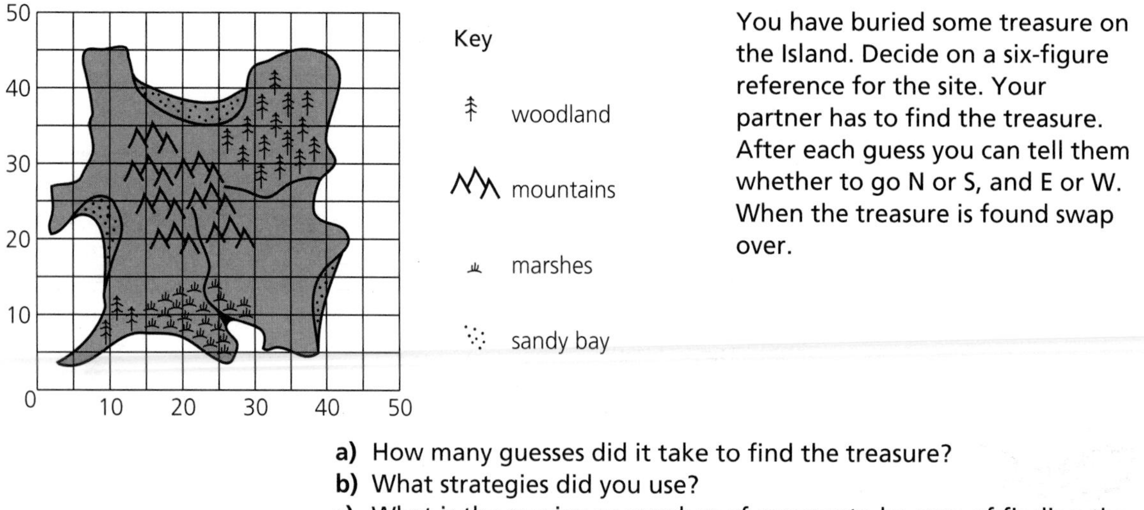

You have buried some treasure on the Island. Decide on a six-figure reference for the site. Your partner has to find the treasure. After each guess you can tell them whether to go N or S, and E or W. When the treasure is found swap over.

a) How many guesses did it take to find the treasure?

b) What strategies did you use?

c) What is the maximum number of guesses to be sure of finding the treasure?

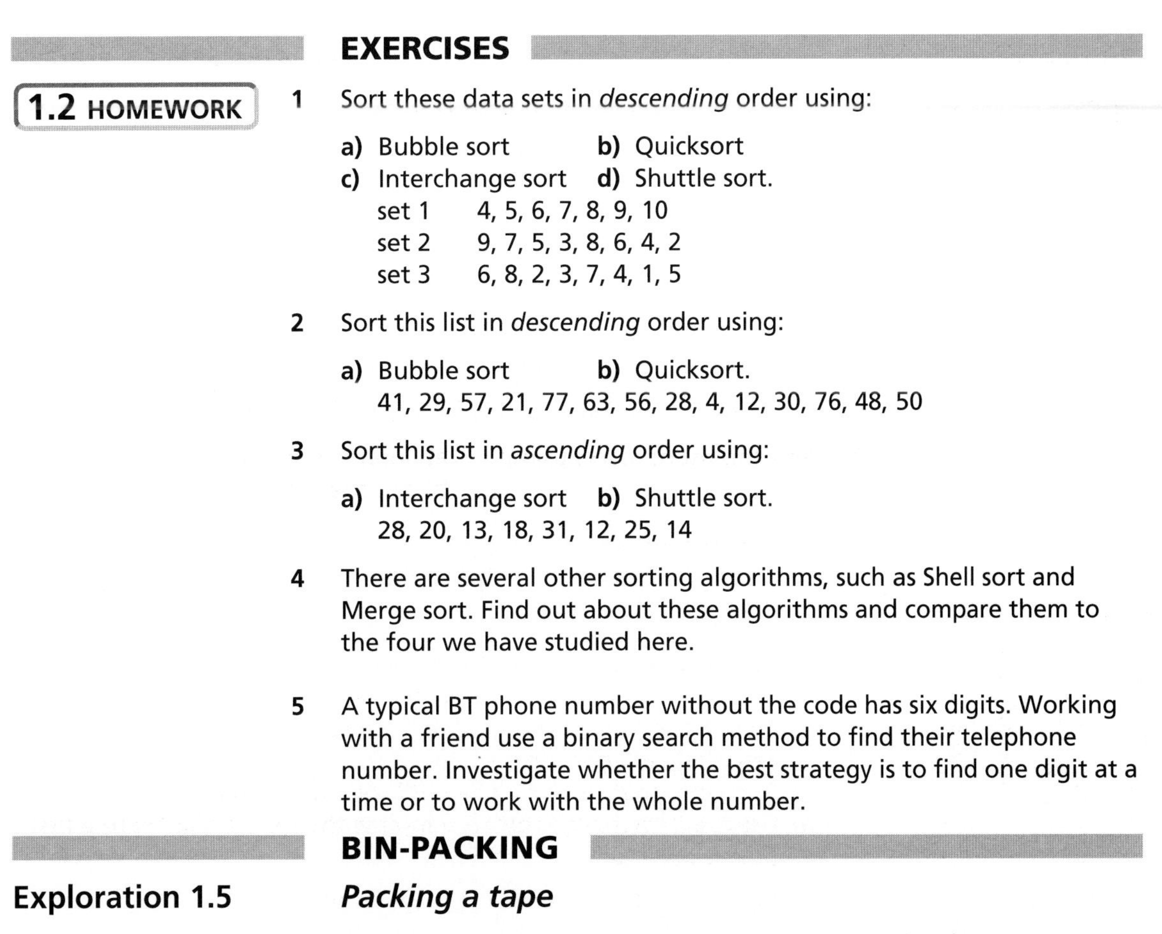

EXERCISES

1.2 HOMEWORK

1 Sort these data sets in *descending* order using:

a) Bubble sort **b)** Quicksort
c) Interchange sort **d)** Shuttle sort.

set 1 4, 5, 6, 7, 8, 9, 10
set 2 9, 7, 5, 3, 8, 6, 4, 2
set 3 6, 8, 2, 3, 7, 4, 1, 5

2 Sort this list in *descending* order using:

a) Bubble sort **b)** Quicksort.

41, 29, 57, 21, 77, 63, 56, 28, 4, 12, 30, 76, 48, 50

3 Sort this list in *ascending* order using:

a) Interchange sort **b)** Shuttle sort.

28, 20, 13, 18, 31, 12, 25, 14

4 There are several other sorting algorithms, such as Shell sort and Merge sort. Find out about these algorithms and compare them to the four we have studied here.

5 A typical BT phone number without the code has six digits. Working with a friend use a binary search method to find their telephone number. Investigate whether the best strategy is to find one digit at a time or to work with the whole number.

BIN-PACKING

Exploration 1.5

Packing a tape

Over a bank holiday weekend you want to record eight programmes on your video recorder. You have four 180-minute tapes. The lengths of the programmes are given in minutes.

A	90	E	120
B	35	F	45
C	105	G	60
D	90	H	100

Can you record all the programmes? If so, how would you do it?

Efficient packing

We are all familiar with the problem of packing too many things into too little time or space. You may have difficulty fitting all the activities you want to do into the time available, or you may have trouble fitting things into your suitcase when you go on holiday. Efficient packing is very important in business and industry, whether it is fitting a number of jobs into the minimum time or packing goods into a warehouse to waste the least amount of space.

As with sorting, these problems need a clear strategy but, unlike sorting, there is no known algorithm which will always produce the best, or **optimal**, solution. **Heuristic algorithms** set out to find good solutions Heuristic methods use logical and often intuitive steps to find a good solution for a particular problem.

The word heuristic *is from the Greek* heuriskein *meaning find.*

In the exploration above, we were trying to pack the programmes into the tapes, which are called the **bins**. We start by establishing the most and least tapes we will need; these are called the **upper and lower bounds**. Since there are eight programmes, we will not use more than eight tapes, so 8 will be an upper bound.

The total time for the programmes is 645 minutes, each tape lasts 180 minutes, so the least amount of tape we could possibly need is $645 \div 180 = 3.58$, so 4 will be a lower bound.

We know that we shall need at least four tapes and no more than eight tapes. We can now consider three algorithms that will give us a better solution. Problems like this are called **bin-packing** problems.

Full-bin algorithm

When there is a relatively small amount of information, we may look for combinations that will fill one bin, then the remaining items are packed in the first available space which is big enough. In the example:

$A + D = 180$

$E + G = 180$

$B + F + H = 180$

leaving C (105) to be placed on a separate tape. This gives us a solution using four tapes.

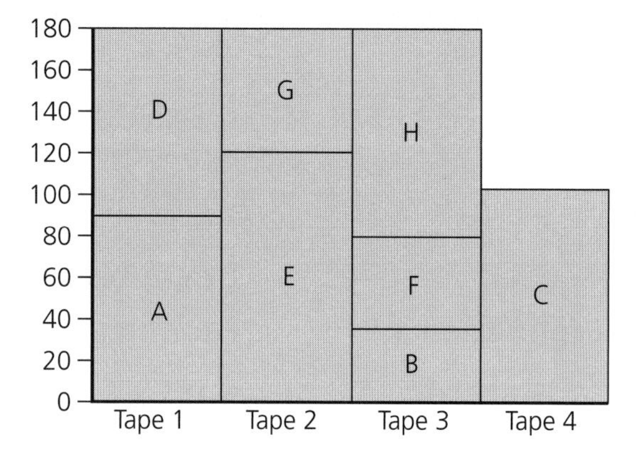

First-fit algorithm

Full-bin combinations work well for simple problems, but if there is a large amount of information or if no full-bin combinations exist, we need a more general method. The first-fit algorithm places each item in the first available bin that has enough room. Then the solution to the tape problem looks like this.

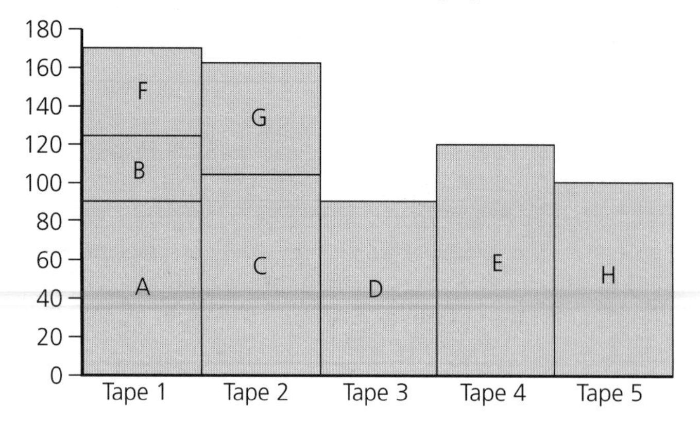

We put programme A on the first tape, using up 90 minutes. Programme B needs 35 minutes and since there is enough space on tape 1 we record it there. Tape 1 now has 55 minutes of space. Programme C needs 105 minutes so we begin Tape 2 and so on.

This algorithm gives a solution requiring five tapes, with more tape left unused that the previous solution. It is less efficient in terms of filling up space, and in fact would not enable us to record all the programmes, if only four tapes are available.

First-fit decreasing algorithm

This algorithm re-orders the information in descending order of size, then follows the pattern of the first-fit algorithm. This puts the largest items first, so that a more efficient solution can be found. For the programmes above the first step of re-ordering gives:

E 120, C 105, H 100, A 90, D 90, G 60, F 45, B 35

The first-fit algorithm then gives the following solution.

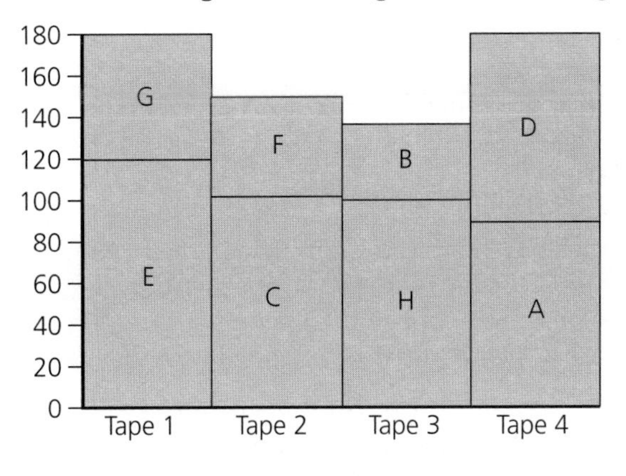

Again we get a solution using four tapes, so there is less space wasted. However, as we must re-order the list first, the algorithm takes rather longer than the first-fit algorithm.

Bin-packing techniques are used to model many practical situations for example:

- How many workers are needed to complete a job in a given amount of time?
- What is the best way to load a ferry, to fit in the maximum number of cars and lorries?
- How does a plumber minimise the number of standard lengths of pipe needed to cut a given number of lengths for a job?
- How many computer disks will be needed to store a selection of programs?
- How does a TV company allocate advertising slots, to minimise the number of breaks in its programmes?

The choice of which 'bin-packing' method to use depends upon the type of problem.

EXERCISES

1.3 CLASSWORK

In this exercise you should try using all three 'bin-packing' methods on each problem you do. Compare the results and see if you can reach any conclusions as to when a particular algorithm will be most effective.

1 Vertical racks in a storage room are designed to take boxes with standard bases but different heights. If the racks are 1 metre high what is the minimum number of racks needed to store the boxes shown?

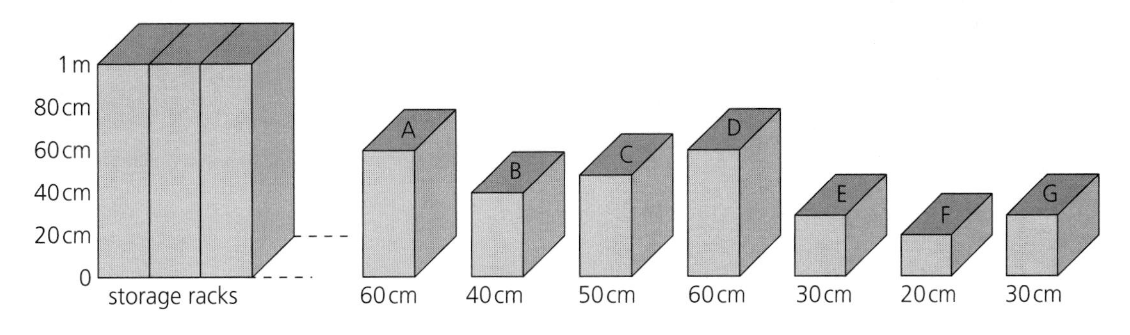

2 A plumber has some 12-foot lengths of copper pipe. For a particular job he needs to cut the sections listed below. What is the best way of cutting the lengths so that he wastes as little pipe as possible?

Section	A	B	C	D	E	F	G	H	I	J	K	L
Length (feet)	2	2	2	3	3	3	4	4	5	6	6	7

3 A computer programmer wishes to fit the following programs onto no more than four 800 kbyte disks.

Program	A	B	C	D	E	F	G	H	I	J	K	L	M	N	O	P
Size (kbytes)	180	150	140	145	200	50	100	130	160	300	170	250	165	200	150	350

Which solution would you recommend? Why?

4 Celestial Television wishes to schedule the following advertisements during the showing of its Saturday night theatre presentation. However, the producer does not want to incorporate more than three breaks, so as not to disturb the artistic flow of the play. A standard length break is 180 seconds.

Is it possible to schedule all the advertisements?

	Product	Time (seconds)
A	Doggibrek	40
B	Zap detergent	30
C	Stukup glue	25
D	Deeliteful dessert	46
E	Carl's Cars	52
F	Mammoth meals	46
G	Trudi dolls	33
H	Deals on Wheels	38
I	Dinosnax	20
J	Bodgit builders	25
K	Soppybabe nappies	40
L	Rekkit DIY	18
M	Miss Ellie's cakes	30
N	Pasta la Vista	20
O	Venus beauty products	40
P	Aardvark carphones	37

State, with reasons, the solution that you would recommend to the producer.

5 A project consists of several activities, as shown in the table below, with their expected duration.

Activity	A	B	C	D	E	F	G	H	I	J	K	L
Duration (days)	8	7	5	6	5	6	9	4	7	9	8	6

Use a suitable packing algorithm to determine the number of workers required to complete the job in twelve days.

EXERCISES

1.3 HOMEWORK

In this exercise you should try using all three 'bin-packing' methods on each problem you do. Compare the results and see if you can reach any conclusions as to when a particular algorithm will be most effective.

1 A storage warehouse has storage racks 2.4 m high. They provide their clients with boxes of three different heights: 60 cm, 80 cm and 120 cm. A client fills three 60 cm boxes, A, B and C, four 80 boxes D, E, F and G and one 120 cm box H. What is the minimum number of racks he needs to store his boxes?

2 Bodgit Brothers are installing a central heating system. They need the following lengths of pipes.

Length (m)	2	2.5	3.0	3.5
No.	4	3	2	3

The piping comes in 6 m lengths. What is the best way to cut the lengths, to minimise wastage?

3 Wendy is backing up the files on the hard disk of her computer onto 1.4 Mbyte floppy disks. Here is a printout of her file details.

	File	Size (kbytes)
A	bridge.tif	409
B	cbl.tif	819
C	chapt1.doc	114
D	chapt2.doc	127
E	chapt3.doc	225
F	chapt4.doc	356
G	chapt5.doc	142
H	ctm.tif	414
I	frame1.dir	48
J	frame2.dir	109
K	frame3.dir	72
L	graph.doc	271
M	intro.doc	134
N	motion.doc	231
O	pipe.doc	130
P	ted.tif	672
Q	wave.tif	532

What is the minimum number of disks she needs?
(1 Mbyte = 1000 kbytes)

4 The following acts are all keen to take part in a variety show. The show is organised in three half-hourly parts. Is it possible for all acts to take part? What schedule would you recommend to the producer?

	Act	Time (minutes)
A	Daves Duo	4
B	Jeremy's Jugglers	5
C	Twinkle Toes dance troupe	7
D	The Bodmin Clog dancers	6
E	The Plymstock Brass Band	10
F	Marvo the Conjurer	6
G	Victor the Ventriloquist Fiddler	5
H	Coral Sheen (Comedienne)	5
I	Maureen Bright's Startling Origami	4
J	Stuart Stardust – the fire eater	7
K	Mike the Magnificent (illusionist)	8
L	Lamplight Lucy (vocalist)	4
M	Rocky Wilde (female impersonator)	4
N	Woofers and Tweeters Jazz Band	7
O	The Caterpillars (band)	5

5 On a trip home from Finland, John and his family want to pack as few suitcases as possible. Each suitcase must have a mass of less than 22 kg. They have parcelled up the things they want to take into twelve bundles.

Bundle	A	B	C	D	E	F	G	H	I	J	K	L
Mass (kg)	8	12	14	6	6	6	6	14	8	4	8	4

How many suitcases are needed?

COURSEWORK INVESTIGATION

Investigation

Newspaper layout

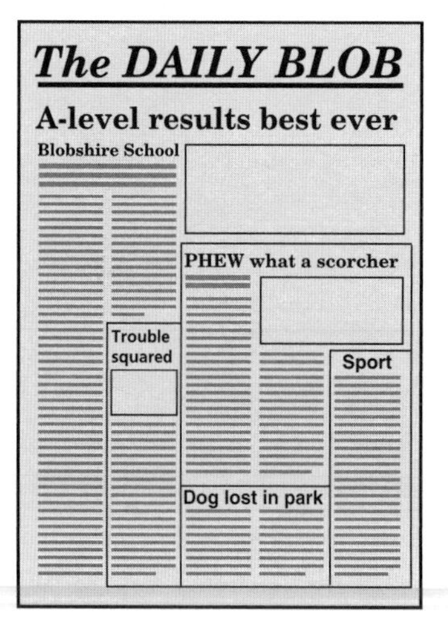

If you look at the way a newspaper is set out you will see that the text is organised in columns, with pictures and headlines often taking up more than one column width. Consider ways in which bin-packing algorithms could be adapted to help with the layout of a newspaper.

CONSOLIDATION EXERCISES FOR CHAPTER 1

1 The following algorithm finds the greatest common divisor of two positive integers (e.g. the greatest common divisor of 24 and 36 is 12).

INPUT A
INPUT B
REPEAT
  Let Q be the largest whole number such that QA is less than or equal to B.
  Let R be $(B–QA)$
  Let the new value of B be A
  Let the new value of A be R.
UNTIL $R = 0$
PRINT B
STOP

i) Demonstrate the algorithm in use on $A = 2520$ and $B = 5940$.
ii) Show what happens if the order of input is reversed, i.e. if $A = 5940$ and $B = 2520$.
iii) In the worst case, a formula giving the approximate number of iterations of this algorithm is:

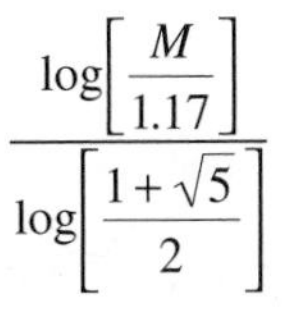

$$\frac{\log\left[\frac{M}{1.17}\right]}{\log\left[\frac{1+\sqrt{5}}{2}\right]}$$

where M is the larger of A and B.

Explain what is meant by the phrase '... in the worst case ...'.

(OCR MEI (Specimen Paper 2000) D & D1)

2 Use the quicksort algorithm to rearrange the following numbers into ascending order, showing the new arrangement at each stage. Take the first number in any list as the pivot.

9, 5, 7, 11, 2, 8, 6, 17

(AQA A Specimen Paper 2000 Discrete 1)

3 The following flowchart defines an algorithm which operates on two inputs, x and y.

a) Run the algorithm with inputs of $x = 3$ and $y = 41$, counting how many times the instructions in box number 3 are repeated.

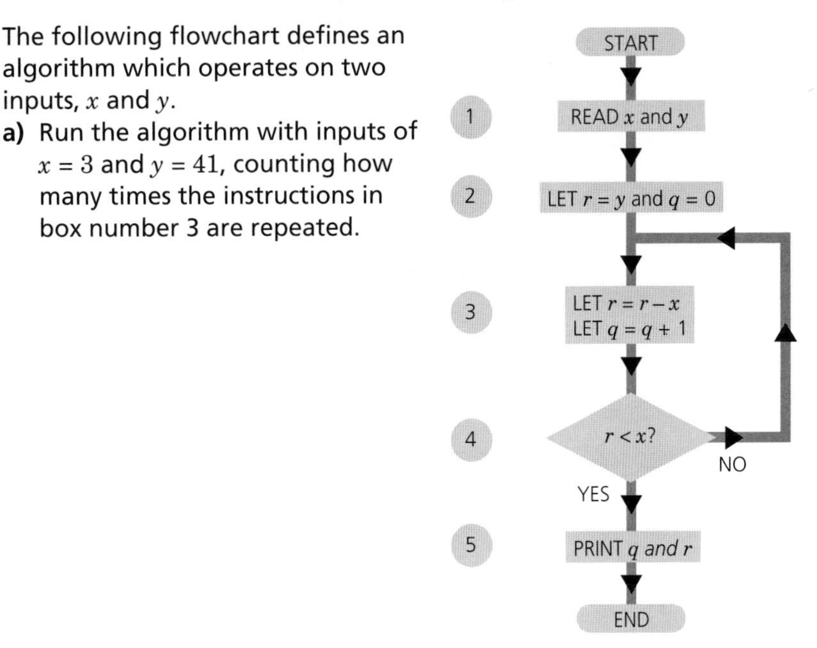

b) Say what the algorithm achieves.

The following flowchart defines an algorithm which operates on three inputs, x, y_1 and y_2.

c) Work through the algorithm with $x = 3$, $y_1 = 4$ and $y_2 = 1$. Record your work by completing the table below. Keep a count in the spaces provided of how many times the instructions in boxes numbers 3 and 6 are repeated.

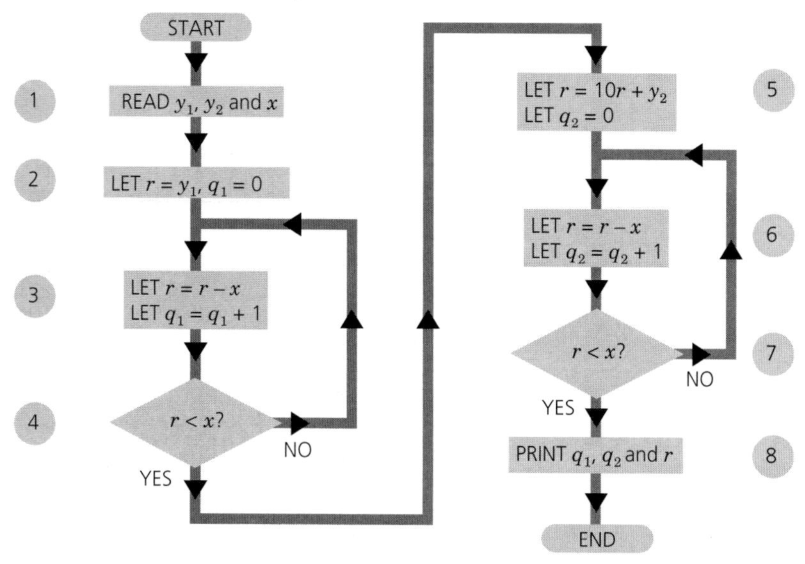

r	
q_1	
q_2	
No. of repetitions of box number 3	
No. of repetitions of box number 6	

d) The second algorithm achieves the same result as the first. Say what you think are the advantages and disadvantages of each.

(O & C MEI Question 3, Decision & Discrete Mathematics Paper 19, January 1995)

4 **a)** The instructions labelled 1 to 7 below describe the steps of a bubble sort algorithm. Apply the bubble sort algorithm to the list of numbers shown after the instructions. Record in the table provided the state of the list after each pass, and record the number of comparisons and the number of swaps that you make in each pass. (The result of the first pass has already been recorded.)

Instructions

1 Let i be equal to 1.
2 Repeat lines 3 to 7, stopping when i becomes 6.
3 Let j equal 1.
4 Repeat lines 5 and 6, stopping when j becomes $7 - i$.
5 If the jth number in the list is bigger than the $(j + 1)$th, then swap them.
6 Let the new value of j be $j + 1$.
7 Let the new value of i be $i + 1$.

List: 7 9 5 1 11 3

Table:

*i*th pass	$i = 1$	$i = 2$	$i = 3$	$i = 4$	$i = 5$
	7				
	5				
	1				
	9				
	3				
	11				
Comparisons	5				
Swaps	3				

b) Suppose now that the list is split into two halves, {7, 9, 5} and {1, 11, 3}, and that the algorithm is applied to each half separately, giving {5, 7, 9}, and {1, 3, 11}. How many comparisons and swaps does this entail altogether?

c) A modified version of the bubble sort algorithm, in which lines 2 and 3 have been changed, is now used to sort the revised list {5, 7, 9, 1, 3, 11}.

Revised instructions

1 Let i equal 1.
2 Repeat lines 3 to 7, stopping when i becomes 4.
3 Let j equal $4 - i$.
4 Repeat lines 5 and 6, stopping when j becomes $7 - i$.
5 If the jth number in the list is bigger than the $(j + 1)$th, then swap them.
6 Let the new value of j be $j + 1$.
7 Let the new value of i be $i + 1$.

Apply the revised algorithm to the list {5, 7, 9, 1, 3, 11}, recording your results in the table provided.

*i*th pass	$i = 1$	$i = 2$	$i = 3$
Comparisons			
Swaps			

Does splitting the list, applying the bubble sort algorithm to each half separately, and then applying the revised algorithm to the revised list, appear to offer any advantages? Justify your answer.

d) Explain why the changes to statements 2 and 3 of the original algorithm were made to produce an algorithm to sort the revised list.

(O & C MEI Question 3, Decision & Discrete Mathematics Paper 19, June 1995)

5 **a)** The following eight lengths of pipe are to be cut from three pieces, each of length 2 m.

1.1 m, 1.2 m, 0.3 m, 0.4 m, 0.3 m, 0.7 m, 0.4 m, 0.2 m

i) Use the first-fit decreasing algorithm to find a plan for cutting the lengths of pipe. Show your plan by marking cuts on the diagram below.

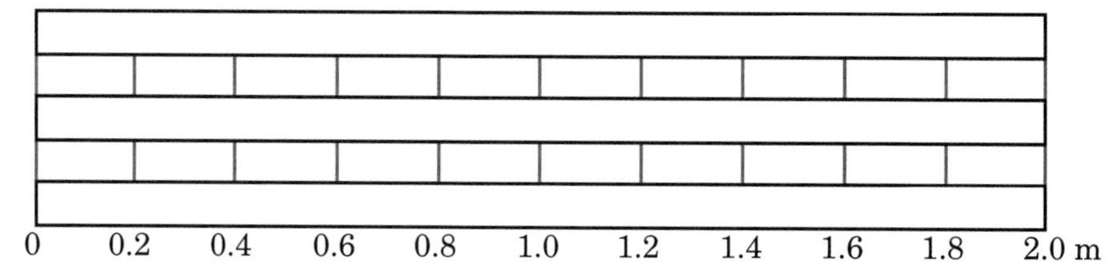

ii) Suppose that it is preferable to be left with a small number of longer pieces of pipe, rather than a larger number of shorter lengths. Use any method to produce an improved cutting plan.

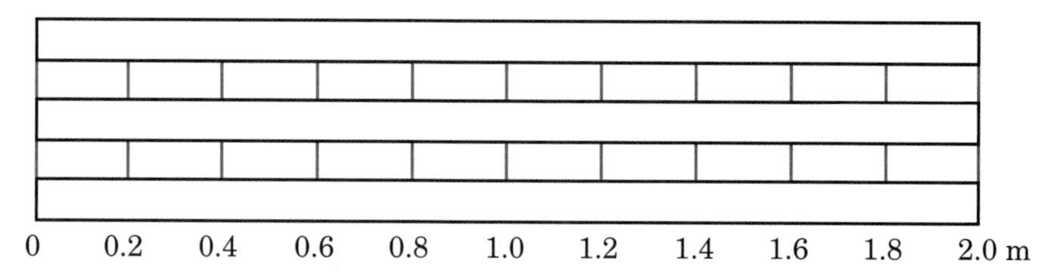

(O & C MEI Question 3 (part), Decision & Discrete Mathematics Paper 19, January 1996)

6 A schoolboy is surprised to discover an old method of multiplication which requires knowledge of just one multiplication table, the two times table. He attempts to explain the method to a friend by writing it out as an algorithm with Q and R as positive integers.

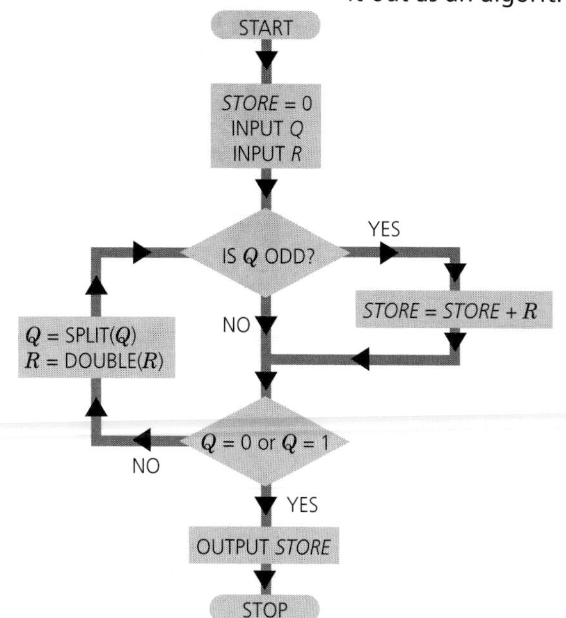

He uses two functions called DOUBLE AND SPLIT.

'DOUBLE' has its usual meaning of 'multiply by two', so DOUBLE(7) = 14.

'SPLIT' means 'divide by two but ignore any halves', so SPLIT(7) = 3, SPLIT(8) = 4.

a) Use the given algorithm to test whether the input of 7, 9 gives the expected output. Show the result of each step.

b) Repeat to find the output of the input 4, 8.

c) Without carrying out the algorithm, explain why the input of 10, 25 would be performed faster than the input of 25,10.

(AQA A Specimen Paper 2000 Discrete 1)

7 Use the binary search algorithm to locate the name GREGORY in the following list.

1 ARCHER
2 BOWEN
3 COUTTS
4 DENYER
5 EATWELL
6 FULLER
7 GRANT
8 GREGORY
9 LEECH
10 PENNY
11 THOMPSON

(Edexcel Specimen Paper 2000 D1)

8 A class is given a task to perform by its mathematics teacher. Each pupil is to design an algorithm to achieve what is required. One pupil's attempt is as follows.

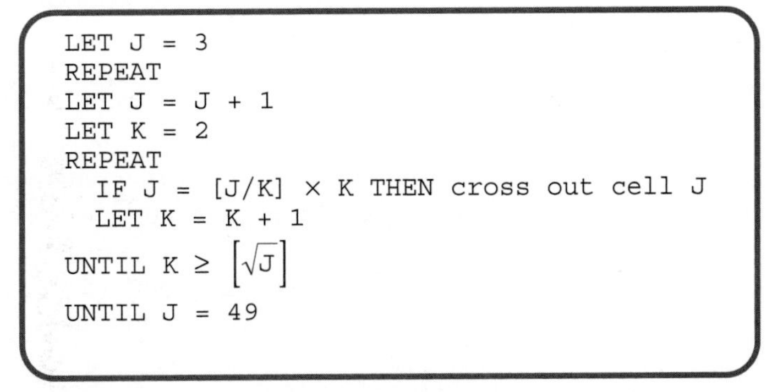

```
LET J = 3
REPEAT
LET J = J + 1
LET K = 2
REPEAT
  IF J = [J/K] × K THEN cross out cell J
  LET K = K + 1
UNTIL K ≥ [√J]
UNTIL J = 49
```

$[x]$ stands for the largest whole number less than or equal to x; e.g. $[3.2] = 3$; $[5] = 5$.

a) Make a copy of the following table representing all the values of J considered by the algorithm. Follow through the algorithm crossing out cells as directed. (You do not need to show more than one cross in any cell.)

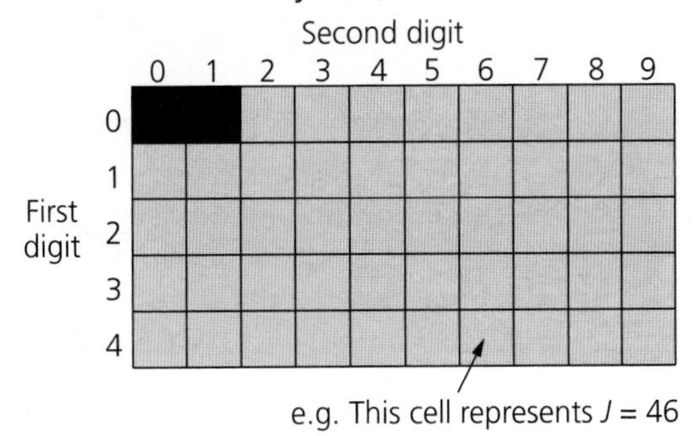

What do you think the task set by the teacher was? Does the algorithm produce the correct result?

A second pupil had the idea of looking at each number in turn, starting with 2, and then crossing out all cells representing a *proper* multiple of the number (i.e. 2 × the number, 3 × the number, 4 × the number, etc.).

b) Make another copy of the table and follow through the second pupil's approach.
Indicate the order in which you cross out cells by putting a number instead of a cross when you *first* eliminate a cell.

c) Produce an algorithm, in similar form to that given for the first pupil, for the second pupil's idea.

d) Describe how the efficiency of the second algorithm could be assessed.

(UODLE Question 10, Decision Mathematics AS, 1994)

9 **a)** In a two-person game one player thinks of a word and invites the other to guess what that word is. In response to an incorrect guess the first player indicates whether the mystery word is before or after the guess in alphabetical (dictionary) order.

i) Given that a dictionary is available, describe a strategy for the second player to follow in trying to find the word in as few guesses as possible.

ii) Given that the dictionary has approximately 100 000 words, find the maximum number of guesses which might be required to find the mystery word using your method from part **i)**. Explain your reasoning in arriving at this number.

b) In a game of 'Battleships' a player chooses a square on a grid to represent a battleship. A second player tries to find that square by making a sequence of guesses. In response to an incorrect guess, the second player is told *either* whether the battleship is to the east or west of the guess (i.e. to the left or the right), *or* whether it is to the north or south (above or below).
The game is being played on a 15×20 grid and a first guess has been made at square K8, as indicated on the diagram.

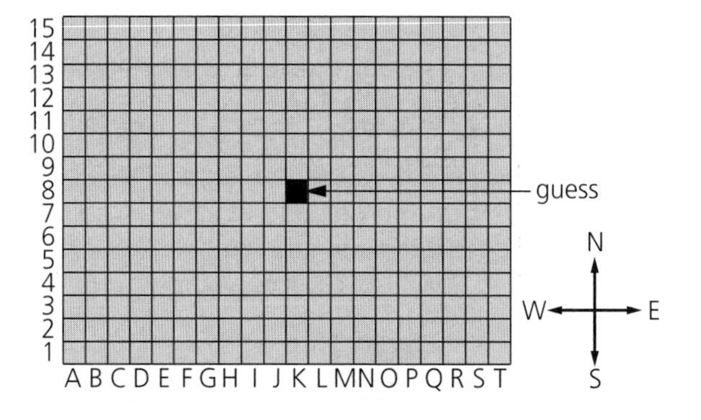

The information provided is that the battleship is to the south of the guess.

i) In a binary search the set of possibilities is, as nearly as possible, halved at each iteration. Given the guess and the information, what would be the next guess?

ii) If the information following that next guess is that the battleship is to the north of the guess, what would be the next guess in a binary search?

iii) What difficulty will be encountered when the information is that the battleship is to the west?

iv) Starting a game afresh, what is the minimum number of guesses that will be needed to be sure of locating the battleship?

c) 'Spaceships' is similar to 'Battleships', but is played on a 3-dimensional grid. Playing a binary search strategy on a $15 \times 15 \times 15$ grid, how many guesses will be needed to be sure of locating the spaceship?

(O & C MEI Question 3, Decision & Discrete Mathematics Paper 19, January 1997)

10 Andy wants to record the following twelve TV programmes onto video tape. Each video tape has space for up to three hours of programmes.

Programme	A	B	C	D	E	F	G	H	I	J	K	L
Length (hours)	$\frac{1}{2}$	$\frac{1}{2}$	$\frac{3}{4}$	1	1	1	1	$1\frac{1}{2}$	$1\frac{1}{2}$	$1\frac{3}{4}$	2	2

i) Suppose that Andy records the programmes in the order A to L using the first fit algorithm. Find the number of tapes needed, and show which programmes are recorded onto which tape.

ii) Suppose instead that Andy is transferring the programmes from previously recorded tapes, so that they can be copied in any order, and that Andy uses the first fit decreasing algorithm. Find the number of tapes needed, and show which programmes are recorded onto which tape.

(OCR Specimen Paper 2000 Discrete D1)

Summary

After working through this chapter you should:

- understand that an algorithm is a systematic process that can be used to give the solution to a problem
- know how an algorithm can be stated or described and be able to follow the steps of a given algorithm
- be able to use Bubble sort, Quicksort, Shuttle sort and Interchange sort algorithms
- know how you can assess the speed and efficiency of these algorithms, by considering the number of comparisons and exchanges made at each iteration
- know how to perform a linear search, a sequential search and a binary search
- model situations using bin-packing algorithms and apply full-bin, first-fit and first-fit decreasing algorithms to a variety of problems.

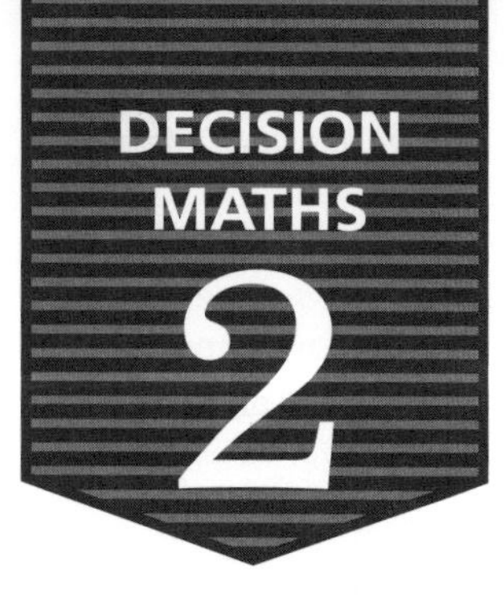

Graphs and networks 1: Shortest path

By the end of this chapter you will:

- *know that we can use networks to model many real-life situations*
- *find that networks are an application of an area of mathematics called graph theory*
- *explore Dijkstra's algorithm which finds the shortest distance between any two points or* ***vertices*** *in a network.*

NETWORK PROBLEMS

Exploration 2.1 *Route maps*

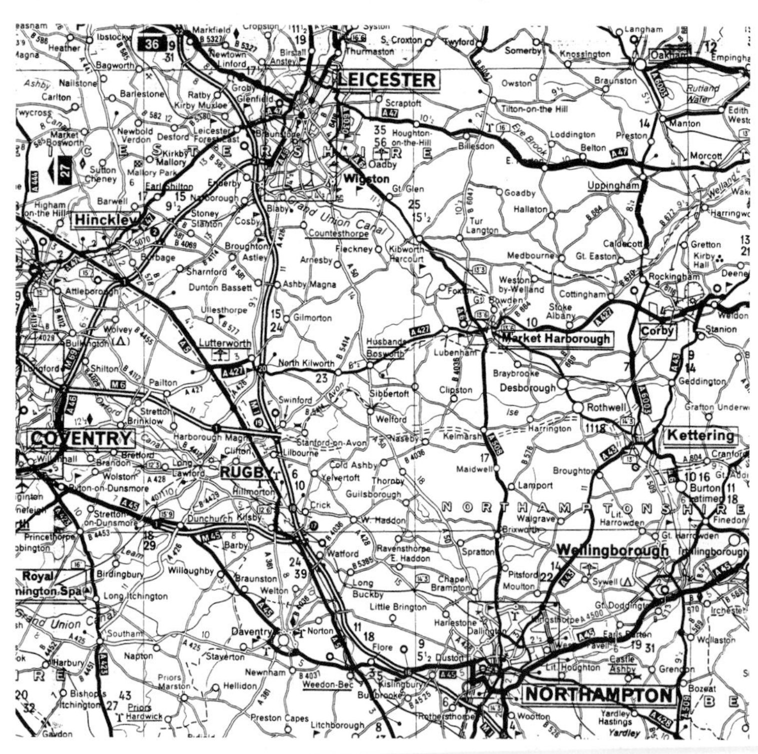

This is a map of the area south of Leicester. It represents many (though not all) of the roads and towns in the area. A conference is being held in Wellingborough. Design a simpler road map of the area to give information on how to get to Wellingborough from nearby towns.

Simplified diagrams

The original map gives far more detail than is needed just for travelling through the region in a car. It would be easier if we could represent it in a simpler form. Such a simplified plan, showing only the major roads, is called a **route map**.

This is a diagrammatic way of showing how places are linked by roads. It does not necessarily give information of the type of road or places of interest in the area. This type of simplified diagram is useful in many other areas.

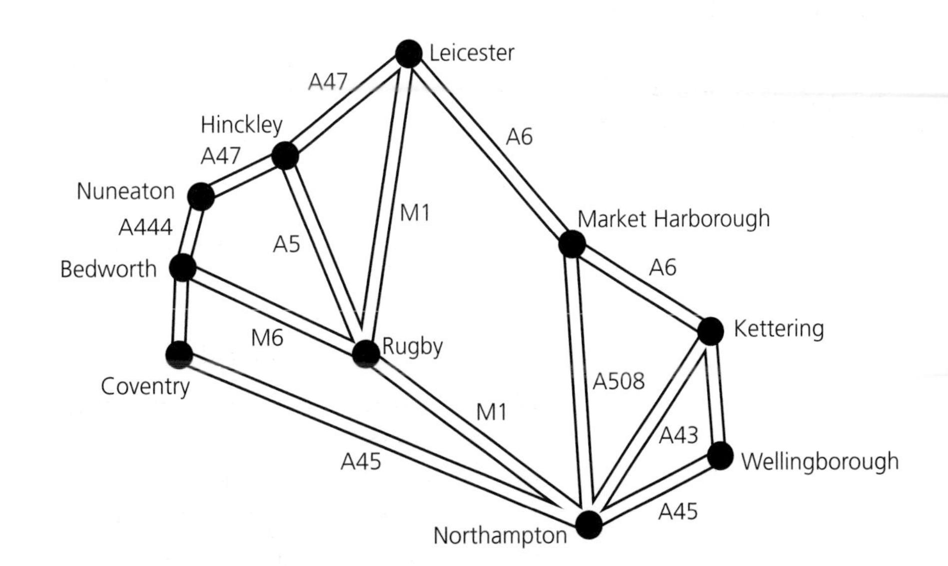

Electrical circuits

This diagram represents an electrical circuit made up of a battery, four ammeters, a resistor and three lamps. This type of diagram shows how components are connected together, although, as with the route map, it does not give much detail about features such as how long or how thick the individual lengths of wire are.

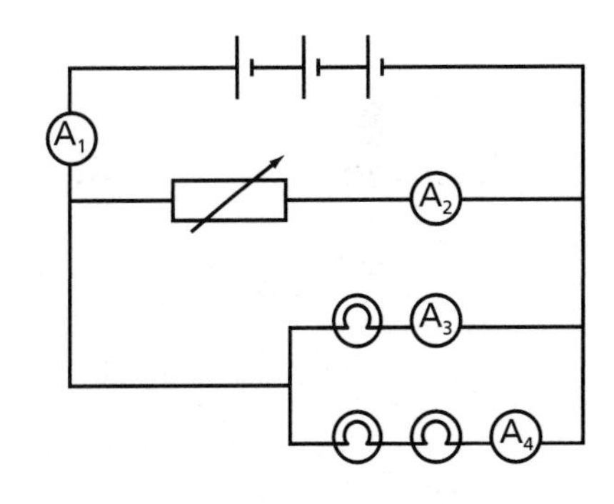

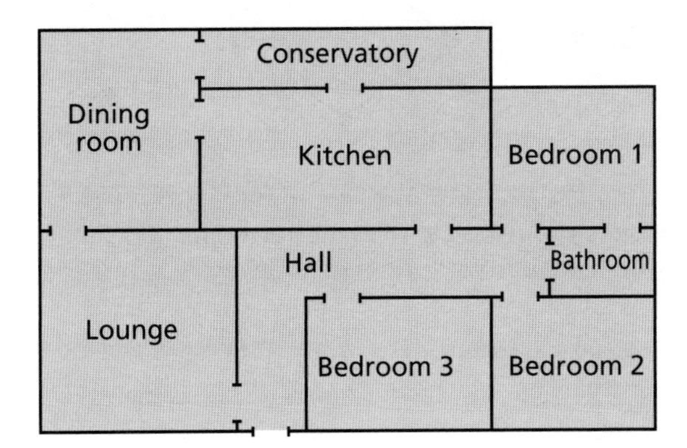

Circulation diagrams

This is a plan of the ground floor of a house. A plan is useful for showing the layout of a small building, but if you were designing a large public building to be used by lots of people, such as a school, a simpler diagram showing the connections between rooms would be useful.

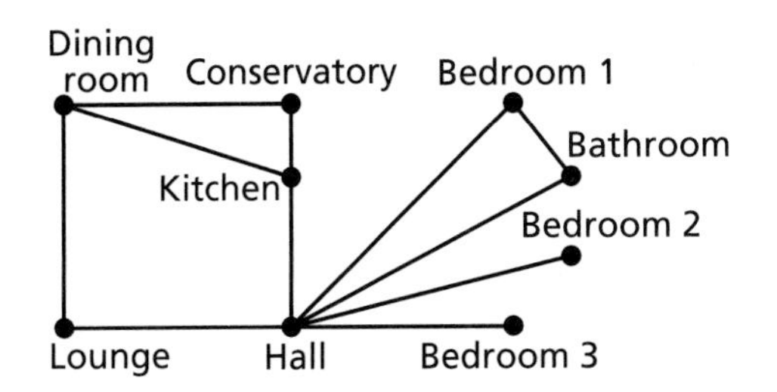

This is a circulation diagram, of the type used by architects. Once again, it gives only limited information; it does not tell us the size or shape of the rooms or the length and width of the corridors.

Structural diagrams

If you are studying Chemistry, you will probably be familiar with these diagrams. They show how atoms link together to give molecules. These diagrams give no information about how atoms are aligned in space: they are a two-dimensional representation of a three-dimensional object.

Introduction to graph theory

*In some books vertices are called **nodes**.*

All the examples above involve the way in which certain objects are related to each other. In each diagram the objects are represented by **points** and the connections are represented by **lines**. This type of diagram is called a **graph**. Each point representing an object is called a **vertex** (plural: **vertices**). Each line connecting two vertices is called an **edge** if it has no direction or an **arc** when it has direction.

A **simple graph** is one in which there is no more than one edge connecting each pair of vertices, and there are no loops.

The **degree** of a vertex is defined as the number of arcs incident with the vertex.

A graph is **connected** if there is at least one route through the graph connecting any given pair of vertices (nodes). In the following diagram the first graph is connected because there is a route along the edges between any pair of vertices e.g. ACDE connects A to E. In the second graph there is no route to vertex E.

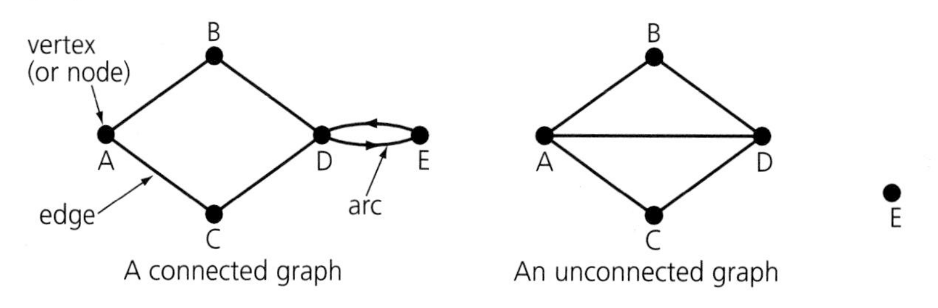

A graph where each vertex is connected to every other vertex is **complete**. A complete graph with n vertices is denoted by K_n.

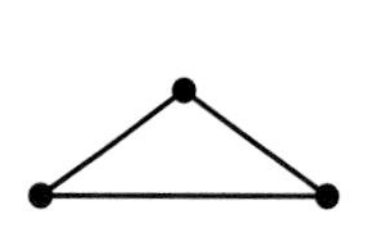

K_3 It has three vertices and three edges.

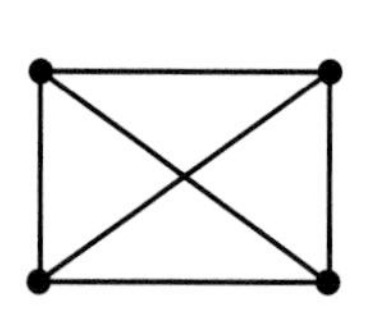

K_4 It has four vertices and six edges.

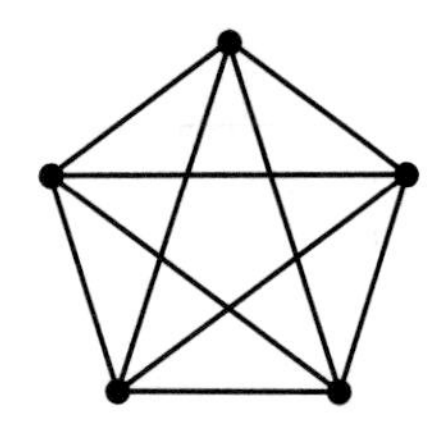

K_5 It has five vertices and ten edges

Subgraphs

A **subgraph** is part of a graph which is itself a graph. A graph can have several subgraphs, not all of which are connected. For example, these are all subgraphs of K_5.

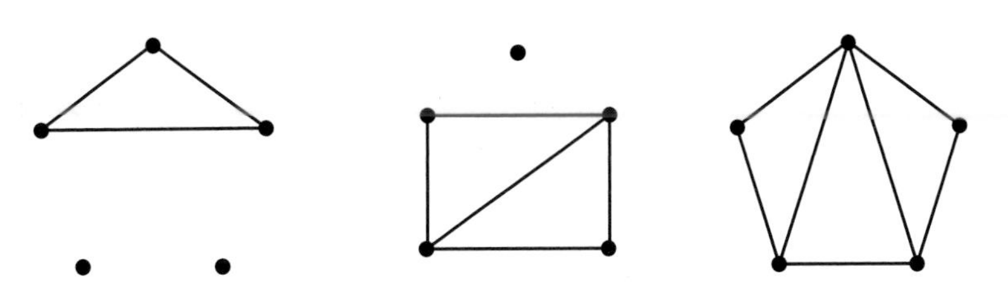

Digraphs

Digraph is short for directed graph.This is a graph in which one or more of the edges has a direction, shown by an arrow.

Adjacency matrices

So far we have represented all the graphs by diagrams that are clear and easy to understand. However it is not always practical to draw this kind of graph, especially if there are large numbers of vertices and edges which would create a very complicated picture, or if we wish to store the information in a computer. Another way in which graphs can be represented is by using **matrices**.

	A	B	C	D
A	0	2	1	1
B	2	0	0	1
C	1	0	0	1
D	1	1	1	2

Above, left, is a graph with four vertices. The matrix on the left here describes the graph. Each row and each column represents a vertex of the graph and the numbers in the matrix give the number of edges joining each pair of vertices, so there are two edges from A to B, one edge from A to C and so on. This is the **adjacency** matrix for the graph. D to D is 2 because you can go in either direction.

Paths

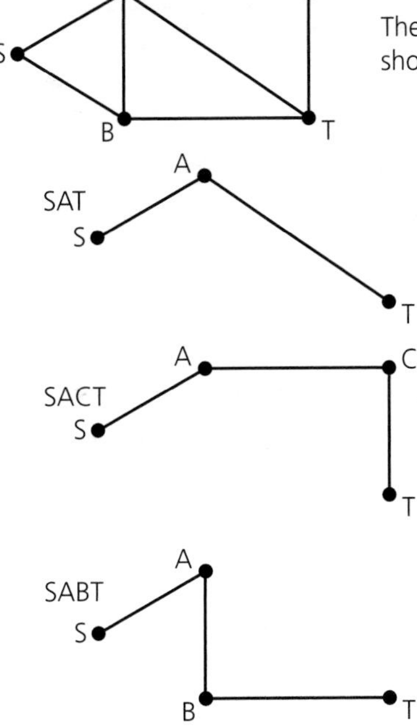

The paths from S to T in this graph are shown below.

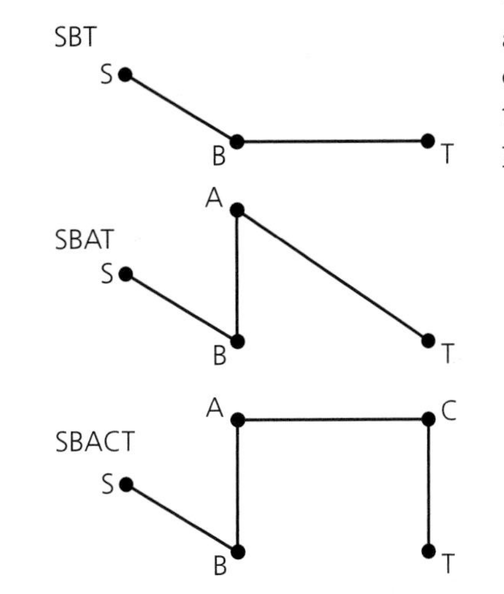

Many of the applications of graph theory are about getting from one vertex to another. If we make a route through a graph which does not go along any edge more than once or visit any vertex more than once, it is called a **path**.

Networks

A **network** is a graph in which each edge (or arc) is given a value called its **weight**. The weight on an edge may represent a distance, a time or a cost: for example, if we add distances to our route map the resulting network would look like this.

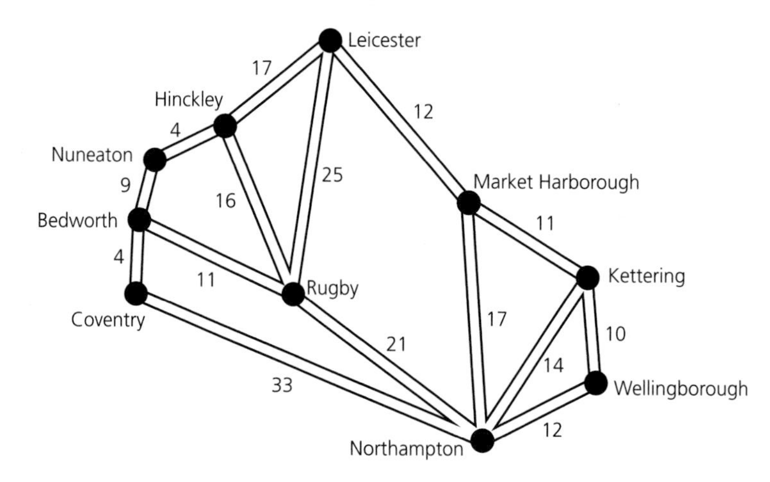

Representing networks using matrices

Just as we can use an adjacency matrix to represent a graph, we can also use matrices to represent networks. These can show:

- the distance between vertices
- the time it takes to travel between vertices
- the cost of moving between vertices, called **cost matrices**.

Here is an example of a distance matrix and its associated network.

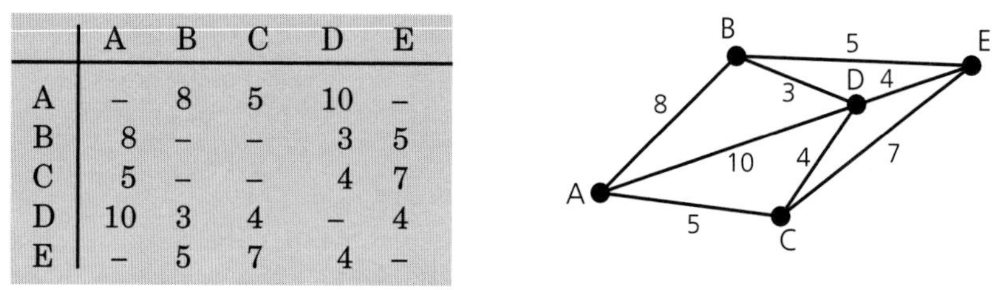

	A	B	C	D	E
A	–	8	5	10	–
B	8	–	–	3	5
C	5	–	–	4	7
D	10	3	4	–	4
E	–	5	7	4	–

EXERCISES

1 For each of the following graphs give the number of vertices, the number of edges and say whether or not the graph is connected.

In each case, suggest a real situation that the graph might represent.

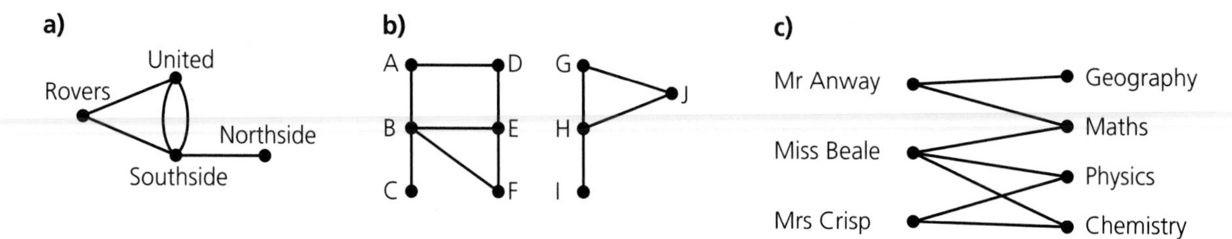

2 **a)** Draw the graph for this adjacency matrix.

	A	B	C	D	E
A	0	1	2	1	1
B	1	0	1	2	1
C	2	1	0	0	1
D	1	2	0	0	0
E	1	1	1	0	2

What is the significance of the 2 here?

b) Draw the adjacency matrix for this graph.

3 List all the paths from A to Z in these graphs.

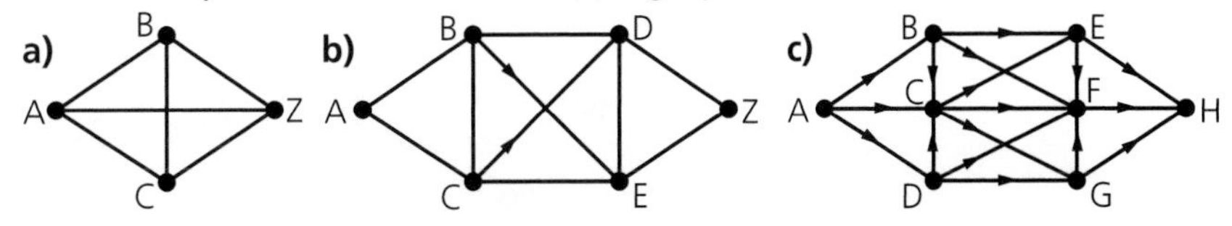

4 Draw a circulation diagram for your house or part of your school. Can you suggest any improvements to the design?

5 Draw the graphs K_6 and K_7. By considering the number of arcs find a general formula for the number of arcs in K_n.

EXERCISES

2.1 HOMEWORK

1 The figure shows different regions of a city with a river running through it. There are six areas of the city labelled A, B, C, D, E and F and bridges connect the areas as shown.

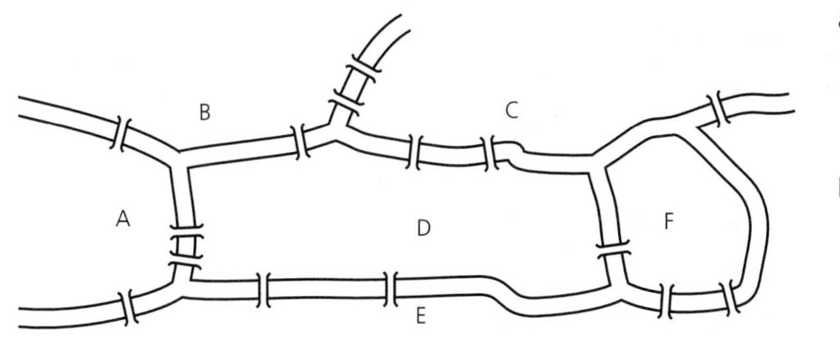

a) Draw a graph to represent this city. Each edge should represent a possible route over a bridge between two areas of the city.

b) Can you find a path starting from any area of the city which passes through each other part and returns to the starting area?

2 Construct the adjacency matrix for the bridge crossings for the city in question **1**.

3 List all the paths from A to S in these graphs.

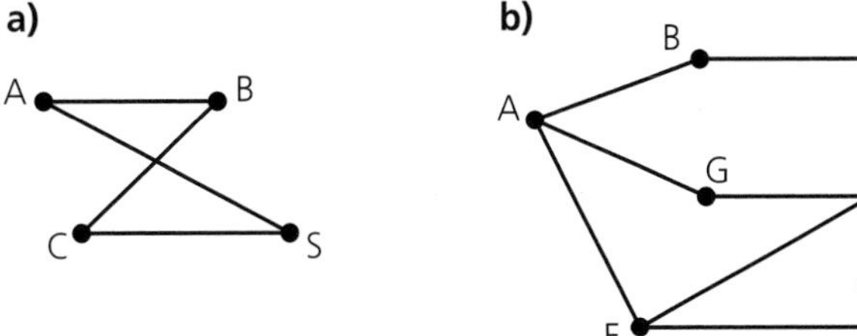

4 What is the degree of each vertex in:

a) K_2 **b)** K_5 **c)** K_n?

We shall meet this again in Chapter 5, *Networks 4: Route inspection problems*.

5 Use a map of the motorway system and main trunk roads of England to find the shortest distances between each pair of teams in the top half of the Premier Soccer league. Represent this information a graph. Is this graph complete?

SHORTEST PATH PROBLEMS

Exploration 2.2

Find the shortest route

Study this network again. Suppose you live in Nuneaton and are attending the conference in Wellingborough. What is the shortest route to take?

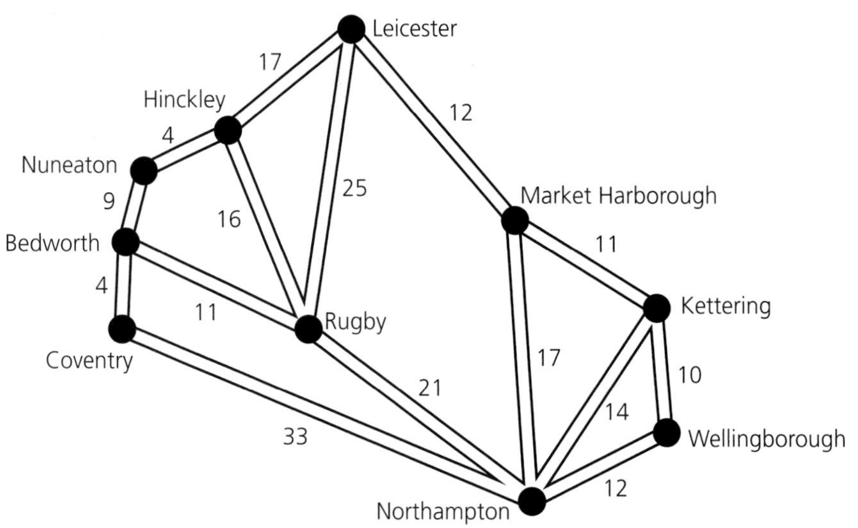

Finding paths

In Exploration 2.2 you probably investigated different routes through the network by means of a trial and error method. Shortest path problems involve finding paths with optimal properties, not necessarily to do with length. They may be **cost paths** (the cheapest way) or **risk paths** (the safest way). They are very important in industry where the shortest or least-cost route is required.

Now suppose that there are many vertices in your network; in question 3 of Exercises 2.1 classwork you will have discovered that as the number of vertices increases, the number of possible routes through a network increases faster. Looking for every possible path and finding its length – a method called **exhaustive search** – will certainly provide the correct solution (provided you do not make any mistakes or overlook any path) but it will be very time-consuming.

It seems reasonable to suppose that it would be possible to devise an algorithm to find the shortest path through a network more efficiently and which could be programmed into a computer to solve large problems. Such an algorithm was developed by E. W. Dijkstra in 1959.

Dijkstra's algorithm

Dijkstra's algorithm is an example of a **greedy** algorithm. Greedy algorithms are generally used to solve optimisation problems. They proceed in steps; at every step the algorithm chooses the best option *at that stage*. Once an item is included in the solution, it will not be changed; it never reconsiders its path, but it always provides an optimal solution if the type of problem is appropriate.

Dijkstra's algorithm finds the shortest route between two vertices in the network. At each stage a label is assigned to a vertex, showing the shortest distance from the starting point to that vertex. The label is either temporary or permanent, depending on whether we have considered all the ways of getting to that vertex. When the label is temporary, we simply write the value next to the vertex like this.

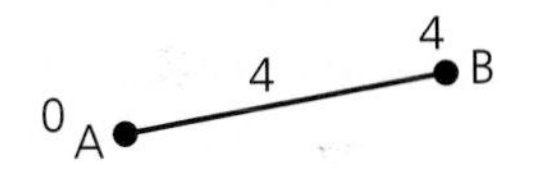

When we have decided that the label is permanent, we put a box round it giving the value and its position in the sequence of permanent labelling, which is also the number of iterations performed so far.

Recording the order in which we assign permanent labels to the vertices is an essential part of the algorithm.

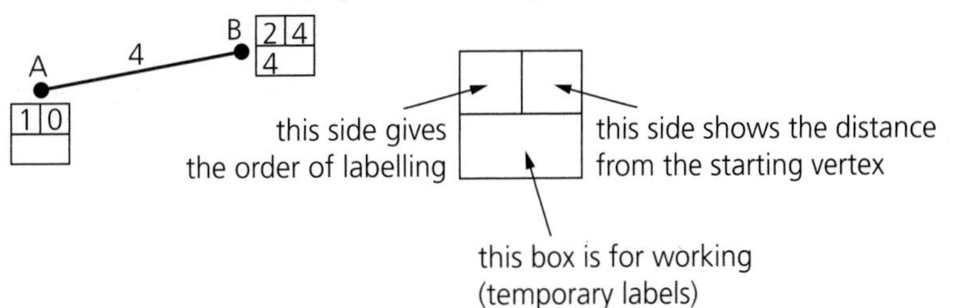

The algorithm gradually changes all the temporary labels into permanent ones. This can be demonstrated on this simple network.

Example 2.1

Find the shortest path from A to F.

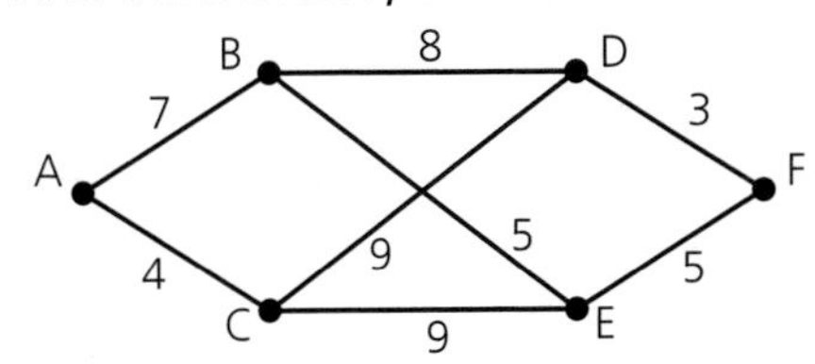

Solution

We start by giving a permanent label 0 to our starting vertex A. Next we consider the vertices which can be reached directly from A and give them temporary labels equal to their distances from A: so B is labelled 7 and C is labelled 4.

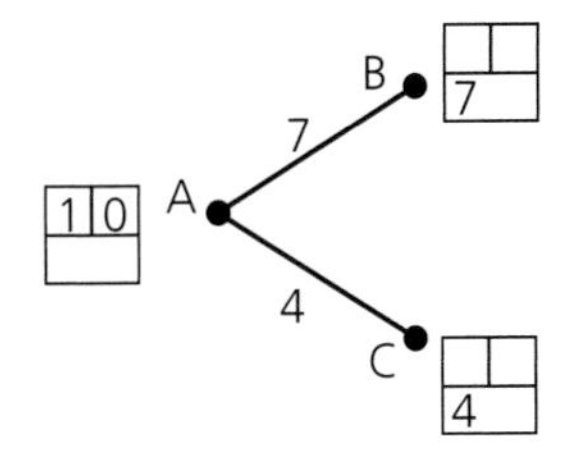

We now look at the vertices that have temporary labels and select the smallest. We assign a permanent label to it. In this case we assign a permanent label 4 to C.

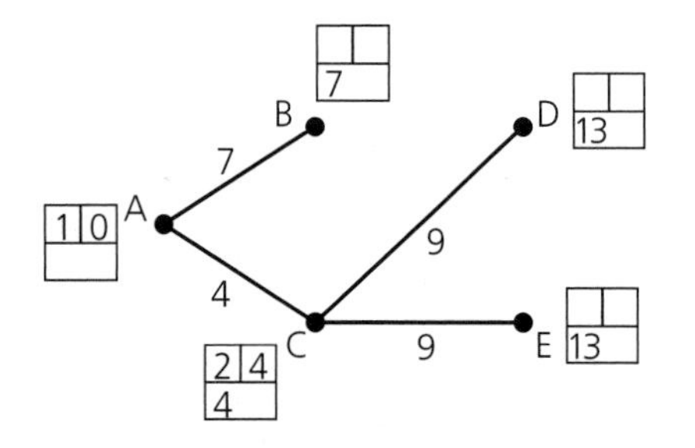

Now we consider all the vertices that are connected directly to our newly-labelled vertex C; they are D and E. We calculate the shortest distance from A to D via C by adding the permanent label at C to the distance from C to D.

$A \to C \to D = 4 + 9 = 13$ so we assign the temporary label 13 to D.

Similarly $A \to C \to E$ is found to be $4 + 9 = 13$ so we assign the temporary label 13 to E.

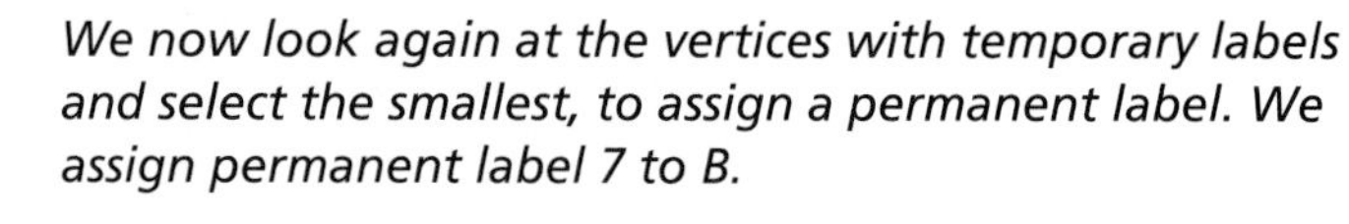

We now look again at the vertices with temporary labels and select the smallest, to assign a permanent label. We assign permanent label 7 to B.

Now we repeat the process, considering all the vertices that can be reached directly from B and give temporary labels to these vertices.

$A \to B \to D = 7 + 8 = 15$	*This is larger than the current temporary label at D so we do not change it.*
$A \to B \to E = 7 + 5 = 12$	*This is smaller than the current temporary label at E so we replace the 13 with 12.*

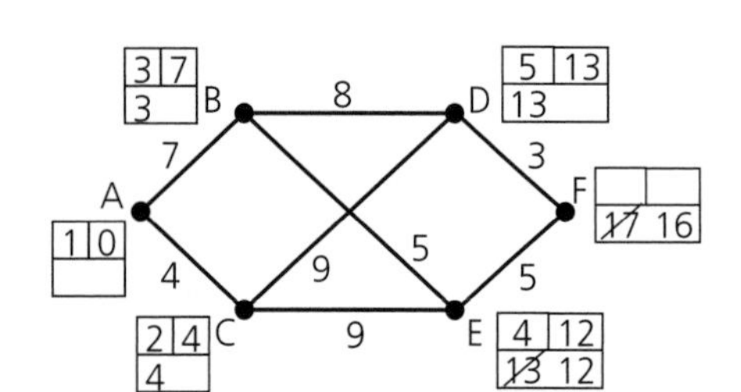

Our smallest temporary label is now 12 at E so we make this a permanent label and we give F a temporary label 17.

The fifth permanent label is then assigned to D with a value of 13. This will lead to the replacement of the temporary labels at F, since DF has weight 3, giving a total of $13 + 3 = 16$ which is smaller than the current temporary label.

Finally we assign permanent label 16 to F. Now all the vertices have permanent labels and we can see that the shortest distance from A to F is 16.

We find the shortest path by working backwards from F. The final iteration brought us to F from D. The value of 13 at D cannot have come from B (since $13 - 8 = 5$, not 7), so it must have come from C. In a similar way, we reached C from A so tracing back through the network: $F \to D \to C \to A$.

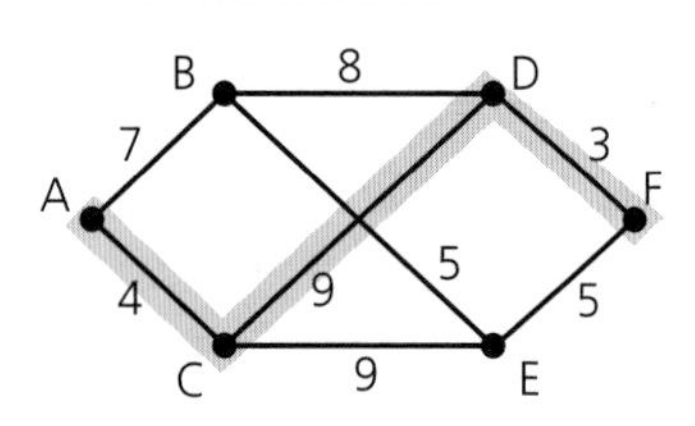

Hence the shortest path from A to F is: $A \to C \to D \to F$ with length 16.

We can now express Dijkstra's algorithm as a set of steps.

> ***Dijkstra's algorithm***
> **Step 1:** Assign the permanent label 0 to the starting vertex.
> **Step 2:** Assign temporary labels to all the vertices that are connected directly to the most-recently permanently labelled vertex.
> **Step 3:** Choose the vertex with the smallest temporary label and assign a permanent label to that vertex.
> **Step 4:** Repeat steps 2 and 3 until all vertices have permanent labels.
> **Step 5:** Find the shortest path by tracing back through the network.

We are now able to use Dijkstra's algorithm to solve the problem given in Exploration 2.2. Here is a solution.

It is now possible to get computer programs for a home PC which will calculate the shortest route between any 2 towns and will print out the route for you. Alternatively they will calculate the quickest route.

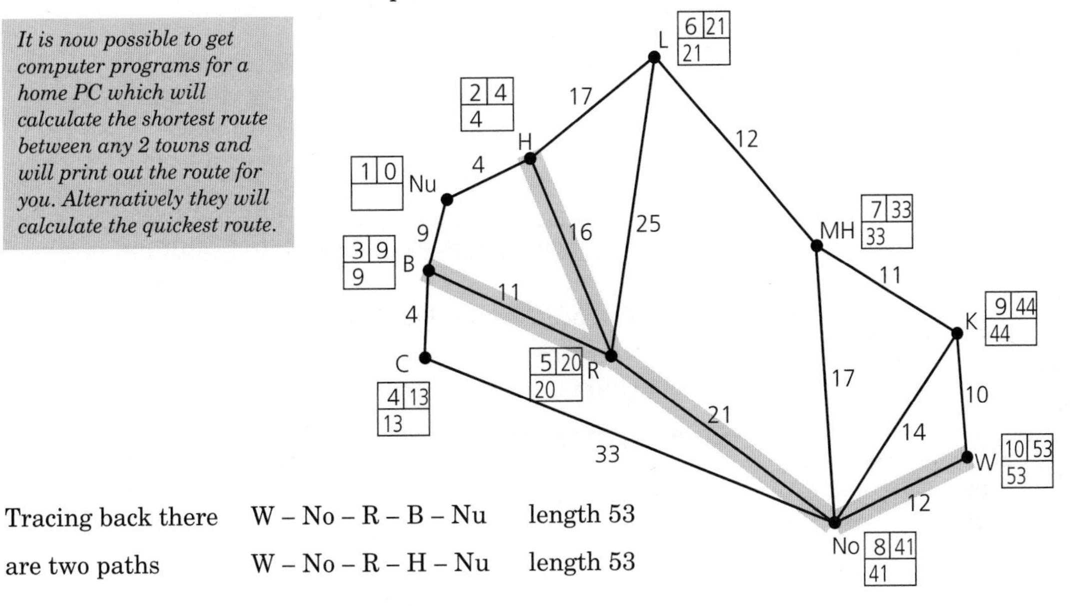

Tracing back there are two paths

W – No – R – B – Nu length 53

W – No – R – H – Nu length 53

Notice that there are two possible solutions, both with length 53 miles:

- Nuneaton, Bedworth, Rugby, Northampton, Wellingborough
- Nuneaton, Hinckley, Rugby, Northampton, Wellingborough.

EXERCISES

1 Find the shortest path from S to T through these networks.

a)

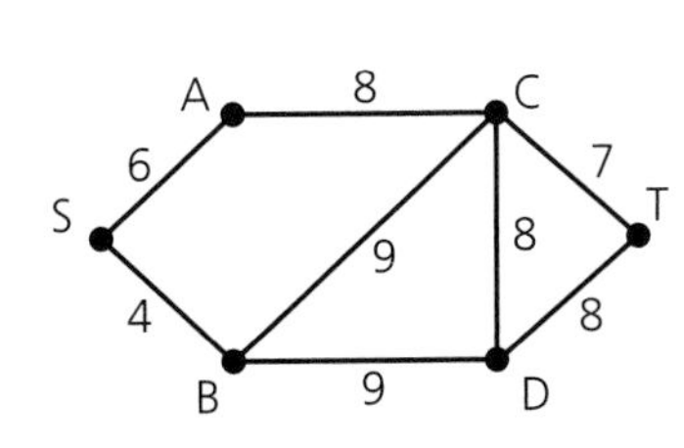

b)

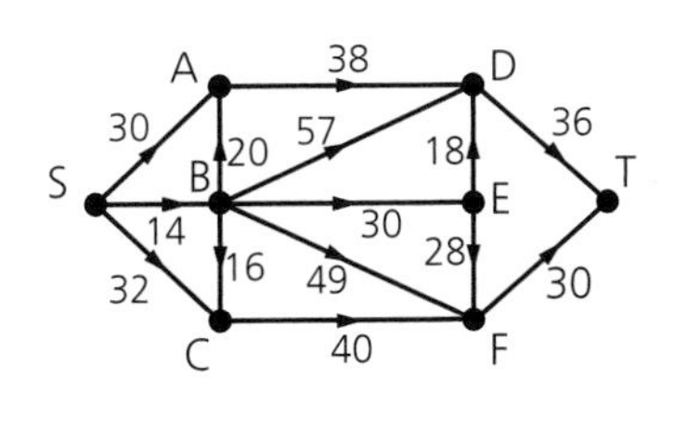

c)

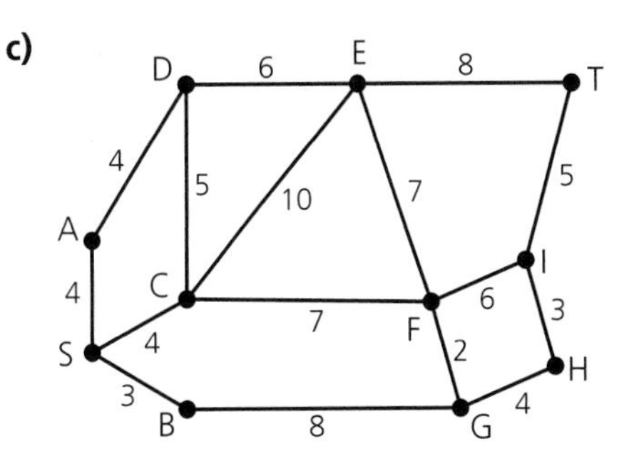

2 Maria travels from her home in Bickleigh to work in Exeter every day. What is her shortest route to work?

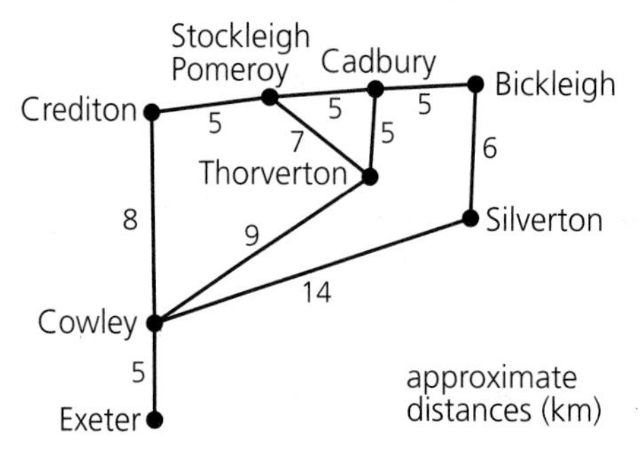

Roadworks cause the road between Thorverton and Cowley to be closed. What is her shortest route to work now?

3 A delivery company operates a service between certain towns in Kent. The charges per kilogram (in £s) for delivery are shown in this network.

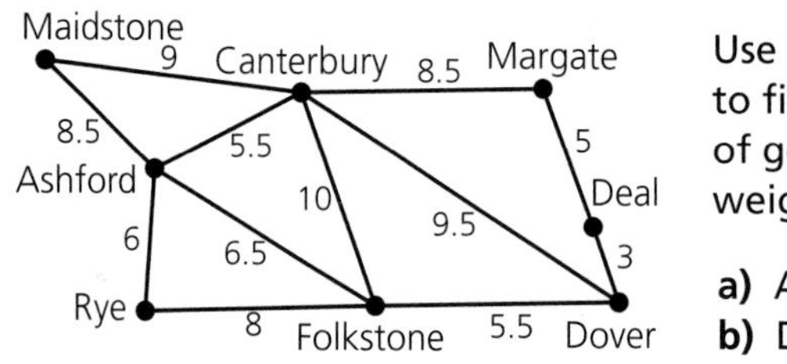

Use Dijkstra's algorithm to find the cheapest way of getting a parcel weighing 4 kg from:

a) Ashford to Dover

b) Deal to Maidstone.

4 Three fire stations are situated at the points marked P, Q and R on this network. A fire breaks out in a factory at X. Which of the fire stations should send their engines to attend the blaze?

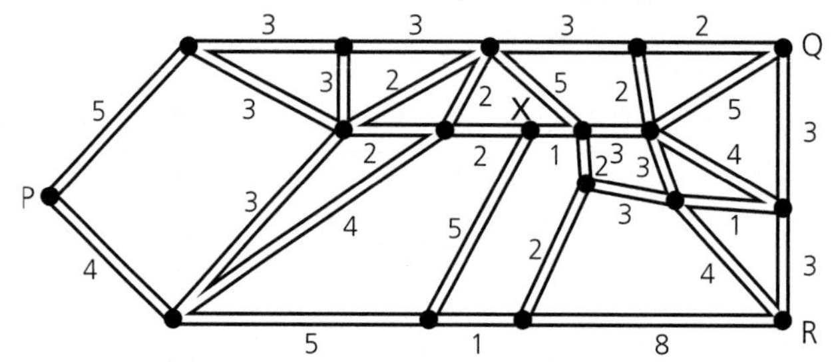

5 A sports centre is to be built that will be used mainly by people in several different villages. It is decided to site the centre in one of the villages. Use the network below to decide in which village it should be sited so as to minimise the average distance travelled by visitors to the centre.

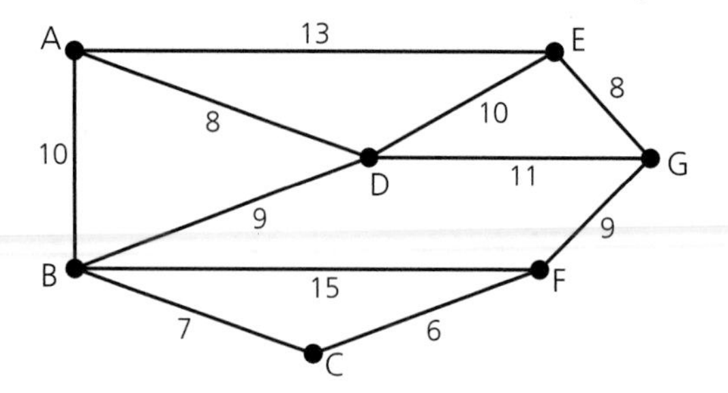

EXERCISES

2.2 HOMEWORK

1 Find the shortest path from S to T through the following networks.

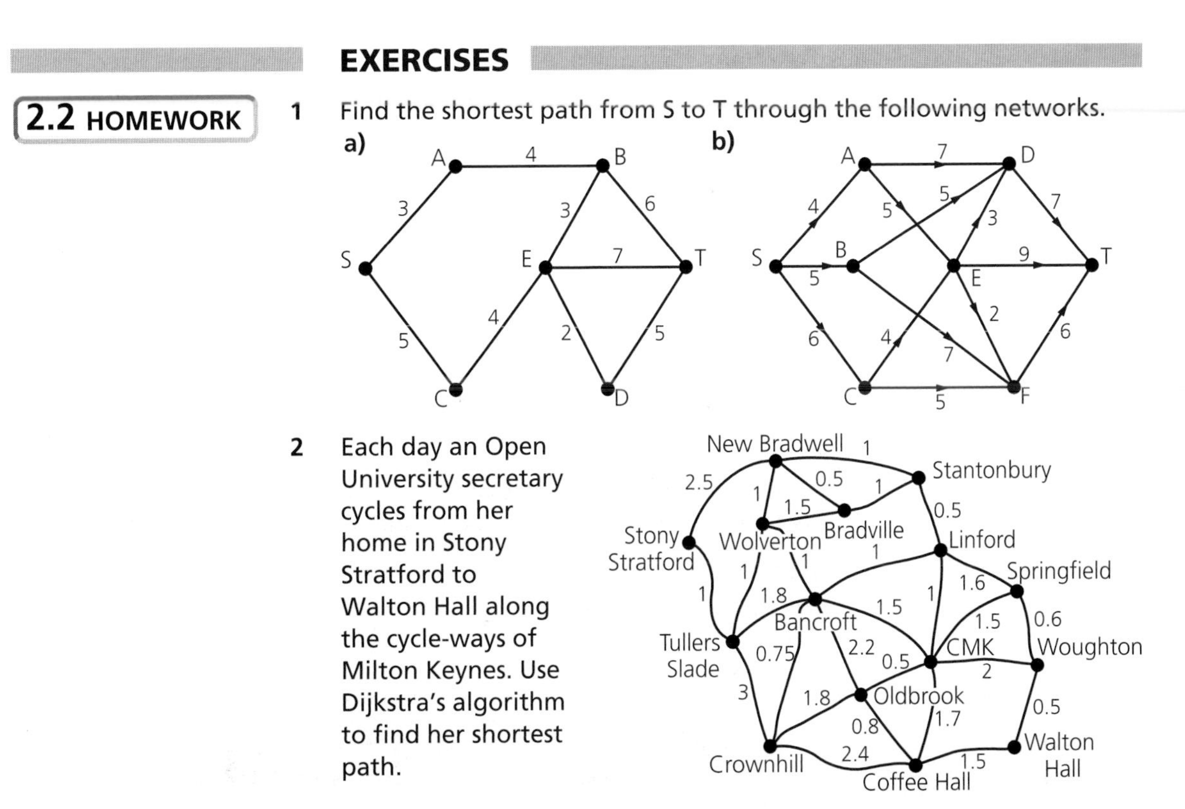

2 Each day an Open University secretary cycles from her home in Stony Stratford to Walton Hall along the cycle-ways of Milton Keynes. Use Dijkstra's algorithm to find her shortest path.

3 The table shows the distances in miles along major routes between cities where schools are using the same mathematics A-level scheme.

	Bristol	Oxford	London	Birmingham	Nottingham	Cambridge
Bristol	–	70	114	89	–	–
Oxford	70	–	56	63	95	55
London	114	56	–	112	160	55
Birmingham	89	63	112	–	49	102
Nottingham	–	95	160	49	–	84
Cambridge	–	55	55	102	84	–

a) Represent this information on a network graph.

b) Books for the scheme are being printed in Bristol. The school in Cambridge finds that it needs an extra set of books urgently. The printing company decide to send books by van.

i) Find the shortest path for the van to travel between Bristol and Cambridge.

ii) Just before leaving the driver is warned of a four-hour delay on the Oxford to Cambridge road. Does the driver need to change the route?

4 A delivery company operates a service between cities in Finland. The charges (in Finnish marks per kilogram) for delivery are shown in the following network.

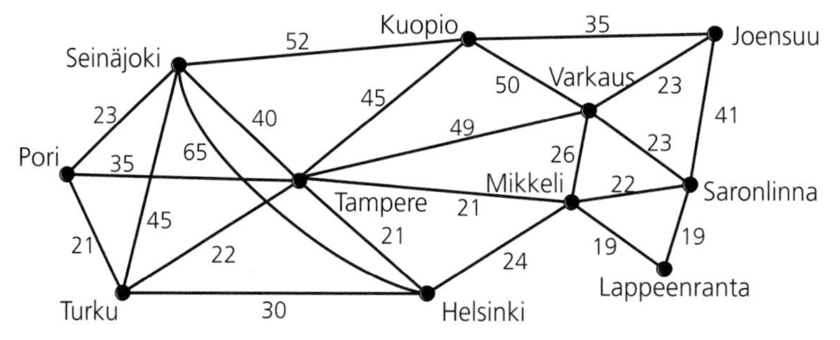

Use Dijkstra's algorithm to find the cheapest way of getting a parcel weighing 5 kg from:

a) Helsinki to Pori **b)** Turku to Joensuu.

5 A salesman in Gothenburg flies regularly using the same airline to the seven cities shown in the table, which also shows the fares (in pounds) for direct flights of the chosen airline between the cities.

	Gothenburg	Helsinki	Amsterdam	Copenhagen	Stockholm	Berlin	Frankfurt	Warsaw
Gothenburg	–	–	68	238	136	–	136	–
Helsinki	–	–	–	102	204	–	–	102
Amsterdam	68	–	–	–	–	–	34	408
Copenhagen	238	102	–	–	68	136	–	–
Stockholm	136	204	–	68	–	–	–	–
Berlin	–	–	–	136	–	–	170	136
Frankfurt	136	–	34	–	–	170	–	–
Warsaw	–	102	408	–	–	136	–	–

Represent this data by a network. Use Dijkstra's algorithm to find the cheapest route between Gothenburg and the other cities.

COMPLEXITY OF ALGORITHMS

In Chapter 1 we looked at the relative efficiency of several sorting algorithms. Another way of comparing efficiency is to find the complexity or order of the algorithm.

Let us consider Dijkstras' algorithm. The difficulty of solving the problem will depend on how many vertices and edges the network has. We will look at the worst case, where all networks are complete.

Consider a complete network with four vertices.

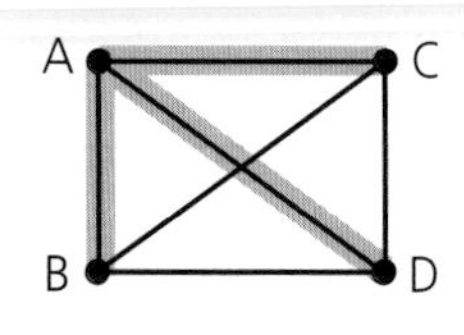

In the first iteration we consider three edges.

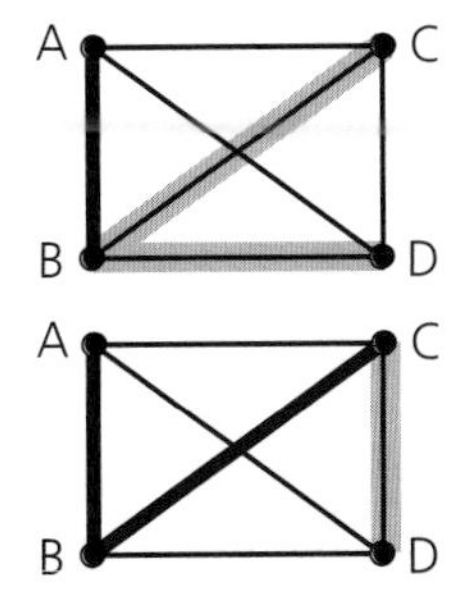

In the second iteration we consider a further two edges connected to the most recently labelled vertex. (B)

In the third iteration we consider the remaining edge.

The results for several sizes of complete network are shown in this table.

Vertices	Maximum number iterations
3	2 + 1 = 3
4	3 + 2 + 1 = 6
5	4 + 3 + 2 + 1 = 10

You will recognise the sequence that is generated as the triangle numbers. The nth term for this sequence is $\frac{1}{2}n\ (n - 1)$.

This is equivalent to $\frac{1}{2}n^2 - \frac{1}{2}n$ which has order 2 which we can write as $O(n^2)$.

Thus we can say that Dijkstra has **quadratic complexity**, implying that if we double the number of vertices it will take four times as much effort to solve the problem.

COURSEWORK INVESTIGATION

Investigation

Fire escapes

In a school, or other public building, there are notices in all the rooms directing people to the nearest fire exit. By considering the plan of such a building, decide on the fire exits which should be used by people in various parts of the building. What other factors need to be considered (width of corridors, other obstructions etc.)? Is shortest distance the most appropriate way to devise a building evacuation plan?

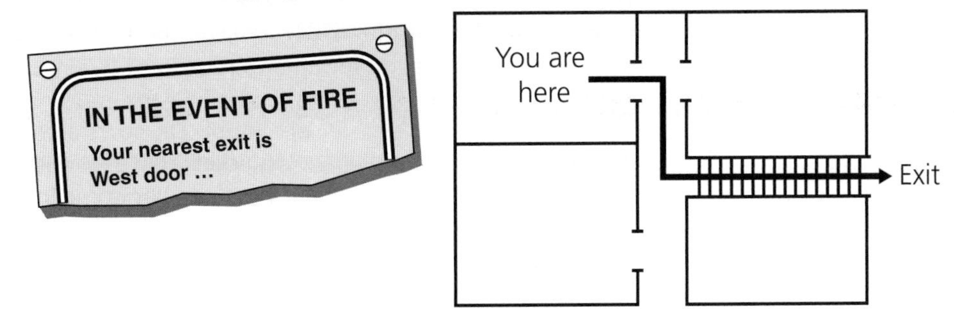

CONSOLIDATION EXERCISES FOR CHAPTER 2

1 G_1 is a connected graph with four vertices and no loops (i.e. every edge connects together two separate vertices, and for any two vertices there is a set of edges forming a path from one vertex to the other).

a) What is the least number of edges that G_1 could have?

b) What is the greatest number of edges that G_1 could have?

G_2 is a connected graph with n vertices and no loops.

c) What is the least number of edges that G_2 could have?

d) What is the greatest number of edges that G_2 could have?

(OCR MEI Specimen Paper 2000 D & D1)

2 Use Djikstra's algorithm to find the shortest distance, and the corresponding shortest path from B to F in the network below. Ensure that you show clearly the steps of the algorithm. Stating the correct answer without showing how you achieved it is not sufficient.

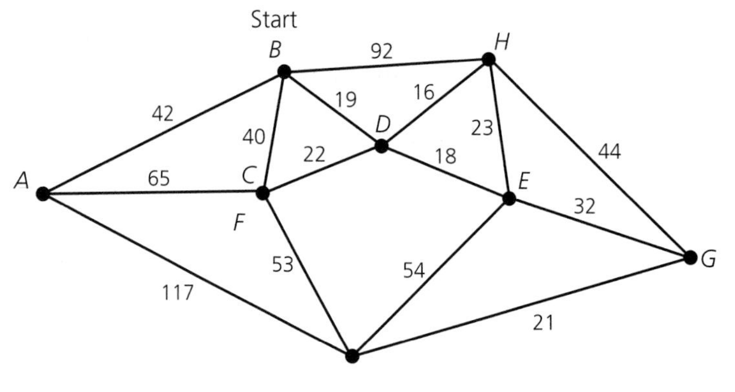

(AQA B Specimen Paper 2000 D1)

3 This network models possible air routes between S and T via certain intermediate cities. The numbers on the arcs indicate the fares in units of £100. The journeys can only be made in the direction indicated. Use the shortest path algorithm to find the route between S and T for which the total fare is a minimum. Your solution should indicate clearly the order in which the nodes (cities) receive their permanent labels.

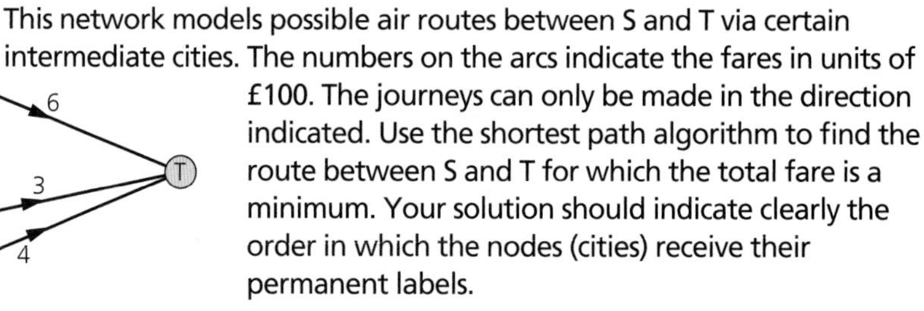

(ULEAC Question 5, Decision Mathematics D1 Specimen Paper)

4 The network shows the road connections between a group of villages. The numbers on the arcs represent the times in minutes that a driver takes to travel along the roads represented by those arcs. The network is set up ready for an application of Dijkstra's algorithm.

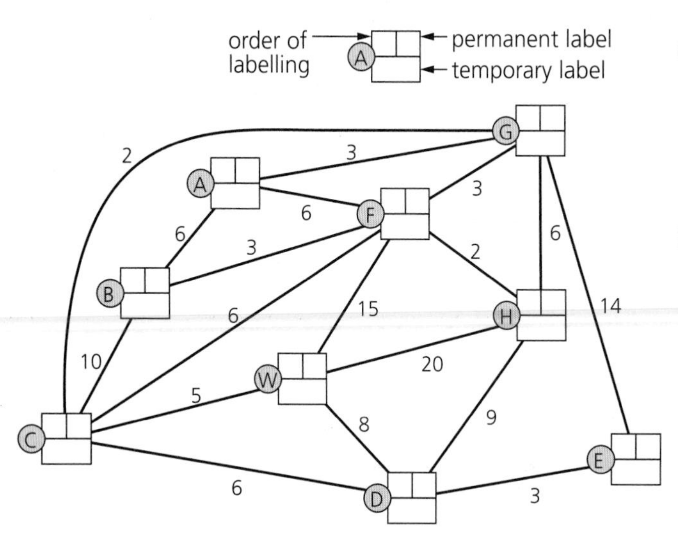

a) Apply Dijkstra's algorithm to find the quickest route for the driver to take between her home in village H and her workplace in village W. Write your temporary labels and permanent labels in the boxes provided.

b) As the driver is about to leave H she hears on the radio that an accident in the centre of village F is causing a delay of 10 minutes to all traffic through that village. Explain briefly how you would find her quickest route under these circumstances. In particular say how you would modify the network in part (a).

(O & C MEI Question 2, Decision & Discrete Mathematics Specimen Paper)

5 A graph has five vertices P, Q, R, S, T, and each vertex is directly connected to every other vertex. Describe how to apply Dijkstra's algorithm to find the shortest path from P to T, and explain why this requires six addition calculations in the worst case.

Show that when Dijkstra's algorithm is used on a graph with six vertices it requires ten addition calculations in the worst case.

The number of additions affects the amount of time that Dijkstra's algorithm takes to run on a computer.

a) Assuming that the problem has already been put into a suitable format, what is the other main factor that would affect the time that Dijkstra's algorithm takes to run on a computer?
b) Dijkstra's algorithm is of quadratic order (order n^2). Explain what this means.

(OCR Specimen Paper 2000 D1)

6

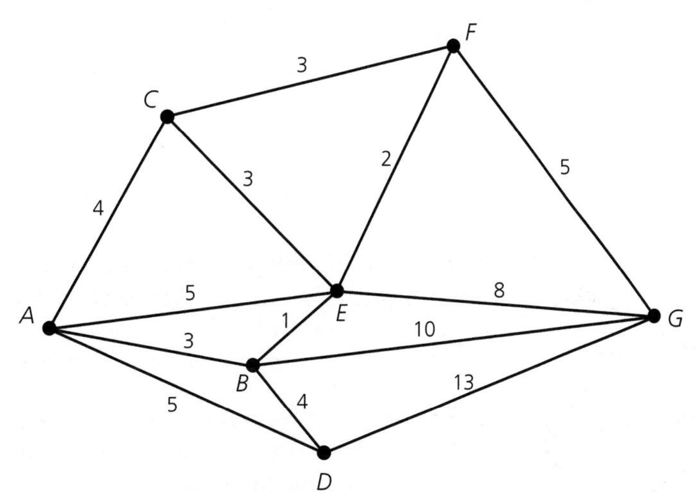

a) Copy the figure and show fully the use of Dijkstra's algorithm to find the shortest path through the graph from *A* to *G*. Display the path clearly and state its length.
b) It can be shown that a trial and improvement solution to the shortest path problem could take more than $(n-2)!$ computations for a network with n vertices. Using Dijkstra's algorithm requires a maximum of $\frac{1}{2}(5n^2 - 3n) + 1$ computations. Evaluate these expressions when $n = 7$ and $n = 70$, and comment on these values.

(AQA A specimen Paper 2000 D1)

7 **a)** An express delivery company operates its service between certain towns in Kent. The charges per kilogram (in £s) are shown on the network.

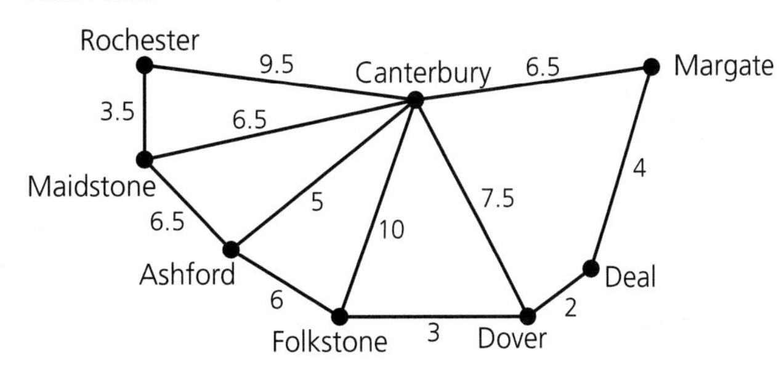

Use Dijkstra's algorithm to find the cheapest way of sending a 4 kg package from:
i) Maidstone to Dover **ii)** Deal to Rochester.

b) A second company sets up in opposition. It can undercut the first company by £1 per kg on the routes between Rochester and Canterbury, Canterbury and Dover, Maidstone and Ashford, and Ashford and Folkestone, but charges 50p more on all the others. Draw the network for the new company. Would it be cheaper to use this company for the packages in part **a)** and if so, would the routes be the same?

8 The diagram shows a small network in which the number on each arc represents the cost of travelling along that arc in the directions indicated.

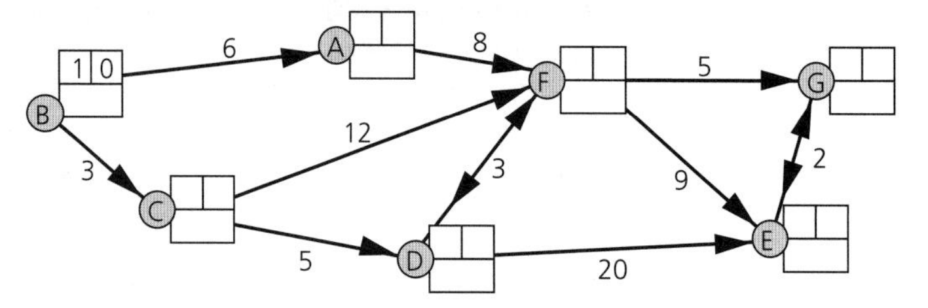

a) Apply Dijkstra's algorithm to find the smallest cost, and the associated route, to get from B to E. (Record the order in which you permanently label nodes, and write your temporary labels and permanent labels, in the boxes provided.)

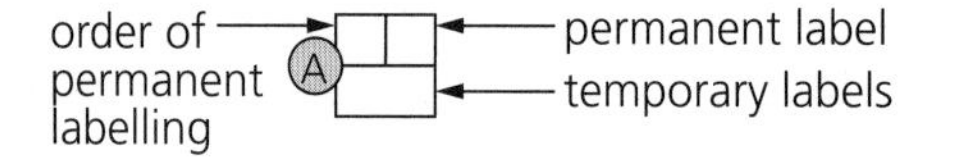

Smallest cost from B to E ...... Cheapest route from B to E ......
Show briefly how the route is found in Dijkstra's algorithm, once the destination vertex has been permanently labelled.

b) Suppose now that a new route is introduced from A to C, and that a profit of 4 may be made by making a delivery from A to C when using that route. This is shown as a negative cost in part of the network reproduced below.

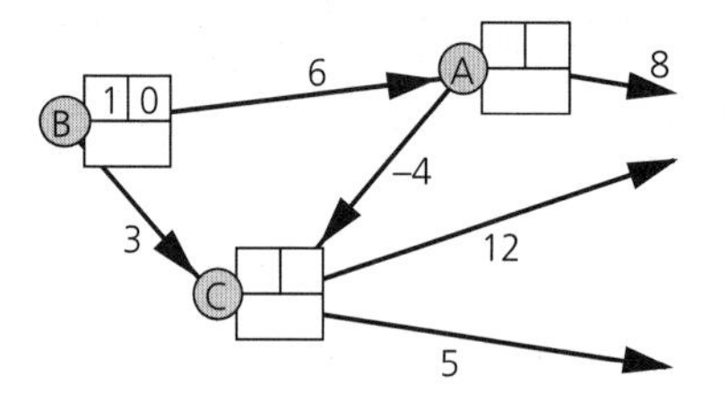

Work through Dijkstra's algorithm for this part of the network with the extra route. Say what is the best route from B to C, and explain why Dijkstra's algorithm fails to find it.

(O & C MEI Question 2, Decision Mathematics, January 1995)

9 The network represents an electronic information network. The vertices represent computer installations. The arcs represent direct links between installations. The weights on the arcs are the rates at which the links can transfer data (in units of 100 000 bits per second). A chain of links may be established between any two installations via any other installations. For such links the rate at which the chain can transfer data is given by the speed of the slowest link in the chain.

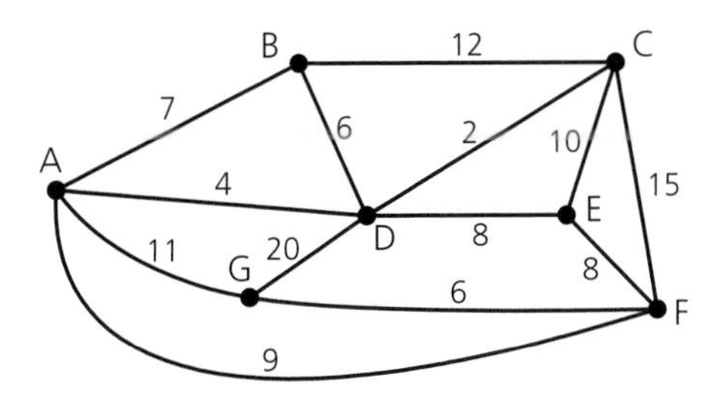

a) i) The fastest chain linking G to B is needed. Devise a way of adapting Dijkstra's algorithm to produce an algorithm suited to this task. Give your adaptations and explain how your algorithm works.

ii) Apply your algorithm, showing all of your steps on a copy of the network.

iii) State the fastest chain and its speed of transferring data.

b) How could you find a slowest chain linking G to B? Give such a chain.

(UODLE Question 22, Paper 4, June 1995)

10 A mathematician is writing a chapter of a book. In this chapter she wishes to start from result A which was proved in a earlier chapter, to prove results B, C, D and E. There are many ways in which she can do this, since she can prove some results from others. She would like to do this as efficiently as possible, so that the total number of lines of proof is as small as possible.
The number of lines of proof required to establish one result directly from another are given in the following table.

		Results to be proved			
		B	C	D	E
Results proved	A	10	9	–	15
	B	–	15	20	7
	C	15	–	11	12
	D	20	11	–	5
	E	7	12	5	–

For example, she can prove B from A in 10 lines.

a) Of the two graphs shown below, one represents a possible solution, the other does not. State, with a reason, which does not represent a solution, and give the number of lines of proof required for the other.

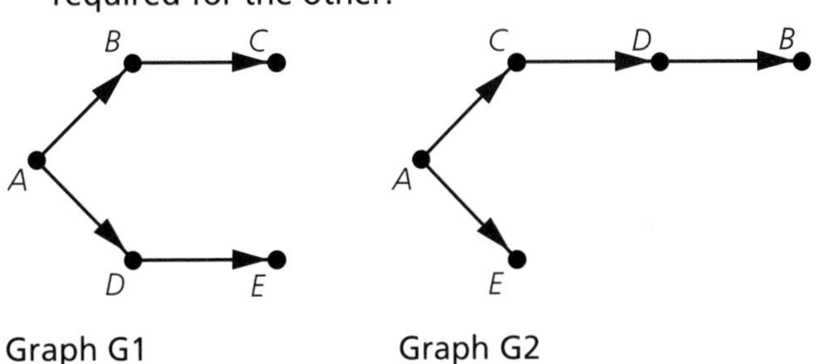

Graph G1 Graph G2

b) Use the following algorithm to find an efficient way for the mathematician to prove all the results B, C, D and E from A. Draw the graph representing your solution and give the total number of lines of proof.

Step 1 Start with A proved.

Step 2 Prove the result which can be proved in the minimum number of lines from a proved result.

Step 3 Repeat step 2 until all results are proved.

c) Suppose that the mathematician subsequently discovers that she can prove C from D in only 8 instead of 11 lines. (It still takes 11 lines to prove D from C, and all other figures are similarly unaffected.)

What difference does this make to the solution?

In this case, does the algorithm succeed in finding the best solution in the corresponding digraph? Justify your answer.

(AQA B Specimen Paper 2000 D1)

Summary

After working through this chapter you should:

- know that:
  a graph is a diagram showing how objects are related to each other
  a vertex or node is a point representing an object in a graph
  an edge is a line joining two vertices, without direction
  an arc is a line joining two vertices in a given direction
  a network is a graph in which each edge (or arc) is given a value or weight
- know that:
  a path is a route through the graph
  a connected graph is one in which there is at least one route through
  a complete graph is one in which each vertex is connected to every other vertex
- be able to draw a graph from its adjacency matrix and a network from its cost matrix
- understand how networks can be used to model problems
- use Dijkstra's algorithm to find the shortest path through a network, and recognise that algorithms that are used to find optimal paths are called greedy algorithms

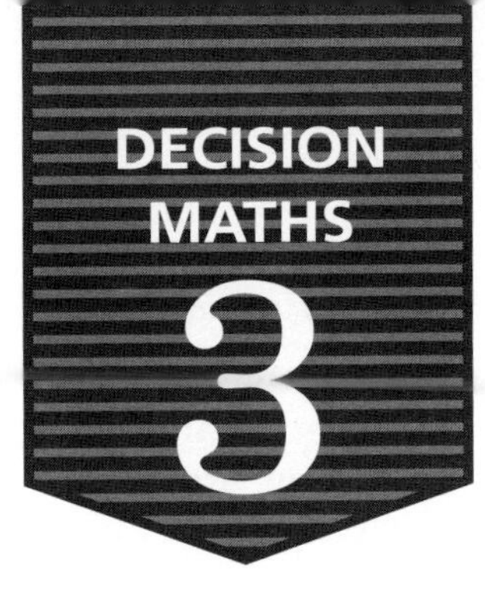

Graphs and networks 2: Minimum-connector

By the end of this chapter you should be familiar with:

- *a spanning tree, which connects all the vertices in a graph with no cycles*
- *the minimum connector, which is the spanning tree of minimum length*

Exploration 3.1

Cable TV: a minimum connector problem

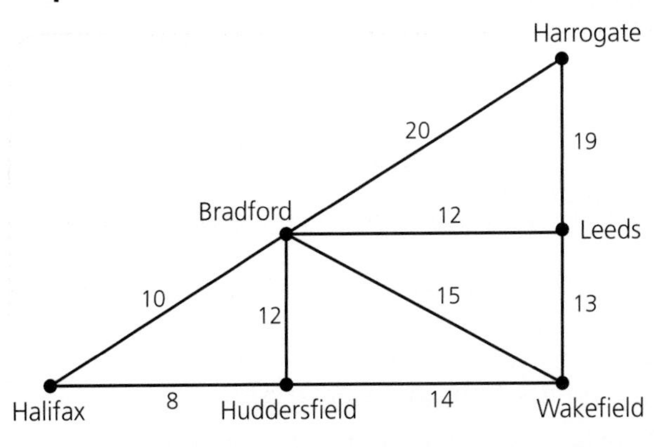

A cable TV company is installing a system of cables to connect all the towns in a region. The numbers in the network show distances in miles.

- Find a layout that will connect all the towns in the region.
- Find a different layout which uses less cable.
- What is the least amount of cable needed?

GRAPH THEORY – TREES

A connected graph which contains no cycles is called a **tree**.

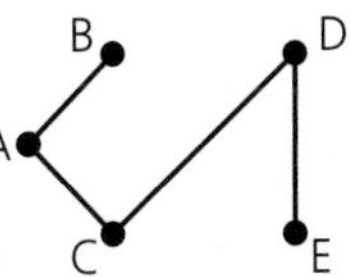

A **spanning tree** is a subgraph that includes all the vertices in the original graph, and is also a tree. A graph will have several trees.

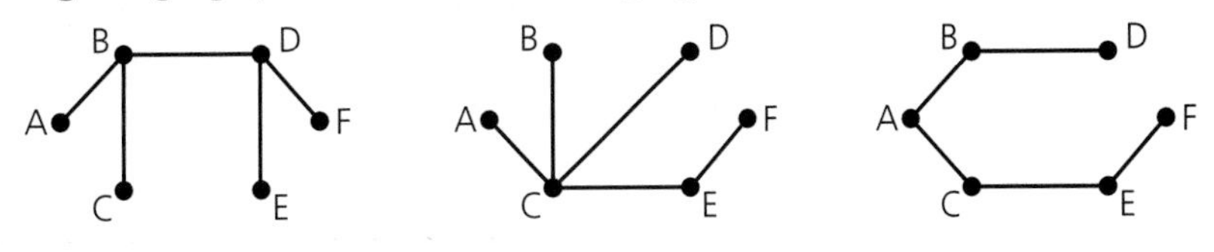

These are all spanning trees of the path ABCDECA above. There are others, can you find them? Confirm that, for a network of n vertices, there will be $n - 1$ edges in the minimum spanning tree.

THE MINIMUM-CONNECTOR PROBLEM

Let us return to the problem in Exploration 3.1. We can now see that to solve it, we need to find the spanning tree of shortest length for the network. This type of problem is a **minimum connector problem**. There are several algorithms for finding the minimum connector for a network; we shall look at two of these.

Kruskal's algorithm

J.B. Kruskal Jr published the algorithm in 1956 in the ***American Mathematical Society proceedings****.*

Kruskal's algorithm is a greedy algorithm, it chooses the smallest available edges, not worrying about any connection to edges that have already been chosen, except that it is careful not to form a cycle. The algorithm can be stated like this.

Step 1: Rank the edges in order of length.
Step 2: Select the shortest edge in the network.
Step 3: Select, from the edges which are not in the solution, the shortest edge which does not form a cycle. (Where two edges have the same weight, select at random.)
Step 4: Repeat Step 3 until all the vertices are in the solution.

Example 3.1

Use Kruskal's algorithm to find the least amount of cable needed to solve the problem in Exploration 3.1.

Solution
First, rank the edges in order of length.

Order of connection				
	1	*Halifax to Huddersfield*	*8*	
	2	*Halifax to Bradford*	*10*	
	3	*Bradford to Leeds*	*12*	*(As these two edges have the same length, the order is arbitrary.)*
This edge forms a loop	→	*Bradford to Huddersfield*	*12*	
	4	*Leeds to Wakefield*	*13*	
This edge forms a loop	→	*Huddersfield to Wakefield*	*14*	
This edge forms a loop	→	*Bradford to Wakefield*	*15*	
	5	*Leeds to Harrogate*	*19*	
This edge forms a loop	→	*Bradford to Harrogate*	*20*	

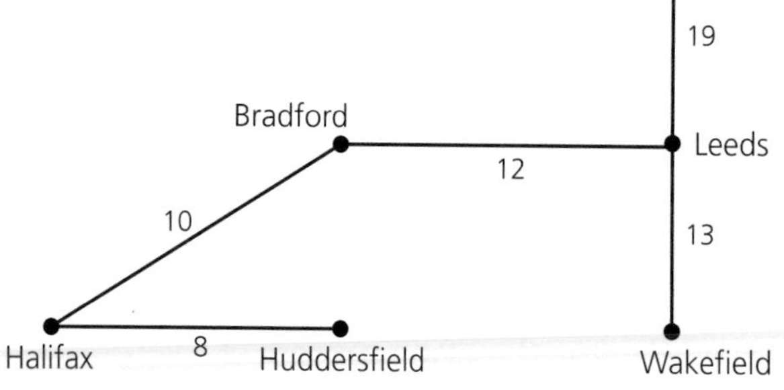

Now we begin to select edges, starting with the smallest.
We have now connected all the vertices into the spanning tree.
The length of our minimum spanning tree is $8 + 10 + 12 + 13 + 19 = 62$ *miles.*

It is not necessary to redraw the network at every iteration. The solution can be shown on one diagram provided you note the order in which the edges were connected into the spanning tree.

Prim's algorithm

Prim's algorithm can also be done onto a network.

When programming a network into a computer, it is generally useful to adopt a matrix format. In Chapter 2, *Networks 1: Shortest path*, we used the idea of an adjacency matrix to describe graphs. It is also possible to describe a network using **distance matrices**. The distance matrix for example 3.1 would look like this.

	L	H	W	B	Ha	Hu
L	∞	19	13	12	∞	∞
H	19	∞	∞	20	∞	∞
W	13	∞	∞	15	∞	14
B	12	20	15	∞	10	12
Ha	∞	∞	∞	10	∞	8
Hu	∞	∞	14	12	8	∞

The symbol ∞ means that there is no direct connection between these vertices. If we use ∞ (of infinite lengths) it ensures these edges will not feature in the solution.

R.C. Prim published this algorithm in the ***Bell System Technical Journal*** *in 1957.*

Prim's algorithm works from a starting point and builds up the spanning tree step by step, connecting edges into the existing solution. It can be applied directly to the distance matrix, as well as to the network itself, so it is more suitable for using with a computer if the network is large. The algorithm can be stated follows.

Step 1: Choose a starting vertex.
Step 2: Join this vertex to the nearest vertex directly connected to it.
Step 3: Join the next nearest vertex, not already in the solution, to any vertex in the solution, provided it does not form a cycle.
Step 4: Repeat until all vertices have been included.

We now apply the algorithm to the distance matrix above, with the solution taken straight from the graph shown alongside.

Matrix method

Network method

Choose a starting vertex, say L. Delete row L. Look for the smallest entry in column L.

	L (1 ↓)	H	W	B	Ha	Hu
~~L~~	~~∞~~	~~19~~	~~13~~	~~12~~	~~∞~~	~~∞~~
H	19	∞	∞	20	∞	∞
W	13	∞	∞	15	∞	14
B	(12)	20	15	∞	10	12
Ha	∞	∞	∞	10	∞	8
Hu	∞	∞	14	12	8	∞

● L

LB is the smallest edge joining L to the other vertices. Put edge LB into the solution. Delete row B. Look for the smallest entry in columns L and B.

	L (1 ↓)	H	W	B (2 ↓)	Ha	Hu
H	19	∞	∞	20	∞	∞
W	13	∞	∞	15	∞	14
~~B~~	~~12~~	~~20~~	~~15~~	~~∞~~	~~10~~	~~12~~
Ha	∞	∞	∞	(10)	∞	8
Hu	∞	∞	4	12	8	∞

BHa is the smallest edge joining L and B to the other vertices. Put edge BHa into the solution. Delete row Ha. Look for the smallest entry in columns L, B and Ha.

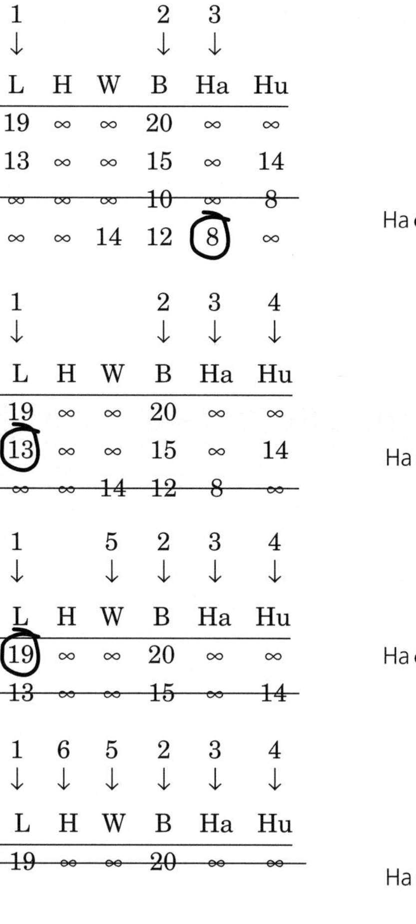

	1 ↓			2 ↓	3 ↓	
	L	H	W	B	Ha	Hu
H	19	∞	∞	20	∞	∞
W	13	∞	∞	15	∞	14
~~Ha~~	~~∞~~	~~∞~~	~~∞~~	~~10~~	~~∞~~	~~8~~
Hu	∞	∞	14	12	(8)	∞

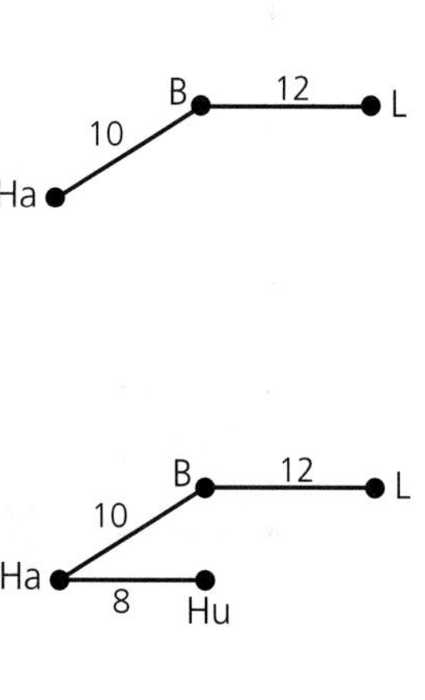

HaHu is the smallest edge joining L, B and Ha to the other vertices. Put edge HaHu into the solution. Delete row Hu. Look for the smallest entry in columns L, B, Ha and Hu.

	1 ↓			2 ↓	3 ↓	4 ↓
	L	H	W	B	Ha	Hu
H	19	∞	∞	20	∞	∞
W	(13)	∞	∞	15	∞	14
~~Hu~~	~~∞~~	~~∞~~	~~14~~	~~12~~	~~8~~	~~∞~~

LW is the smallest edge joining L, B, Ha and Hu to the other vertices. Put edge LW into the solution. Delete row W. Look for the smallest entry in columns L, B, Ha, Hu and W.

	1 ↓		5 ↓	2 ↓	3 ↓	4 ↓
	L	H	W	B	Ha	Hu
H	(19)	∞	∞	20	∞	∞
~~W~~	~~13~~	~~∞~~	~~∞~~	~~15~~	~~∞~~	~~14~~

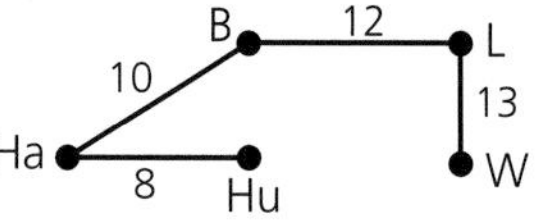

LH is the smallest edge joining L, B, Ha, Hu and W to the other vertices. Put LH into the solution.

	1 ↓	6 ↓	5 ↓	2 ↓	3 ↓	4 ↓
	L	H	W	B	Ha	Hu
~~H~~	~~19~~	~~∞~~	~~∞~~	~~20~~	~~∞~~	~~∞~~

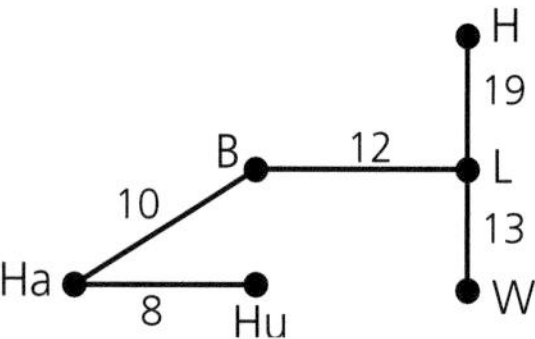

We have now connected all the vertices into the spanning tree. Can you see that we have arrived at the same solution as with Kruskal's algorithm?

In reality, we would not rewrite the matrix at every iteration, but would work from one matrix, deleting the lines as we progress through the algorithm, and making a record of the order in which we connected the vertices along the top of the matrix, so the solution would look like this.

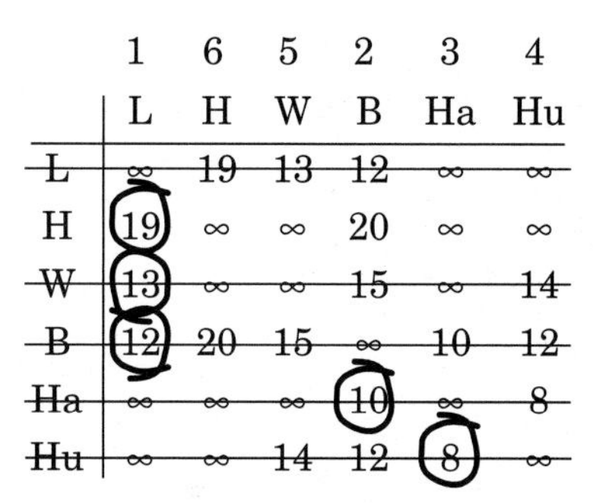

	1	6	5	2	3	4
	L	H	W	B	Ha	Hu
~~L~~	~~∞~~	~~19~~	~~13~~	~~12~~	~~∞~~	~~∞~~
H	(19)	∞	∞	20	∞	∞
~~W~~	~~(13)~~	~~∞~~	~~∞~~	~~15~~	~~∞~~	~~14~~
~~B~~	~~(12)~~	~~20~~	~~15~~	~~∞~~	~~10~~	~~12~~
~~Ha~~	~~∞~~	~~∞~~	~~∞~~	~~(10)~~	~~∞~~	~~8~~
~~Hu~~	~~∞~~	~~∞~~	~~14~~	~~12~~	~~(8)~~	~~∞~~

length 19 + 13 + 12 + 10 + 8 = 62 miles

COMPLEXITY OF MINIMUM CONNECTOR ALGORITHMS

Kruskal's algorithm

The first step is to find the shortest edge, then the second shortest, and so on until the tree is complete. We can consider the amount of effort to be proportional to the number of comparisons made, as this example shows.

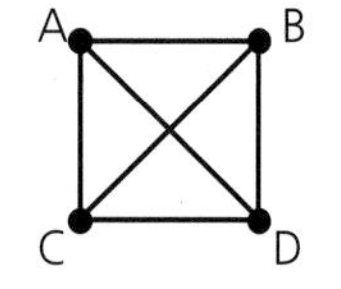

Compare AB, AC, AD, BC, BD and CD (five comparisons) and select the smallest.

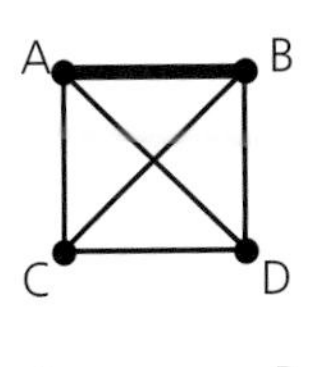

Compare the remaining five edges (four comparisons) and select the smallest.

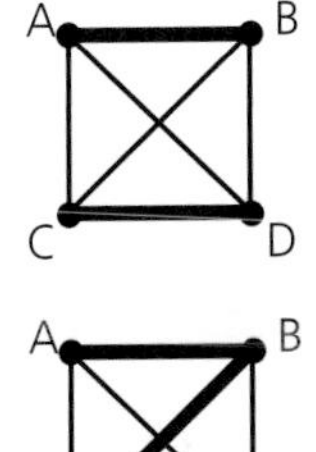

Compare the remaining four edges (three comparisons) and select the smallest.

The spanning tree is now complete as it contains three edges. The results for several sizes of complete network are shown in this table

Vertices	Comparisons
3	$2 + 1 = 3$
4	$5 + 4 + 3 = 12$
5	$9 + 8 + 7 + 6 = 30$
6	$14 + 13 + 12 + 11 + 10 = 60$

We can generalise by saying that a complete network with n vertices will have $\frac{1}{2}n(n-1)$ edges in total and the spanning tree will contain $(n-1)$ edges. The first step will therefore involve $[\frac{1}{2}n(n-1)-1]$ comparisons, the next step will involve one fewer edge (since one edge has already been placed in the spanning tree), so the number of comparisons will be $[\frac{1}{2}n(n-1)-2]$ and so on until the tree has $(n-1)$ edges. Hence we obtain the sequence:

$$[\tfrac{1}{2}n(n-1)-1] + [\tfrac{1}{2}n(n-1)-2] + [\tfrac{1}{2}n(n-1)-3] + \ldots + [\tfrac{1}{2}n(n-1)-(n-1)]$$

The sum of this sequence is $\frac{1}{2}n(n-1)(n-2)$ which is a cubic expression, so the algorithm has order 3 or $O(n^3)$. This means that doubling the number of vertices requires 8 times as much effort to solve the problem.

You can find the general term of the sequence 3, 12, 30, 60, ... using the method of differences and working out the coefficients by using simultaneous equations. There are programmes for solving sets of equations on most graphic calculators. Check this formula to see whether it is correct.

Prim's algorithm

If we treat Prim's algorithm in the same way, we can consider the number of comparisons like this:

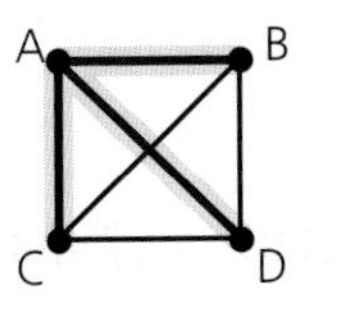

Each vertex has three free edges connected to it, so there are two comparisons to make. If we start at A, we must compare AB, AC and AD.

Each of the two connected vertices has two free edges connected to it, so there are three comparisons to make in total.

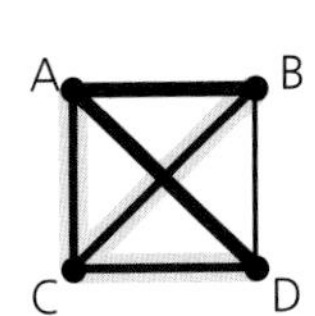

Each of the three connected vertices has one free edge connected to it, so there are two comparisons to make in total.

The results for networks of different sizes are summarised in this table.

Vertices	Comparisons
3	[2 – 1] + [2(1) – 1] = 2
4	[3 – 1] + [2(2) – 1] + [3(1) – 1] = 7
5	[4 – 1] + [2(3) – 1] + [3(2) –1] + [4(1) – 1] = 16
6	[5 – 1] + [2(4) – 1] + [3(3) –1] + [4(2) – 1] + [5(1) – 1] = 30
n	$(n - 2) + [2(n - 2) - 1] + [3(n - 3) -1] + ...+ [(n - 1) - 1]$

You could check the result for Prim in the same way as we did for the result for Kruskal.

The sum of this sequence is $\frac{1}{2}n(n - 1)(n - 2)$ which is also cubic, so Prim's algorithm has order 3 or $O(n^3)$.

Both of these solutions assume the worst case where a network is complete. They also take no account of the possibility of a loop occurring during the process; this would make the problem considerably more difficult to solve.

EXERCISES

3.1 CLASSWORK

1 Use Kruskal's algorithm to find the minimum spanning tree for the network on the right.

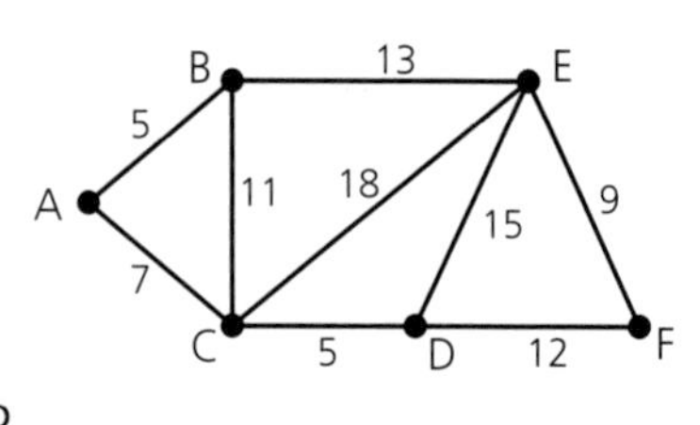

2 Use Prim's algorithm to find the minimum spanning tree for the network on the right. (You will need to put the information into a table first.)

3 The table below gives the distances in miles between some cities in the north of England and Scotland. Use Prim's algorithm to find the shortest route connecting them.

	Aberdeen	**Carlisle**	**Edinburgh**	**Glasgow**	**Inverness**
Aberdeen	∞	221	125	145	105
Carlisle	221	∞	100	95	263
Edinburgh	125	100	∞	40	160
Glasgow	145	95	40	∞	168
Inverness	105	263	160	168	∞

From the information in the table, draw a network and use Kruskal's algorithm to find the shortest route connecting the cities. Comment on the relative efficiency of the two algorithms for a network of this size. Do you think your conclusion would be different if the network had twice as many vertices?

4 This distance chart gives the distances, in miles, between towns in Ireland. Adapt the matrix so that you can use Prim's algorithm to find a minimum spanning tree for these towns.

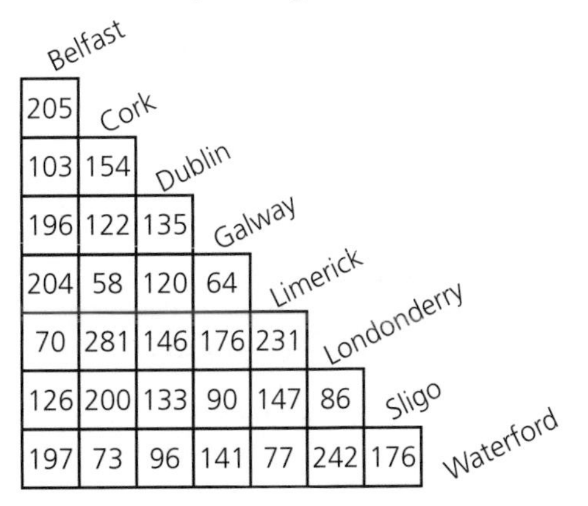

Belfast							
205	Cork						
103	154	Dublin					
196	122	135	Galway				
204	58	120	64	Limerick			
70	281	146	176	231	Londonderry		
126	200	133	90	147	86	Sligo	
197	73	96	141	77	242	176	Waterford

5 Using the information in the distance chart above, construct a network and use it to find the minimum spanning tree using Kruskal's algorithm.

EXERCISES

3.1 HOMEWORK

1 **a)** Use Kruskal's algorithm to find and draw the minimum spanning tree for this network.

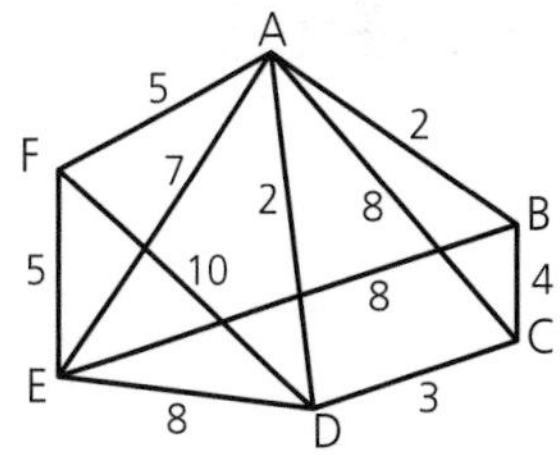

b) Put the information in the network in a table and use Prim's algorithm to verify your answer to part **a)**.

c) Discuss the differences between Kruskal's algorithm and Prim's algorithm. Which method do you think would be faster to use on a computer?

2 To generate a maximum spanning tree i.e. spanning trees with the greatest possible weight, we generate a new weighted graph with the same vertices and edges as the one given but where each weight W_L is replaced by $M - W_L$ where M is any number greater than maximum weight w_i, and proceed as before. Determine the maximum spanning tree for the network above and verify your answer using Prim's algorithm.

3 The table below gives the distances in miles between some cities in the north of England and Scotland.

	Aberdeen	Carlisle	Edinburgh	Glasgow	Inverness	Perth
Aberdeen	∞	221	125	145	105	81
Carlisle	221	∞	100	95	263	144
Edinburgh	125	100	∞	40	160	42
Glasgow	145	95	40	∞	168	60
Inverness	105	263	160	168	∞	115
Perth	81	144	42	60	115	∞

Determine, by any appropriate method, the shortest route connecting the centres. Compare the answers with that for question 3 of Exercises 3.1 CLASSWORK and comment.

4 Find *all* the minimum spanning trees for this weighted graph.

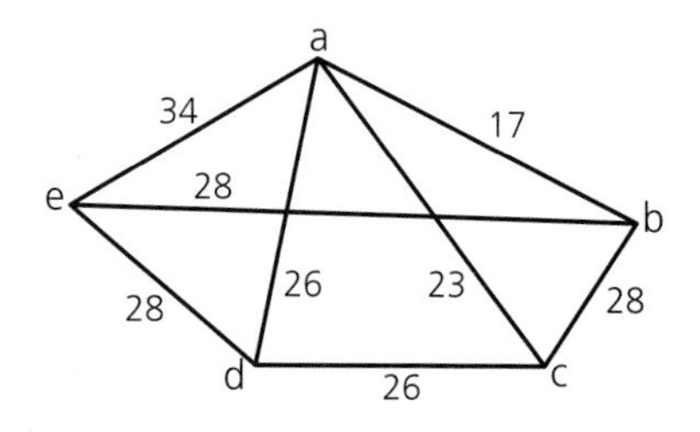

5 **a)** The table below gives the distances in miles, between the first eight centres, listed alphabetically, in my road atlas. Construct a graph to illustrate the data. How many distances would be given if
i) 10 centres
ii) n centres were to be used?

By considering a completed rectangle for the figure justify your conjecture in part **ii)**.

b) Use the information in the table to construct a matrix and hence find the minimum spanning tree by applying Kruskal's algorithm.

c) Verify your result in part **ii)** by adapting the matrix so that Prim's algorithm may be applied to the data.

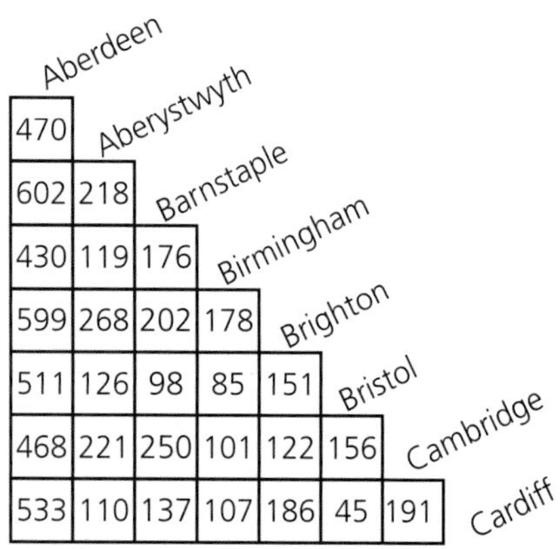

	Aberdeen	Aberystwyth	Barnstaple	Birmingham	Brighton	Bristol	Cambridge
Aberystwyth	470						
Barnstaple	602	218					
Birmingham	430	119	176				
Brighton	599	268	202	178			
Bristol	511	126	98	85	151		
Cambridge	468	221	250	101	122	156	
Cardiff	533	110	137	107	186	45	191

CONSOLIDATION EXERCISES FOR CHAPTER 3

1 The figure shows some places in the south of England. The numbers on the arcs are the distances, in miles, between them. A cable TV company based in Salisbury wishes to link all the places using a minimum length of cable.

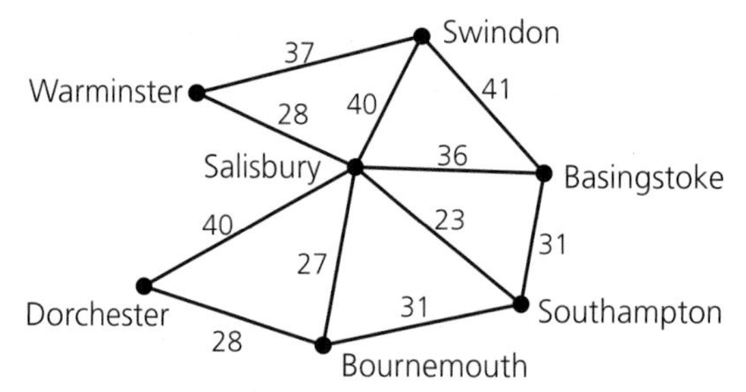

Use the greedy algorithm (Kruskal's algorithm) to find this minimum length of cable. In your solution you should indicate clearly the order in which the places are added to the minimum spanning tree.

(ULEAC Question 2, Specimen Paper, 1996)

2 Five new houses labelled A, B, C, D and E on the diagram are to be connected to a drainage system. Each house is to be connected to the sewer (point S on the diagram), either directly or via another house. Alternatively houses may be connected to an intermediate manhole at M, directly or via another house. This manhole must in turn be connected to S, either directly or via another house.

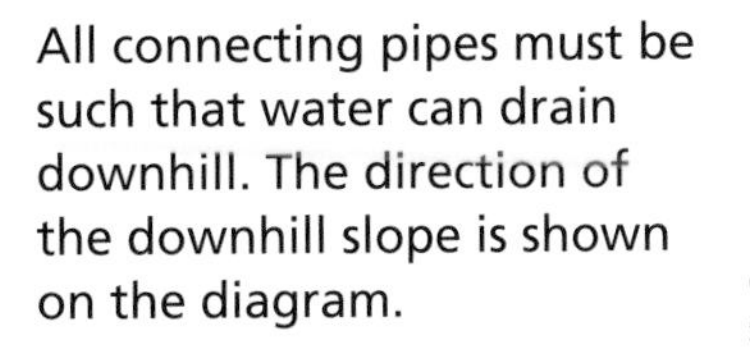

All connecting pipes must be such that water can drain downhill. The direction of the downhill slope is shown on the diagram.

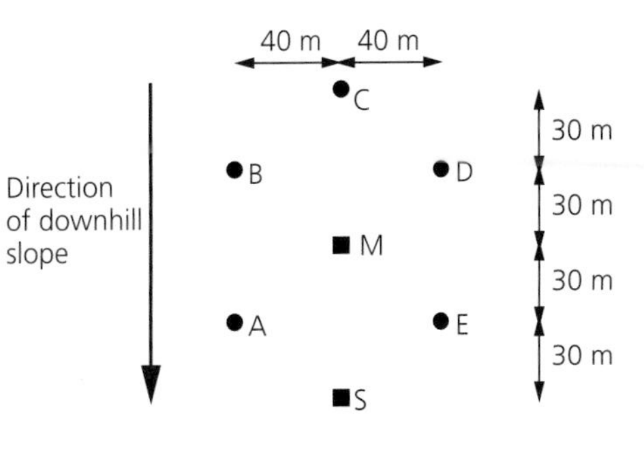

a) Copy and complete the matrix below, showing the lengths of all 16 possible connecting pipes.

To

From	A	B	C	D	E	M	S
A	–	–	–	–	–	–	50
B	60	–	–	–	–	50	
C							
D							
E							
M							
S							

b) Starting from S, use Prim's algorithm to find a minimum connector for A, B, C, D, E, M and S. Show which pipes are in your connector, indicating the order in which they were included, and give their total length.

Explain whether or not your connector represents a system which drains correctly.

(UODLE Question 9, Paper 4, 1993)

3 **a)** The five points on this graph may be connected in pairs by straight lines. Use a greedy algorithm to find a connector of minimum length for the five points. Show your answer on a copy of the diagram, explain your method, and give the length of your minimum connector.

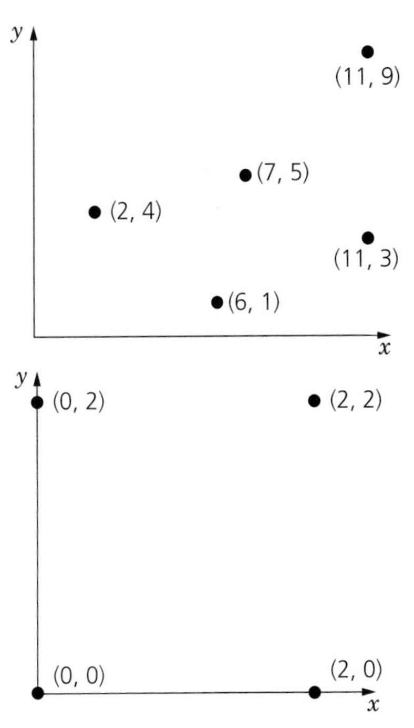

b) The points on this graph may also be connected in pairs by straight lines.

i) Find a minimum connector for the four points.

ii) Find a fifth point so that the minimum connector for the five points is shorter than the minimum connector for the original four points. Show your minimum connector and give its length.

(UODLE Question 7, Paper 4, 1996)

4 The map below shows six campsites, A, B, C, D, E and F, in an area of rough country, together with paths connecting them. The numbers by the paths show lengths, in miles, of sections of path.

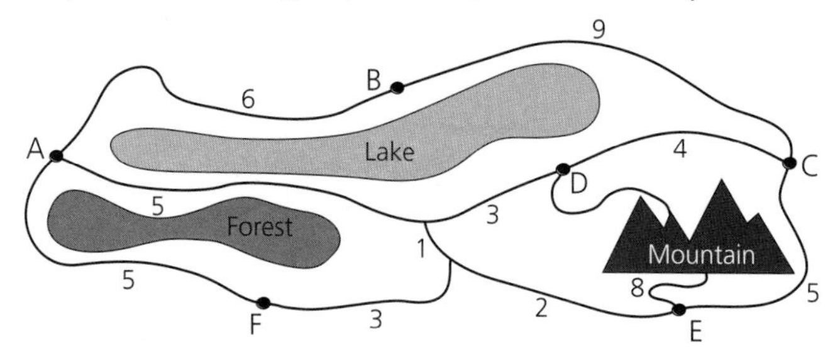

a) Draw a network showing the shortest direct distances between youth hostels. For example, the shortest direct distance between E and F is 5 miles. Between B and D there is no direct link. So you must find the shortest route (you may do this by inspection).

b) Use Prim's algorithm to find a minimum spanning tree for your network. Describe each step in your use of the algorithm, and draw your minimum spanning tree.

(UODLE Question 7, Paper 4, 1990)

5 A company has offices in six towns, A, B, C, D, E and F. The costs in £, of travelling between these towns are shown in the table.

Town	A	B	C	D	E	F
A	–	15	26	13	14	25
B	15	–	16	16	25	13
C	26	16	–	38	16	15
D	13	16	38	–	15	19
E	14	25	16	15	–	14
F	25	13	15	19	14	–

Use Prim's algorithm, starting by deleting row A, to find the cheapest way of visiting the six towns. You should show all your working and indicate the order in which the towns were included.

(OCR Specimen Paper 2000 D1)

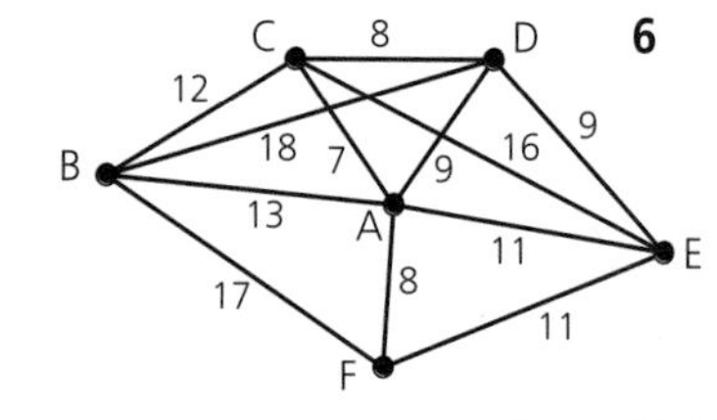

6 This network represents the major routes joining six towns. The distances beside the arcs are the distances in kilometres between the towns.

The council wish to keep the towns connected during cold weather. To this end they wish to determine a minimum connector tree (minimum spanning tree) in order to decide which roads to salt and grit. Use a greedy algorithm to determine a suitable tree for them. Begin building your tree from town A. You should state the algorithm that you use carefully, especially the criteria that you use to include arcs as you build your tree. What is the minimum length of road that the council need to salt and grit?

(UODLE Question 20, Paper 4, 1989)

7 The following matrix represents the distances (in miles) between the six nodes of a network.

	A	**B**	**C**	**D**	**E**	**F**
A	0	11	21	16	9	10
B	11	0	10	31	7	23
C	21	10	0	14	12	20
D	16	31	14	0	11	9
E	9	7	12	11	0	12
F	10	23	20	9	12	0

Use Prim's algorithm to find the minimum connector for the network – start from A and indicate the order in which you include vertices. State which arcs are in your minimum connector and give its total length.

(UODLE Question 10 (part), Paper 4, 1994)

8 The matrix below represents the distances (in miles) of direct roads between six towns.

	A	B	C	D	E	F
A	–	12	3	–	9	10
B	12	–	6	6	18	–
C	3	6	–	6	12	–
D	–	6	6	–	9	–
E	9	18	12	9	–	12
F	10	–	–	–	12	–

a) Starting at A, use Prim's algorithm to find a solution to the minimum connector for the six towns.

b) How many solutions are there? Find them all.

9 The diagram below shows the distance (in miles) between nine towns in SW England. Use Kruskal's algorithm to find the minimum connector for the towns.

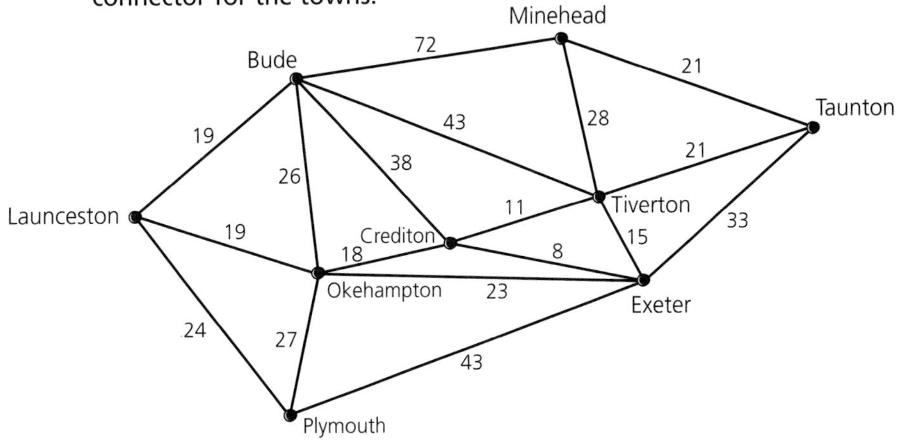

10 The diagram below represents a set of components on a circuit board which need to be joined together as cheaply as possible. Find the minimum length of connector needed to ensure that each component is supplied with power.

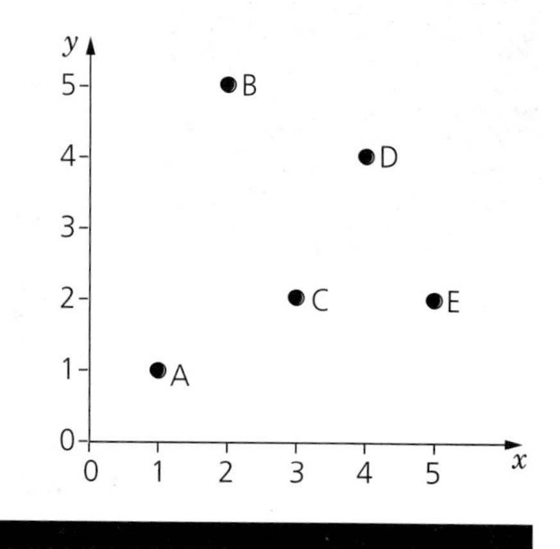

Summary

After working through this chapter you should be familiar with:

- a spanning tree, which connects all the vertices in a graph or network with no cycles
- the minimum connector, which is a spanning tree of minimum length
- Kruskal's and Prim's algorithms for finding the minimum spanning tree in a network.

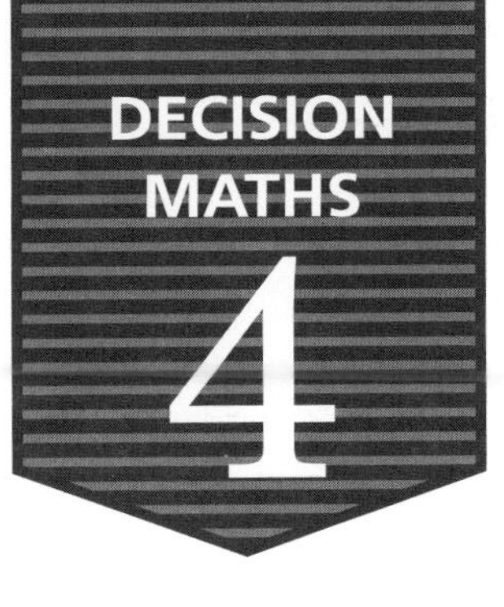

Graphs and networks 3: The travelling salesman problem

By the end of this chapter you should understand that:

- *a cycle is a loop which returns to its starting point*
- *the travelling salesman problem attempts to find a cycle of minimum length that visits all the vertices in a network.*

Exploration 4.1 Cable TV

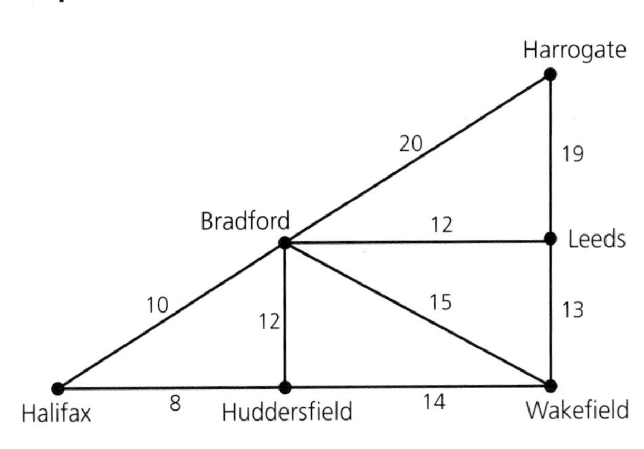

Look back at Exploration 3.1.

Once the cable for the TV network is installed an inspector, based in Leeds, visits all the towns to ensure that the connection has been successful. Find a route for her to visit every town. Is this the shortest possible route?

You are being asked to find a route which visits every town and returns to the starting point, such a cycle being referred to as a **tour**.

TSP is the accepted abbreviation for the travelling salesman problem.

This type of problem, known as the 'travelling salesman problem' (TSP), is well known because it is faced regularly in industry by people who need to minimise distance, time or cost, for example the salesman after whom the problem was named. It is also interesting because there is no exact method for solving it. Mathematicians are trying to find a suitable method that guarantees an optimal solution, but a simple algorithm has not yet been devised. It therefore falls into the group of problems which we classified as **construction problems** in Chapter 1, *Algorithms*. We know that a solution exists, but an algorithm which provides an optimal solution for all networks has not yet been constructed.

Exploration 4.2

Find the lengths of cycles

Find the lengths of all the cycles in this network, beginning and ending at A, and including every vertex at least once. Identify the cycles of minimum length.

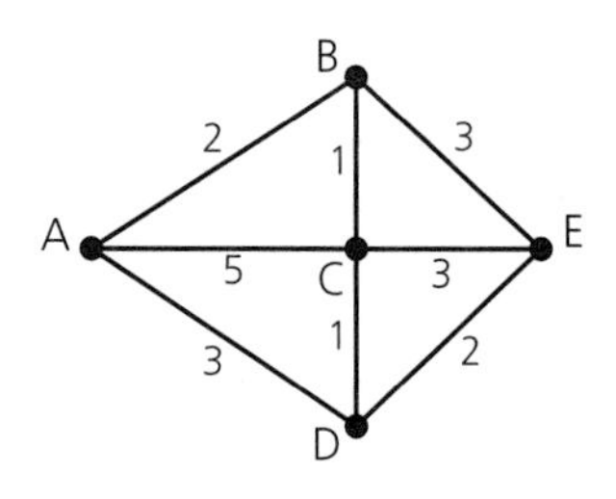

Finding the optimal solution

Carrying out an exhaustive search will give the optimal solution, but it is tedious and time-consuming. It can be shown that for a complete graph with n vertices, there are $\frac{1}{2}(n-1)!$ possible cycles, so in a network with fifteen vertices there would be roughly 4.36×10^{10} cycles; a computer capable of checking 1000 cycles every second would take over a year to check them all! Checking all possible cycles is obviously not a practical proposition, even using a computer. However there is a method of finding a solution that is close to the optimal solution.

GRAPH THEORY – CYCLES

In Chapter 2, *Graphs and Networks 1: Shortest path* we discussed the idea of a path through a graph or network. A **cycle** is a path that completes a loop and returns to its starting point.

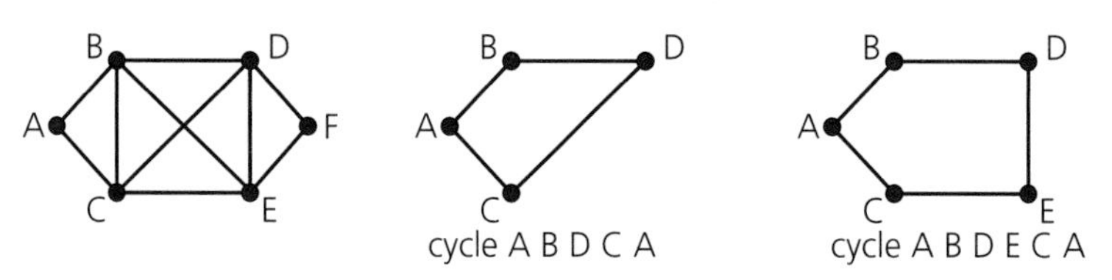

There are two types of cycle which are of interest in solving real-life problems.

Sir William Rowan Hamilton (1805–65) was one of the leading mathematicians of his time. He became Astronomer Royal of Ireland at 22 and was knighted at 30. He did work on algebra, dynamics and optics.

Hamiltonian cycles are cycles that pass through every vertex of the graph exactly once, and only once, and return to the starting point. This is the type of cycle that would be needed to obtain a solution to Exploration 4.1. The path ABDFECA is an example of a Hamiltonian cycle.

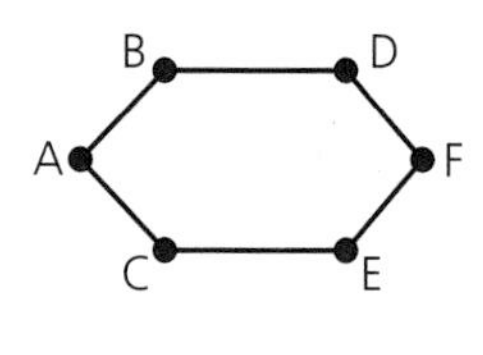

Eulerian cycles are cycles that include every edge of a graph exactly once. The path ABCDBEDFECA is an example of an Eulerian cycle. These cycles are often simply called **Euler cycles**.

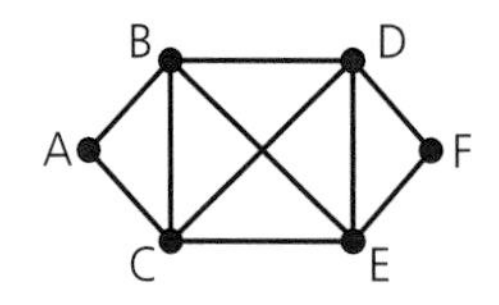

We shall meet this kind of cycle again in Chapter 5, *Networks 4: Route inspection problems*.

Leonhard Euler (1707–83)

In Exploration 4.2 there are many cycles, but the four shown below are of particular importance.

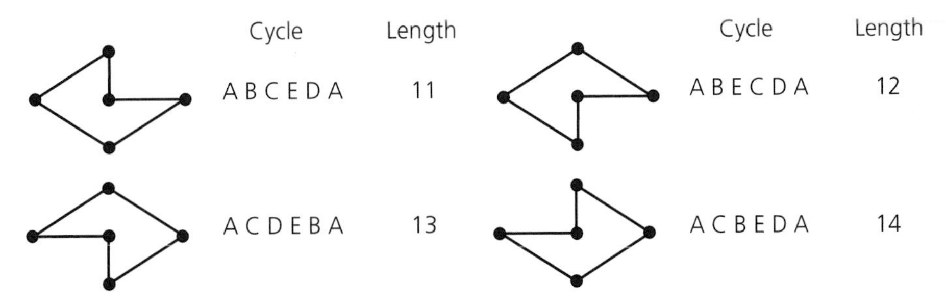

__Note:__ The order of listing the vertices could be reversed, but the cycle would be effectively the same e.g. ABCEDA ≡ ADECBA.

These cycles are important because they visit each vertex once and once only, in other words, they are Hamiltonian cycles. The shortest cycle in the exploration is ABCEDA, length 11. The TSP can be considered as an attempt to find a Hamiltonian cycle of minimum length in the network, if such a cycle exists.

EXERCISES

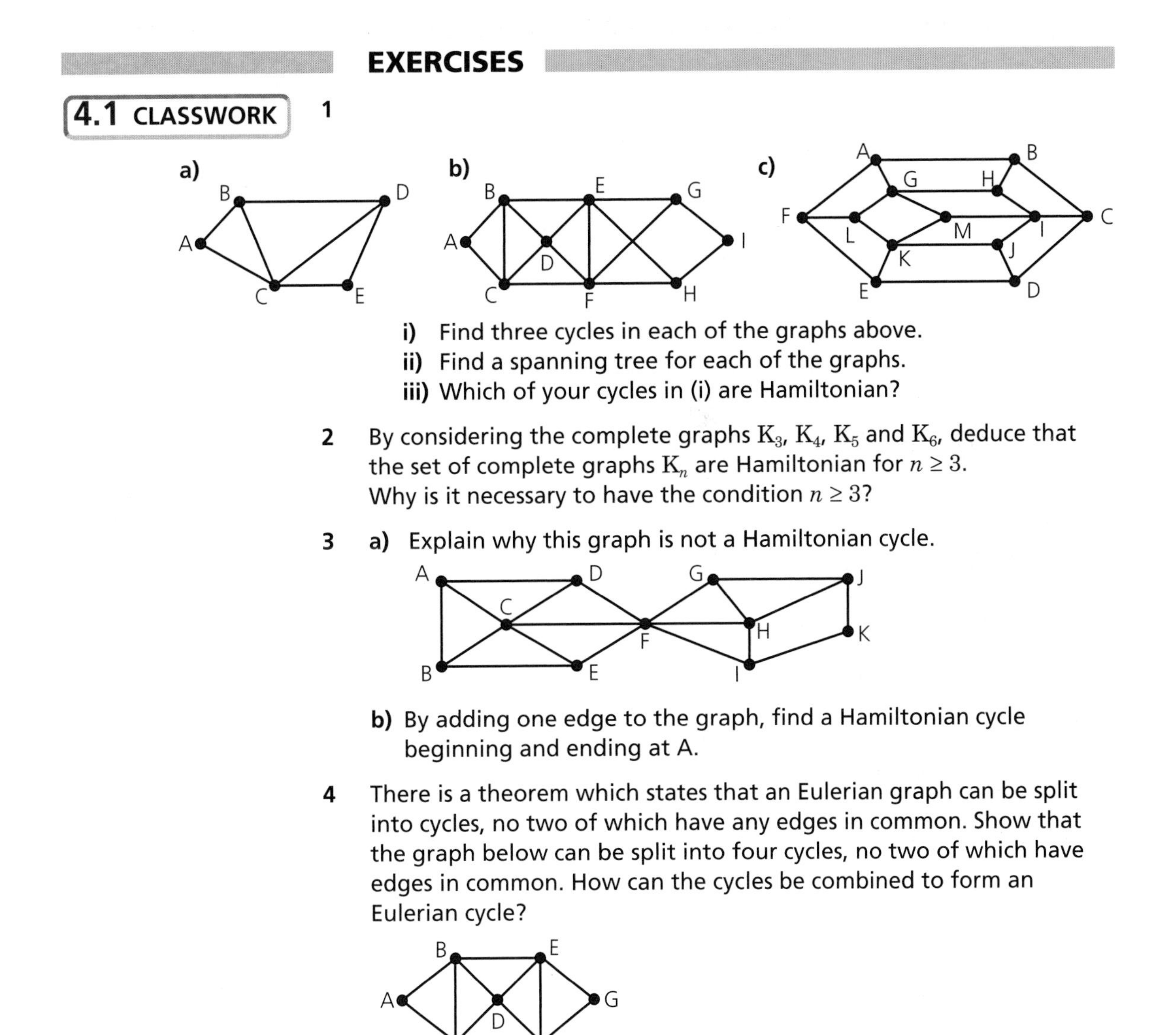

4.1 CLASSWORK

1 **a)** **b)** **c)**

i) Find three cycles in each of the graphs above.
ii) Find a spanning tree for each of the graphs.
iii) Which of your cycles in (i) are Hamiltonian?

2 By considering the complete graphs K_3, K_4, K_5 and K_6, deduce that the set of complete graphs K_n are Hamiltonian for $n \geq 3$.
Why is it necessary to have the condition $n \geq 3$?

3 **a)** Explain why this graph is not a Hamiltonian cycle.

b) By adding one edge to the graph, find a Hamiltonian cycle beginning and ending at A.

4 There is a theorem which states that an Eulerian graph can be split into cycles, no two of which have any edges in common. Show that the graph below can be split into four cycles, no two of which have edges in common. How can the cycles be combined to form an Eulerian cycle?

5 William Hamilton invented the Icosian game, which is played on a board like the one below. The idea of the game is to find a Hamiltonian cycle starting with five given letters. Find two Hamiltonian cycles beginning with the letters BGHJK.

Hamilton sold the game to a dealer in 1859 for £25.00 It was not a commercial success, so the dealer got himself a poor bargain!

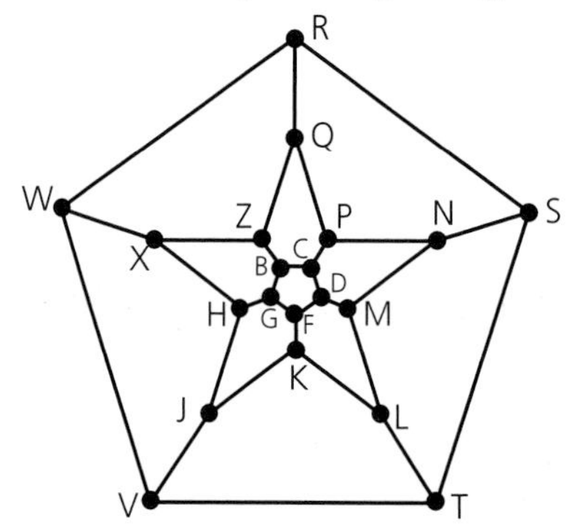

EXERCISES

4.1 HOMEWORK

1 Sketch a graph of the following.

a) Four vertices, one adjacent to all the others but no others adjacent. Is this a tree?

b) Five vertices, with cycles of length 2, 3, 4 and 5.

2 Study graph **G** shown below. Find cycles of lengths, 5, 6, 8 and 9. Do cycles of length 7 exist?

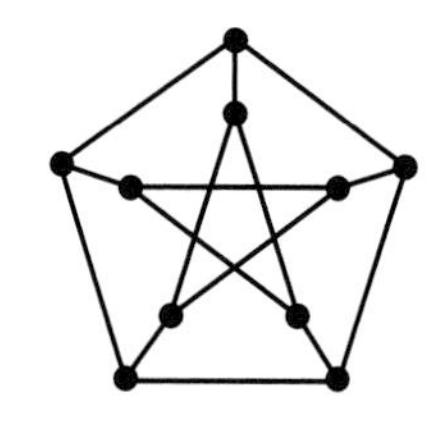

3 Consider the graphs $\mathbf{G_1}$, $\mathbf{G_2}$ and $\mathbf{G_3}$ below.

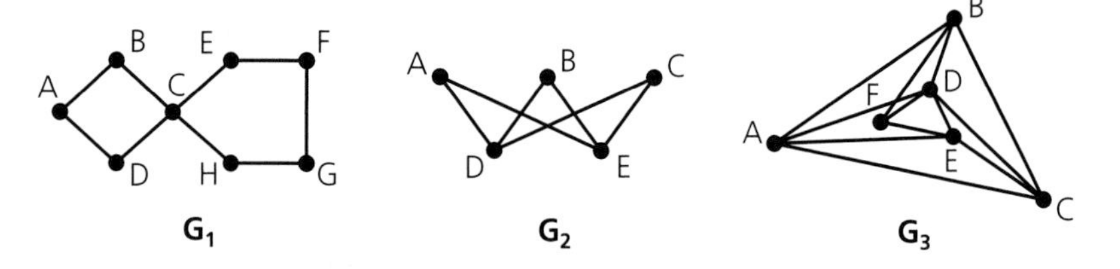

In each case, determine whether the graph is:

a) Hamiltonian

b) Eulerian.

For those graphs which exhibit the appropriate property, identify a cycle.

4 Give an example, with at most seven vertices, which you have not already encountered in this chapter, of:

a) a cycle that is both Hamiltonian and Eulerian

b) an Eulerian cycle that is not Hamiltonian

c) a Hamiltonian cycle that is not Eulerian

d) a cycle that is neither Hamiltonian nor Eulerian.

In each case:

e) find a spanning tree for your graph

f) find cycles for your graph.

5 Suggest conjectures for:
a) the number of edges of any tree with n vertices
b) the number of edges in a complete graph of n vertices.
Justify your conjectures.

Note:
The graph in question 2 in Exercises 4.1 HOMEWORK is known as the **Petersen graph**.

Exploration 4.3 *Is there a Hamiltonian solution?*

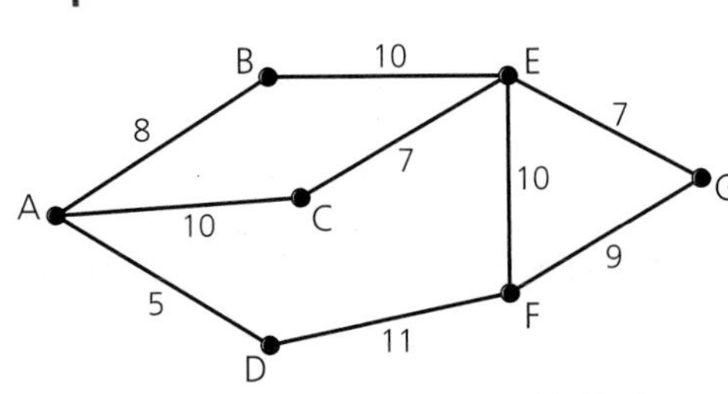

Find a cycle of minimum length, starting and finishing at A.

Is it possible to find a cycle which visits each vertex once, and only once?

Solving the classical problem

In fact it is impossible to find a Hamiltonian cycle in the network in Exploration 4.3, so the cycle must revisit a vertex, to get home. In a practical problem this would be quite acceptable, but in the 'classical' problem of finding Hamiltonian cycles it is not. From the network preceding the exploration, it can also be seen that even if Hamiltonian cycles exist, there may be several of different lengths. Despite this, we can use mathematical modelling to enable us to use the classical problem as a method for finding a cycle which will give an upper bound to the TSP.

Upper bounds

Question 2 of Exercises 4.1 CLASSWORK established that any complete network is Hamiltonian and will contain cycles which visit every vertex within themselves once and only once before returning to the starting point. In this method, the first step is to convert the network into a complete network. We do this by joining each vertex to every other, by the shortest path in each case.

Consider again the network in Exploration 4.3.

Vertex A is joined directly to vertices B, C and D. To make the network complete, we need to add edges joining A directly to E, F and G. In a small network like this one we can find the shortest distances, (minimum weights) by observation.

To reach E from A, we could take:

ABE length 18
ACE length 17

As ACE has the minimum weight, edge AE would have weight 17, giving this result.
If we repeat the same process for all pairs of vertices we will get this complete network.

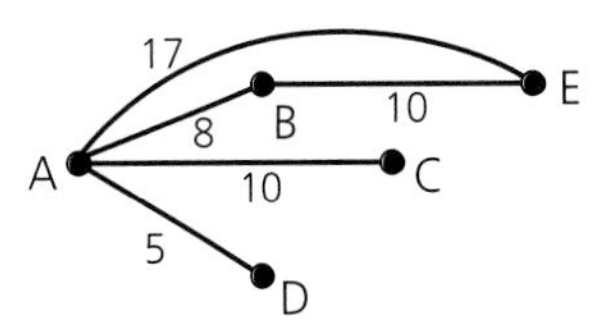

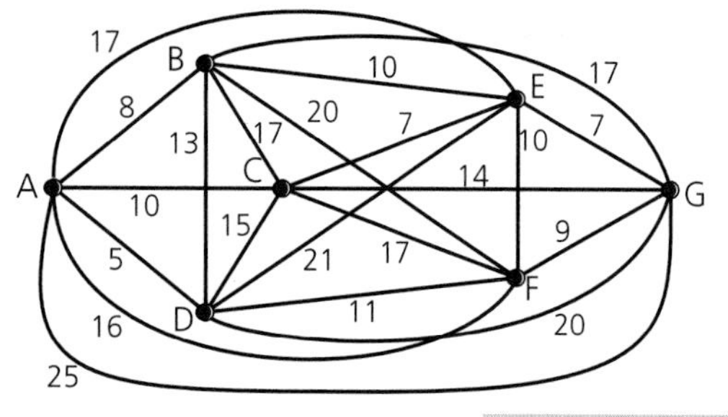

Drawn in this way, the network looks rather cluttered and untidy. It would be difficult to work directly on a diagram like this. To avoid such an untidy network, we could show the information on a distance matrix as we did for Prim's algorithm. This gives the following matrix for the original network.

	A	B	C	D	E	F	G
A	∞	8	10	5	∞	∞	∞
B	8	∞	∞	∞	10	∞	∞
C	10	∞	∞	∞	7	∞	∞
D	5	∞	∞	∞	∞	11	∞
E	∞	10	7	∞	∞	10	7
F	∞	∞	∞	11	10	∞	9
G	∞	∞	∞	∞	7	9	∞

This matrix shows the complete network above.

	A	B	C	D	E	F	G
A	∞	8	10	5	*17*	*16*	*25*
B	8	∞	*17*	*13*	10	*20*	*17*
C	10	*17*	∞	*15*	7	*17*	*14*
D	5	*13*	*15*	∞	*21*	11	*20*
E	*17*	10	7	*21*	∞	10	7
F	*16*	*20*	*17*	11	10	∞	9
G	*25*	*17*	*14*	*20*	7	9	∞

As we know that this complete network will contain Hamiltonian cycles, we now set about building a tour by looking for a cycle with a low weight. For this we will use the **Nearest neighbour algorithm**. This is another greedy algorithm which selects the shortest available edge at each stage, provided that the edge does not form a cycle (except for the last edge which completes the tour). The algorithm works by visiting the nearest vertex which has not yet been visited, until it returns to the starting point. Applying it to this problem gives the following matrix.

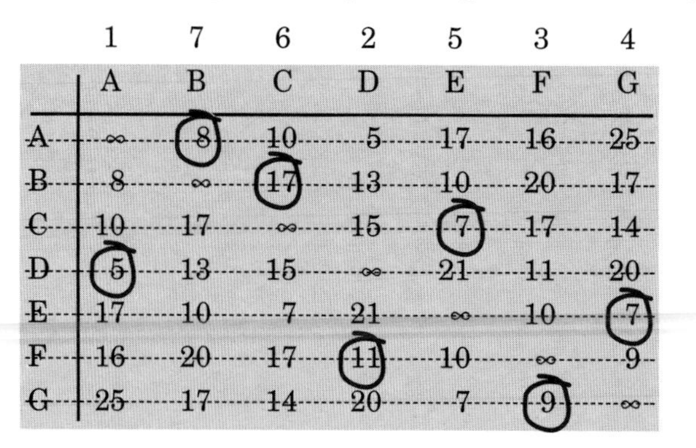

	1	7	6	2	5	3	4
	A	B	C	D	E	F	G
A	∞	(8)	10	5	17	16	25
B	8	∞	(17)	13	10	20	17
C	10	17	∞	15	(7)	17	14
D	(5)	13	15	∞	21	11	20
E	17	10	7	21	∞	10	(7)
F	16	20	17	(11)	10	∞	9
G	25	17	14	20	7	(9)	∞

Start at vertex A. Delete Row A. Choose the smallest entry in column A (AD, 5) and put edge AD in the solution.

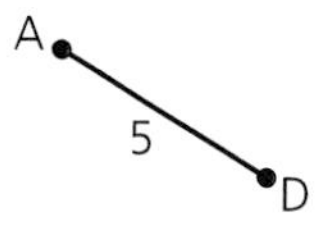

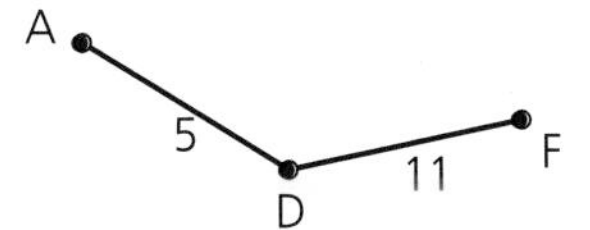

Delete Row D. Choose the smallest entry in column D (DF, 11) and put edge DF in the solution.

Delete Row F. Choose the smallest entry in column F (FG, 9) and put edge FG in the solution.

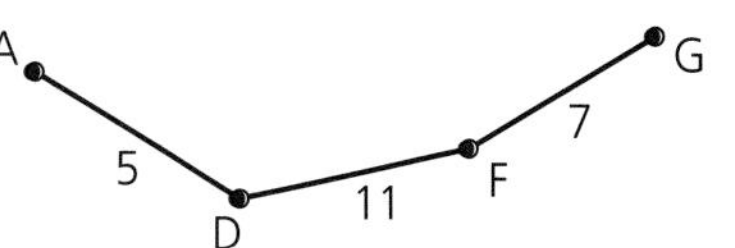

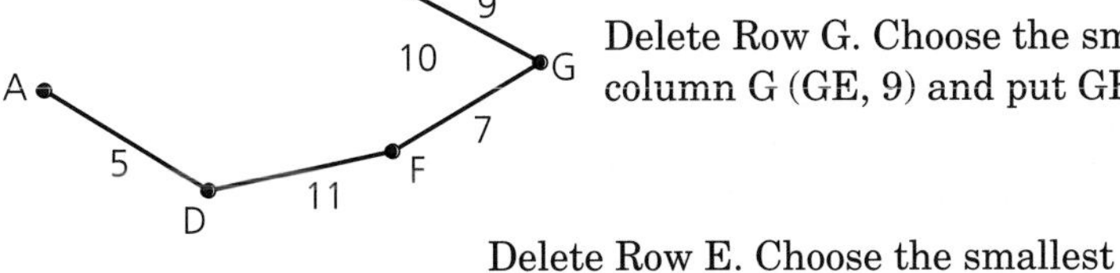

Delete Row G. Choose the smallest entry in column G (GE, 9) and put GE in the solution.

Delete Row E. Choose the smallest entry in column E (EC, 7) and put EC in the solution.

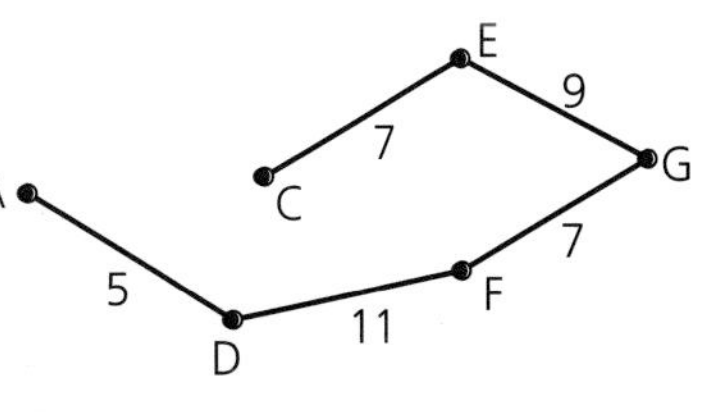

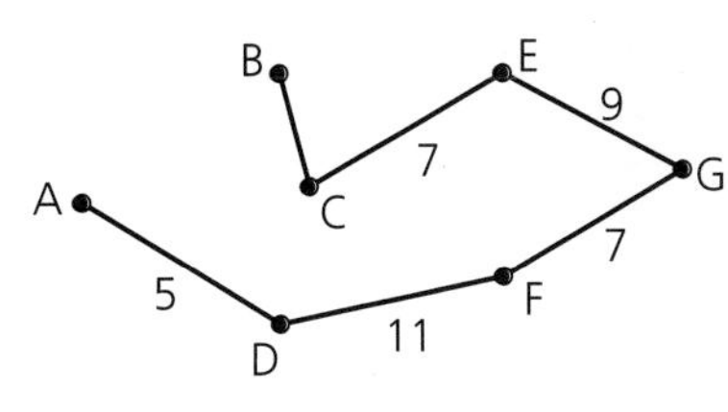

Delete Row C. Choose the smallest entry in column C (the only remaining entry is CB, 17) and put CB in the solution.

We have now connected all the vertices into the solution, so all that remains to do is connect B back to our starting point A, giving the Hamiltonian cycle ADFGECBA, length 64.

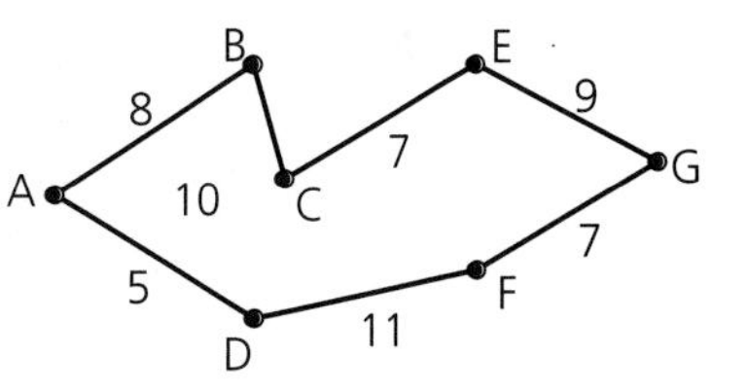

We now need to go back to our original network and interpret this solution in the context of the real problem.

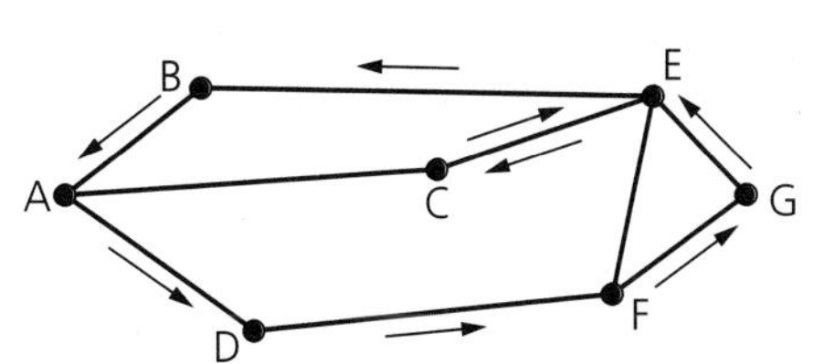

Edges AD, DF, FG, GE, EC and BA were in the network. However, there is no direct edge CB. This was added to make the complete network by taking the path CEB.

Hence our tour is really ADFGECEBA, as shown in the diagram.

This cycle provides us with an upper bound for the optimal length of a tour in the network.

So we can say:

length of optimal tour ≤ 64

The main drawback of the Nearest neighbour algorithm is that it does not consider the weight of the final link until the end. This could prove to be excessively long, although this is not the case in this particular example.

Now we can apply this technique to the problem posed in Exploration 4.1. First, we transform the network into a complete network, so that we can find a Hamiltonian cycle. This is shown in the matrix below.

	Leeds	Harrogate	Wakefield	Bradford	Halifax	Huddersfield
Leeds	∞	19	13	12	20	24
Harrogate	19	∞	32	20	30	32
Wakefield	13	32	∞	15	22	14
Bradford	12	20	15	∞	10	12
Halifax	20	30	22	10	∞	8
Huddersfield	24	32	14	12	8	∞

Using the Nearest neighbour algorithm we can find a tour starting and ending at Leeds.

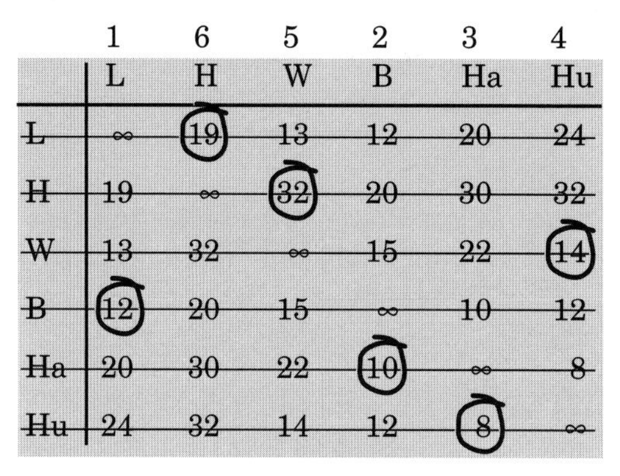

	1	6	5	2	3	4
	L	H	W	B	Ha	Hu
L	∞	(19)	13	12	20	24
H	19	∞	(32)	20	30	32
W	13	32	∞	15	22	(14)
B	(12)	20	15	∞	10	12
Ha	20	30	22	(10)	∞	8
Hu	24	32	14	12	(8)	∞

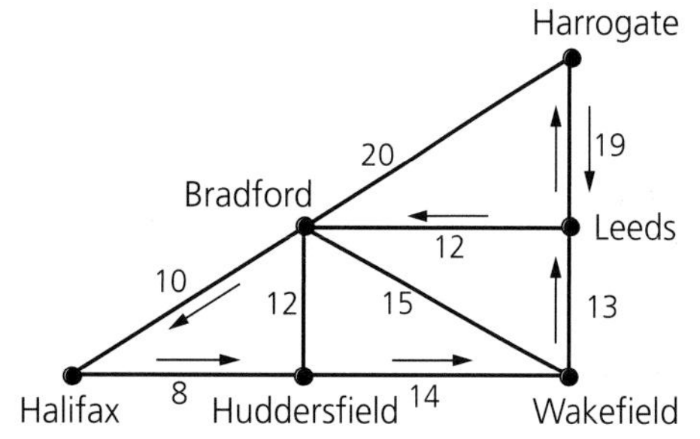

L, B, Ha, Hu, W, H, L:
length = 12 + 10 + 8 + 14 + 32 + 19 = 95 miles

As before, we must now go back to the original problem and interpret our solution in this context.

The tour is:
Leeds, Bradford, Halifax, Huddersfield, Wakefield

Leeds, Harrogate, Leeds

since the edge WH in the complete network goes via Leeds.

The best lower bound is the one closest to the optimal tour – that is the greatest lower bound.

Lower bounds

Having established a tour which provides an upper bound for the solution to the problem, we shall now look at a method for determining a lower bound. We could simply take the minimum connector as being a lower bound, but it would not be a very good one.

A common technique involves:

- deleting a vertex *and* the edges incident on it from the network
- finding a minimum connector for the remaining network
- then reconnecting the deleted vertex by the two shortest edges.

This is acceptable provided the remaining network is connected, but consider the following network.

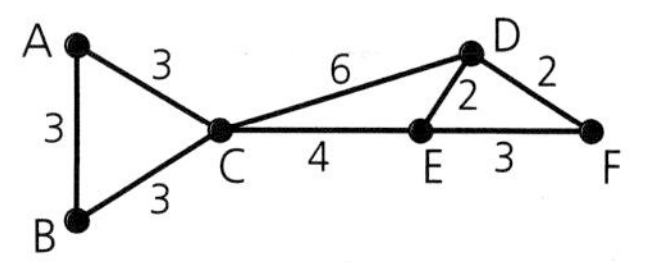

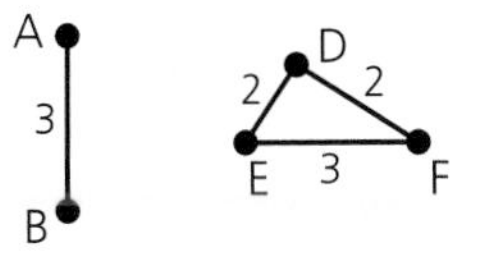

If we delete vertex C, we leave a disconnected network for which no minimum connector can be found.

Once again we return to the classical problem of finding Hamiltonian cycles. If we replace our network by a complete network, the deletion of C does not create a disconnected network and we may proceed with the method.

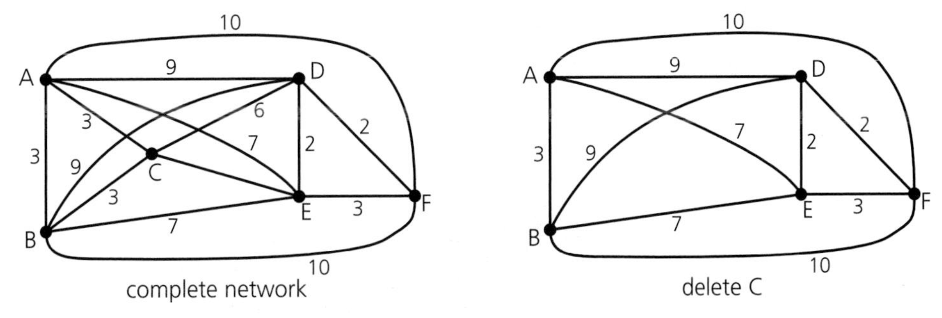

Example 4.1

Use the matrix for the complete network to find a lower bound for the inspector in Exploration 4.1.

Solution

First, delete vertex L, and find the minimum connector for the remaining network.

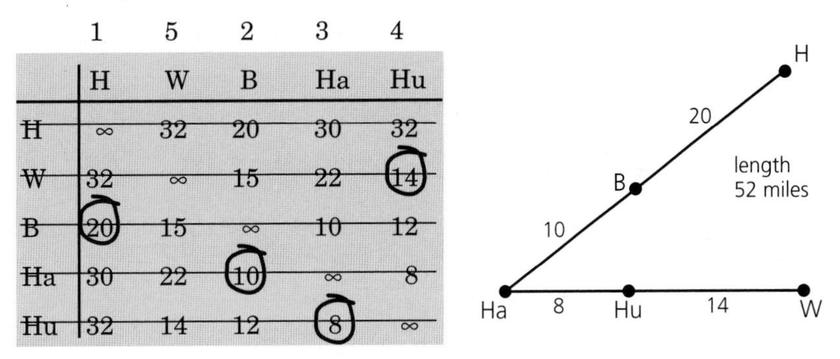

	1	5	2	3	4
	H	W	B	Ha	Hu
H	∞	32	20	30	32
W	32	∞	15	22	(14)
B	(20)	15	∞	10	12
Ha	30	22	(10)	∞	8
Hu	32	14	12	(8)	∞

This gives a lower bound for a TSP on the part of the network not containing Leeds.

However, any solution would have to involve travelling out of Leeds and back again, so we take the two shortest edges leading to Leeds and assume that we travel out of town on one of them and back into town on the other. The two shortest edges are LB (12) and LW (13), which provides a lower bound for a TSP on the part of the network containing only Leeds.

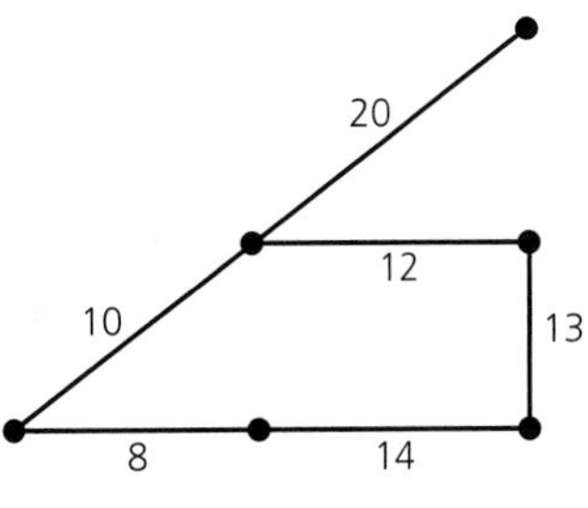

By adding these two together, we obtain a lower bound for the whole network.
Length 25 + 52 = 77 miles

We know that this is a lower bound since it is made up of two lower bounds for subdivisions of the network. We can see that it is a better lower bound than simply using the minimum spanning tree for the network, but is it the best we can do? To answer this question we would need to delete every vertex in turn and compare the results.

This is shown in the table below.

Vertex deleted	Lower bound for deleted vertex	Lower bound for remaining network	Lower bound for TSP
Leeds	25	52	77
Bradford	22	54	76
Harrogate	39	43	82
Wakefield	27	49	76
Halifax	18	56	74
Huddersfield	20	54	74

From the table we can see that the maximum lower bound is 82 miles and is obtained by deleting Harrogate. Thus we can say:

$82 \leq \text{length of route} \leq 95$

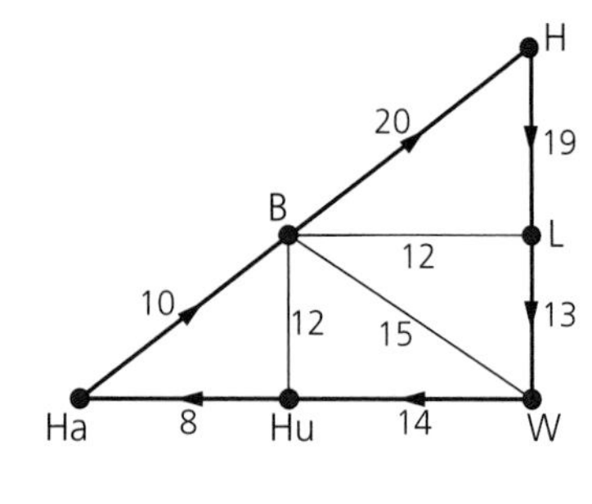

Now we can find a route which will lie within this range. One such is Leeds, Wakefield, Huddersfield, Halifax, Bradford, Harrogate, Leeds with length 84 miles.

Taking into account the practical considerations, this method provides a solution which is good enough. In the inspector's situation, in Exploration 3.1, there would be many other factors which might influence the optimal route on any given day, such as length of journey time, traffic congestion, roadworks and so on, which cannot be accounted for in a mathematical model of this kind. The approximate solution is quite adequate for most TSP problems.

EXERCISES

4.2 CLASSWORK

1 A delivery vehicle leaves the warehouse at A to supply goods to the shops marked B, C, D, E and F on the network, before returning to the warehouse.

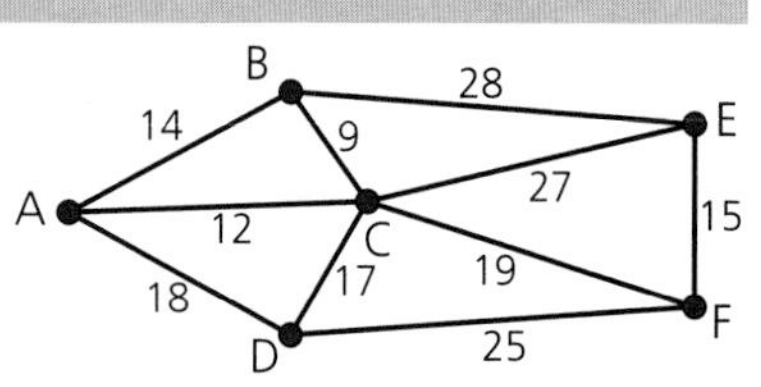

a) By transforming the network into a complete network, find an upper bound for the distance the vehicle will have to travel.

b) Find the best lower bound by deleting each vertex in turn.

2 In Exercises 3.1 CLASSWORK, you solved the minimum connector problem for these cities.

	Aberdeen	Carlisle	Edinburgh	Glasgow	Inverness
Aberdeen	∞	221	125	145	105
Carlisle	221	∞	100	95	263
Edinburgh	125	100	∞	40	160
Glasgow	145	96	40	∞	168
Inverness	105	263	160	168	∞

A buyer, based in Glasgow, needs to visit branches of a chainstore in each of these cities. Obtain an estimate for the upper bound and one for the lower bound by deleting Aberdeen.

3 Marco sells ice cream each day during the summer season, at the places shown in this network.

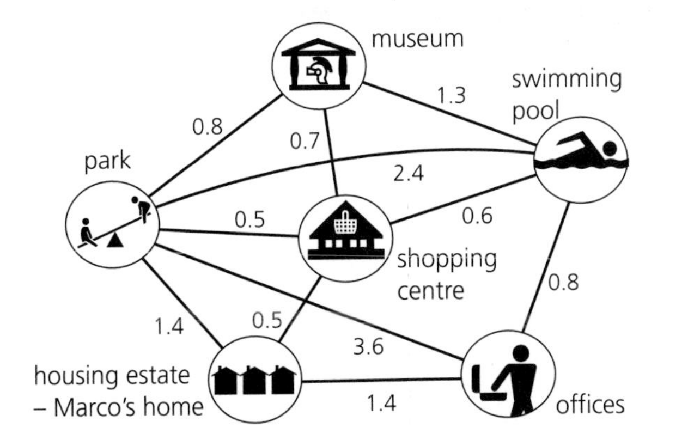

a) Find the shortest route that takes him to each place, starting and finishing at his home, and not visiting any place twice.

b) During the summer holidays, he sells more ice creams at the park than anywhere else. Design a route that enables him to visit the park twice each day.

4 The Island of Guernsey is a popular holiday destination. From the map, draw a network showing the minimum distances between places of interest. Plan a tour of minimum length which visits all the tourist attractions and starts and ends in St Peter Port.

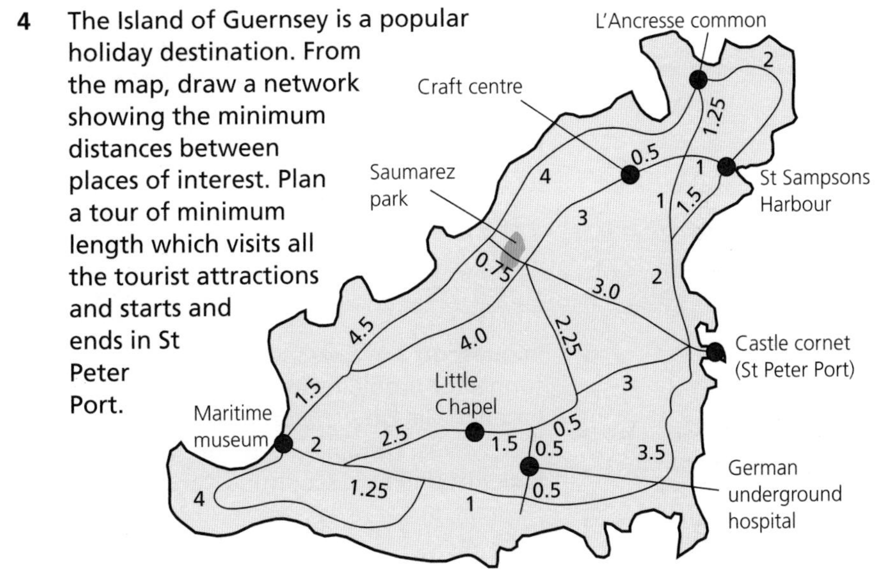

5 In a biscuit factory four types of cookie are made each day by one machine. After each run of one type, the machine must be cleaned, the cleaning time being dependent on the types of cookie being made. The product manageress wishes to find a manufacturing sequence, starting with butter cookies and returning to butter cookies, that will enable each type of cookie to be made whilst minimising the cleaning times.

last made \ to be made	butter	ginger	chocolate	almond
butter	–	15	18	16
ginger	25	–	21	23
chocolate	22	20	–	24
almond	18	12	16	–

a) Explain how this problem can be modelled as a TSP.

b) Use the Nearest neighbour algorithm to establish an upper bound for the cleaning time in one complete cycle.

EXERCISES

4.2 HOMEWORK

1 By redrawing this network as a complete network, find a Hamiltonian cycle that will provide an upper bound for a tour starting and ending at vertex A.

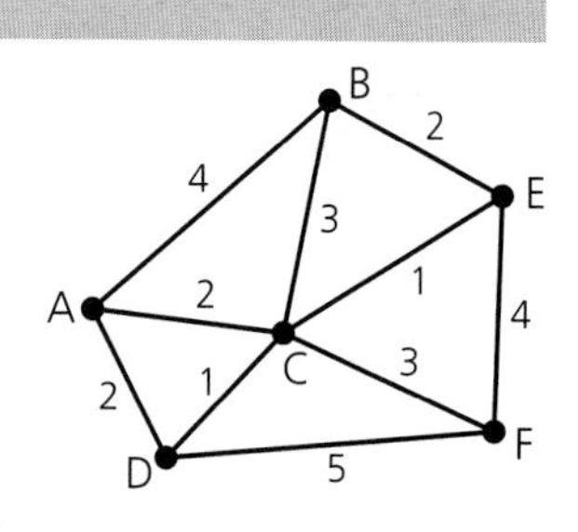

2 Five computer programs have to be run in sequence. Each program needs its own resources, for instance main memory, drives etc. and changing from one set to another uses up valuable time. The table below gives the times for converting resources between programs. Find an ordering of the programs which if run in this order should result in a comparatively small total conversion time.

Program conversion time

	a	b	c	d	e
a	–	80	10	30	20
b	80	–	60	20	20
c	10	60	–	40	40
d	30	20	40	–	70
e	20	20	40	70	–

3 In question 3 of Exercises 3.1 HOMEWORK you solved the minimum connector problems for the augmented Northern centres. A buyer, based in Glasgow, needs to visit branches of a chain store in each of the centres. Obtain an estimate for the upper bound and one for the lower bound by discounting Perth.

4 In question 2 of Exercises 4.2 CLASSWORK the buyer must travel directly between Carlisle and Edinburgh. Describe briefly the method of obtaining a set of lower bounds and obtain three of these.

5 The distances between towns in a particular country are given in the graph below. Because a river flows through the country, there is just one road connecting two towns on either side of the river.

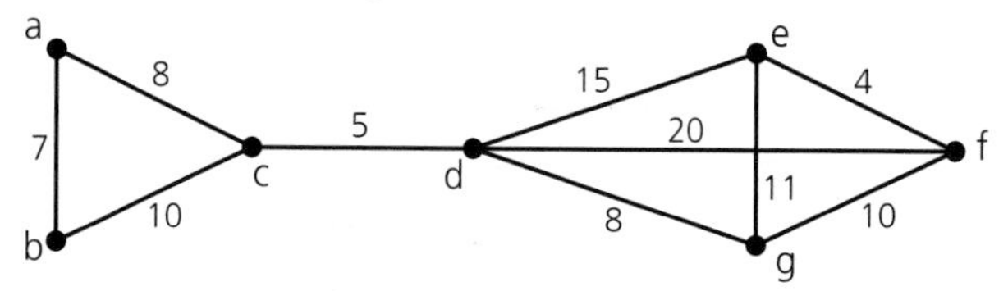

How does this simplify your task of obtaining a good lower bound for the TSP for these towns? Use this fact to obtain two lower bound estimates.

COURSEWORK INVESTIGATIONS

Investigation 1

Cycles

- The table below shows the number of vertices (n), edges (e) and 3-cycles (c_n) in the complete graphs K_1, K_2, K_3 and K_4.

n	1	2	3	4
e_n	0	1	3	6
c_n	0	0	1	4

Extend this table to include K_5 and K_6.

- How many 3-cycles has K_n? Can you complete the row for 4-cycles?
- A graph G_n is obtained from the complete graph K_n by deleting just one 3-cycle, i.e. deleting the three edges which make up the 3-cycle.

Draw examples of G_4 and G_5. Explain why the choice of 3-cycle from K_n would not affect the resulting graph G_n if the vertices were unlabelled.

- If e_n* and c_n* denote the number of edges and 3-cycles respectively in G_n complete this table.

n	3	4	5	6	7
e_n*	0	3	7		
c_n*	0	0			

- Obtain a formula for e_n* in terms of n.

Determine and explain a relationship between e_n* and c_n*. Obtain a formula for c_n* in terms of n and c_n.

COURSEWORK INVESTIGATIONS

Investigation 2

Tour of the building

Plan a tour for a visitor to your school, ensuring that they visit all the communal areas such as the sports facilities, and at least some of the subject areas, as well as a stop for coffee in the staff room. Suppose the visitor is in a wheelchair. How would you amend the tour? Would it be necessary to adapt the algorithms to take account of ramps and lifts?

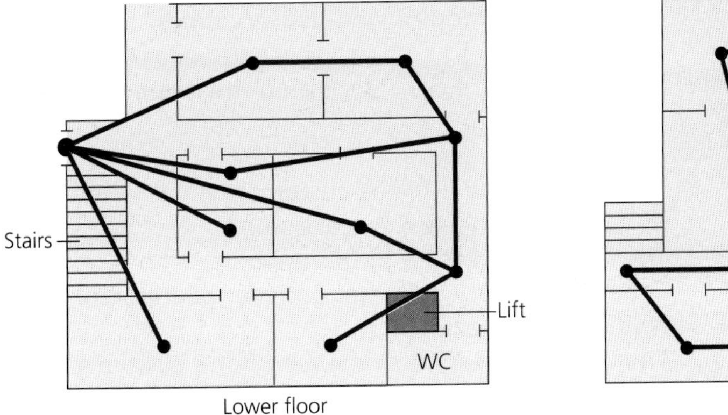

Lower floor

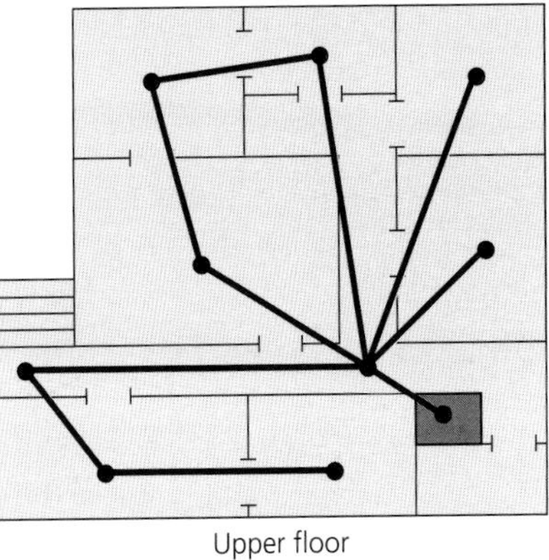

Upper floor

CONSOLIDATION EXERCISES FOR CHAPTER 4

1 **a)** A, B, C and D are the vertices of the complete graph, K_4. List all the paths from A to B (i.e. routes passing through particular vertices at most once).

b) How many paths are there from A to B in the complete graph on the vertices (A, B, C, D, E)?

c) Which of the graphs in parts (a) and (b) are Eulerian, and why?

(AQA B Specimen Paper 2000 D1)

2 An express delivery pizza company promises to deliver pizzas within 30 minutes of an order being telephoned in. Four customers telephone in orders at the same time. While the pizzas are cooking, which takes 10 minutes, a delivery route must be planned for the person who will deliver all four pizzas. The diagram below shows the pizza company P and the four customers A, B, C and D, and the accompanying table shows the travel times for each possible leg of a journey between them.

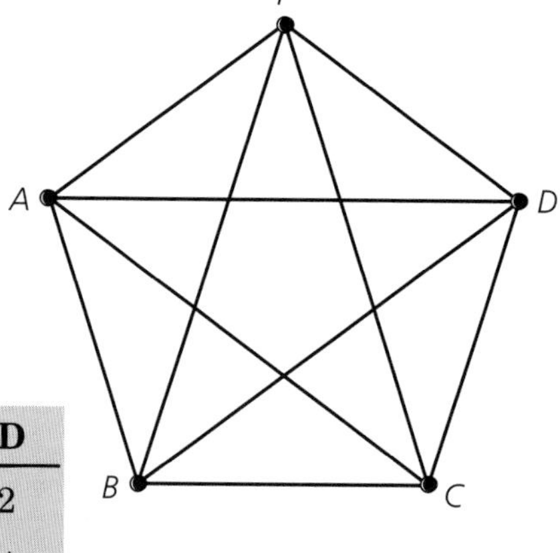

	P	A	B	C	D
P	–	5	3	4	2
A	5	–	1	4	4
B	3	1	–	3	5
C	4	4	3	–	7
D	2	4	5	7	–

The travel times shown exclude the time taken to stop at a customer's house and deliver the pizza; this stopping and delivery time is $2\frac{1}{2}$ minutes.

i) Explain why a travelling salesperson solution taking less than 13 minutes guarantees that all four customers will get their pizzas (including stopping and delivery times) within 30 minutes of their telephone call.

ii) The pizzas are delivered using the nearest neighbour algorithm, as follows.

The first delivery is to the customer who is nearest to P (in the sense of having the shortest travel time).
The second delivery is to the customer who is nearest to the first one.
The third delivery is to the customer nearest the second who is still waiting for a pizza.
The fourth delivery is to the remaining customer.

Write down the order in which the pizzas are delivered using this algorithm, and calculate how long the fourth customer has to wait for their pizza (including the stopping and delivery time).

(OCR D1 Specimen Paper 2000)

3

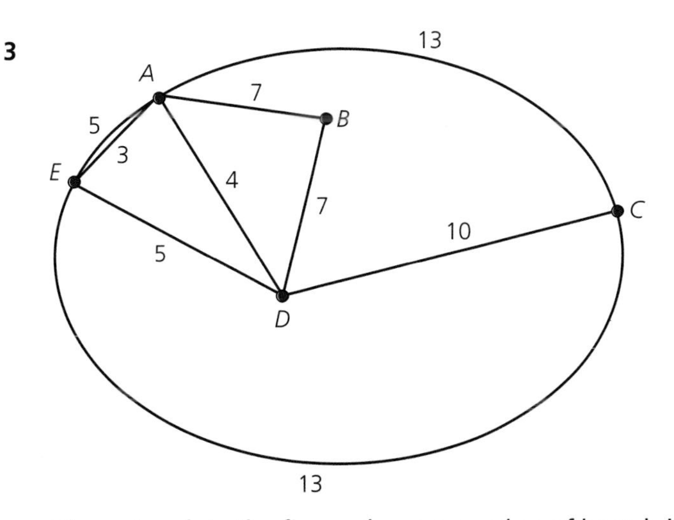

The network in the figure shows a number of hostels in a national park and the possible paths joining them. The numbers on the edges give the lengths, in km, of the paths.

a) Draw a complete network showing the shortest distances between the hostels. (You may do this by inspection. The application of an algorithm is not required.)
b) Use the nearest neighbour algorithm on the complete network to obtain an upper bound to the length of a tour in this network which starts at A and visits each hostel exactly once.
c) Interpret your result in part (b) in terms of the original network.

(EDEXCEL D2 Specimen Paper 2000)

4 A small engineering firm has a production line which is used to make batches of six different aero-engine components, A, B, C, D, E and F. The changeover times in minutes from the line being set up for one component to its being set up for another are given in the table.

	A	B	C	D	E	F
A	—	60	100	40	50	70
B	60	—	90	50	40	40
C	100	90	—	80	60	80
D	40	50	80	—	70	50
E	50	40	60	70	—	70
F	70	40	80	50	70	—

a) Each month, the production line is used to manufacture components, A, B, C, D, E and F in succession. It is then set up ready to produce component A again at the start of the following month. How much time is taken up by the changeovers? Explain why your answer can be considered to be an *upper bound* to a particular travelling salesperson problem. By using a greedy algorithm, or otherwise, show how the engineering firm can reduce the time spent on changeovers. Show sufficient working to make your method clear. You should aim to obtain a time of at most 6 hours.

b) By *initially ignoring component E*, find a lower bound for the time taken up by the changeovers in one month's complete production cycle.

(AQA A Discrete 1 Specimen Paper 2000)

5 The following matrix gives the costs of flight tickets, in £, for direct flights between six connected cities.

To

From		A	B	C	D	E	F
	A	—	45	60	58	90	145
	B	45	—	67	25	83	100
From	**C**	60	70	—	50	70	320
	D	50	25	50	—	35	210
	E	100	80	70	35	—	72
	F	145	110	300	175	80	—

a) A tour is a journey from a city, visiting each other city once and only once, and returning to the starting city. Use a greedy algorithm to find a tour, starting and finishing at A, with a low associated cost. Show that the algorithm has not produced the minimum cost tour.

b) How many different tours which start and finish at A are there altogether?

c) Suppose that, in addition to the cost of tickets, airport taxes must be paid on leaving an airport according to the following table.

A	B	C	D	E	F
20	30	20	40	10	20

Thus the flight from A to B will cost £45 for the ticket and £20 tax, a total of £65. Complete the following matrix of costs, i.e. the total of fares and taxes.

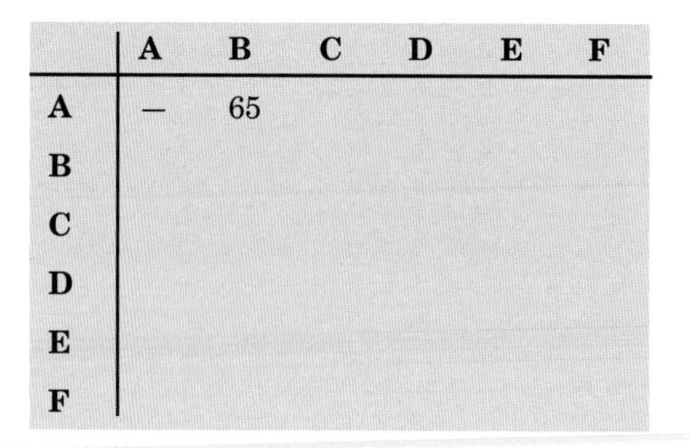

	A	B	C	D	E	F
A	—	65				
B						
C						
D						
E						
F						

d) Give an example where the cheapest route from one city to another differs when taxes are taken into account from that when taxes are not taken into account. Show your two routes and give their costs.

e) Do airport taxes have an effect on the problem of finding the cheapest tour, starting and finishing at A? Why?

(O & C MEI Question 1, Decision & Discrete Mathematics Paper 19, June 1995)

6 The following matrix represents the distances (in miles) between the six nodes of a network.

You may have done the first part of this question in Consolidation Exercise 3.

	A	B	C	D	E	F
A	5	11	21	16	9	10
B	11	0.5	10	31	7	23
C	21	10	57	14	12	20
D	16	31	14	11	11	9
E	9	7	12	11	18	12
F	10	23	20	9	12	43

a) Delete vertex A from the network and use Prim's algorithm to find the minimum connector for the remaining five vertices – start from B and indicate the order in which you include vertices. State which arcs are in your minimum connector and give its total length.

b) By considering the connections between A and the other five vertices, construct a lower bound for the optimal solution of the travelling salesperson problem for the network.

c) Explain why your answer to **b)** is a lower bound for the length of the optimal solution of the travelling salesperson problem for the network.

d) Use the greedy algorithm (that is, go to the nearest vertex not yet visited) to find a solution to the travelling salesperson problem which starts and finishes at A. Calculate its length and compare it with the lower bound which you found in **b)**.

(UODLE Question 10, Paper 4, 1994)

7

	A	B	C	D	E	F	G
A	–	103	89	42	54	143	153
B	103	–	60	98	56	99	59
C	89	60	–	65	38	58	77
D	42	98	65	–	45	111	139
E	54	56	38	45	–	95	100
F	143	99	58	111	95	–	75
G	153	59	77	139	100	75	–

A computer supplier has outlets in seven cities A, B, C, D, E, F and G. The table shows the distances, in km, between the seven cities. Joan lives in City A and has to visit each city to advise on displays. She wishes to plan a route, starting and finishing at A, visiting each city once and covering a minimum distance.

a) Obtain a minimum spanning tree for the network and draw this tree. Start with A and state the order in which the vertices are added.

For three vertices A,B and C, the triangle inequality states that length AB ≤ length AC + length CB.

Given that the network representing this problem is complete and satisfies the triangle inequality:

b) determine an initial upper bound for the length of the route travelled by Joan

c) starting from your initial upper bound for the length of the route, and using an appropriate method, find an upper bound which is less than 430 km

d) by deleting city A, determine a lower bound for the length of Joan's route.

(EDEXCEL D2 Specimen Paper 2000)

8 A manufacturing process consists of five stages, which can be done in any order, before returning to the start again. The costs at each stage are dependent on which stage of the process has been completed immediately before the current one and are shown in the matrix below.

Stage completed (columns); Stage to be done next (rows); in thousands of pounds

	A	B	C	D	E
A	–	5	6	3	8
B	8	–	7	4	4
C	3	7	–	8	6
D	4	5	6	–	3
E	5	6	4	5	–

a) Assuming that the process starts each time with stage A, find the order which will incur the minimum cost.

b) If stage C must be done directly after stage B, adapt the method to find a new least cost plan.

9 The company accountant for the Quasar group of computer stores is planning an inspection. The locations are shown on this network.

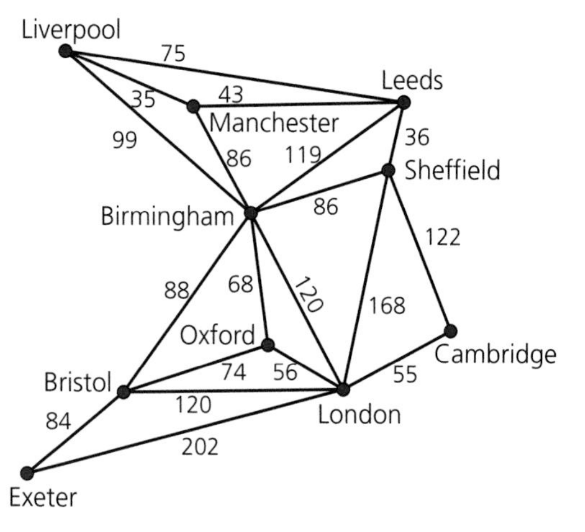

a) Plan a tour for him, starting and finishing at London, which is as short as possible.

b) If he spends 3 hours at each store and allows 1 hour for each 40 miles he travels, how many hours will the trip take?

c) If he can only work at the stores from 8.30 a.m. until 6 p.m., and does not wish to drive between 9 p.m. and 7 a.m., how many days will he need to be away and where should he book overnight hotel accommodation?

10 Plan a tour of minimum length around places of interest in your area.

Summary

After working through this chapter you should be familiar with:

- a cycle, which is a loop which returns to its starting point
- the travelling salesman problem, which attempts to find a cycle of minimum length which visits all the vertices in a network
- Hamiltonian and Eulerian cycles in a simple graph
- finding upper and lower bounds for the travelling salesman problem and appreciate that this method will only provide an approximate solution.

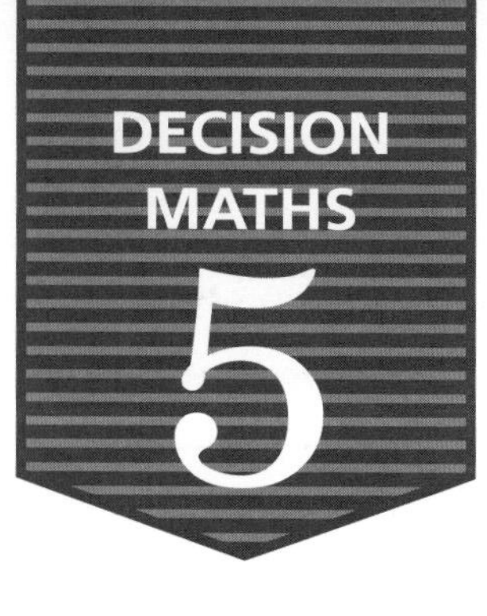

Graphs and networks 4: Route inspection problems

In this chapter we shall:

- *discuss planar graphs, which are graphs that can be drawn in the plane with no edges crossing, and their special properties*
- *discuss isomorphic graphs, which are graphs that have the same number of vertices and the degrees of corresponding pairs of vertices are the same*
- *investigate the degree of a vertex, which is the number of edges incident on the vertex*
- *consider route inspection problems and find the shortest route around a network, travelling along every edge at least once and ending where you started.*

Exploration 5.1 *Postal routes*

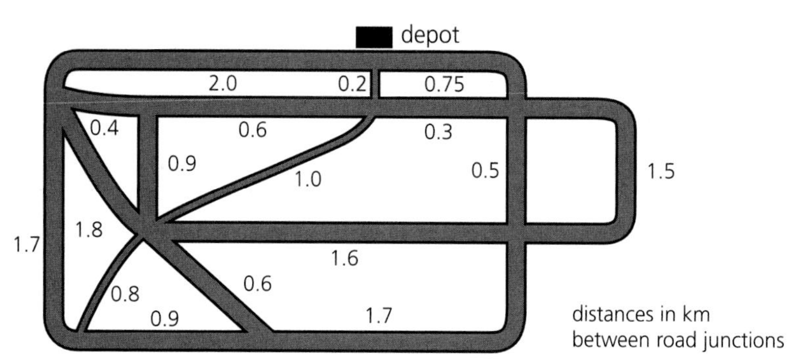

A postman leaves the depot and has to deliver along all the streets in the area shown. What is the shortest distance he must travel?

GRAPH THEORY

Planar graphs

*A **simple** graph is one in which there are no loops and any pair of vertices is connected by no more than one edge.*

A graph that can be drawn in the plane with no edges crossing is called a **planar graph**.

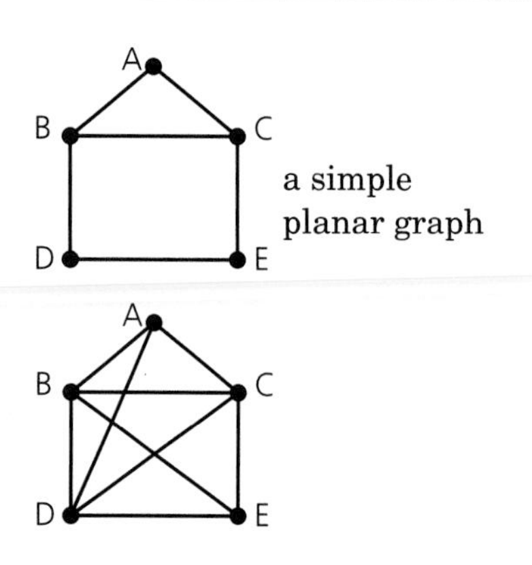

a simple planar graph

It is not always immediately obvious if a graph is planar, so it may be necessary to redraw it to check. Is this graph planar?

Yes, since it can be drawn with no edges crossing by taking edges AD and CD outside.

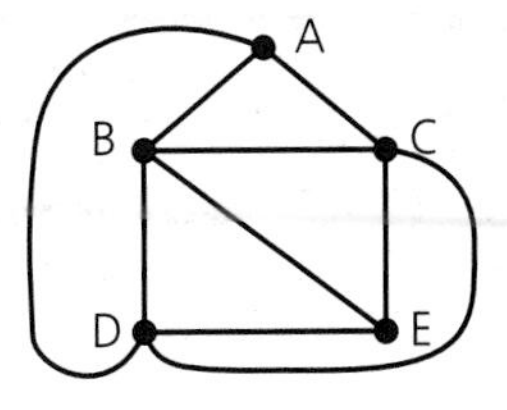

Euler's formula

A planar graph divides the plane into regions which are called **faces**. If we reconsider the graph above we can see that this graph has six faces, one of which, f_6, has no boundary and is called the **infinite** face.

Euler found a simple relationship between the number of vertices (v), edges (e) and faces (f) in a connected planar graph.

$$v - e + f = 2$$

Euler originally developed this formula to apply to the **platonic** solids.

For the graph above this gives:

$$v = 5, e = 9, f = 6 \quad \Rightarrow \quad v - e + f = 5 - 9 + 6 = 2$$

To test whether or not a graph is planar we can use the fact that, if it is, it must satisfy Euler's formula.

Example 5.1

Is the graph K_5 planar?

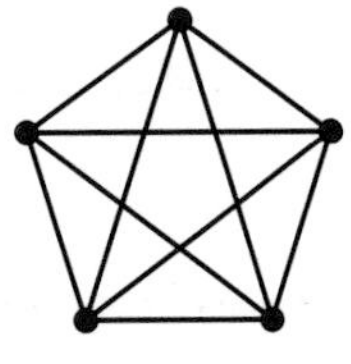

Solution

If K_5 is planar it must satisfy Euler's formula: $v - e + f = 2$.

$$v = 5, e = 10 \quad \Rightarrow \quad 5 - 10 + f = 2 \Rightarrow f = 7$$

So in order to satisfy Euler's formula, the graph must have seven faces. Now each face must be bounded by at least three edges, and since each edge forms a boundary to two faces we can write:

$$2e \geq 3f$$

since $e = 10$, then $20 \geq 3f \quad \Rightarrow \quad \frac{20}{3} \geq f$

Note: *Even the infinite face is bounded by at least three edges.*

However, this contradicts the fact that $f = 7$ for the graph to be planar and hence K_5 must be non-planar.

Degree of a vertex

The **degree** of a vertex is the number of edges incident at that vertex. Consider this graph again.

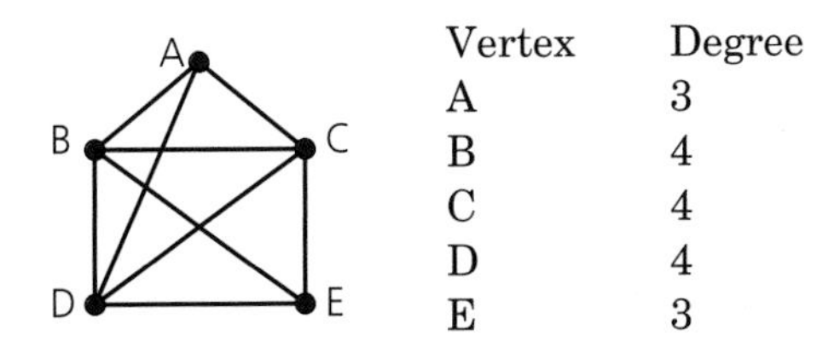

Vertex	Degree
A	3
B	4
C	4
D	4
E	3

If you find the sum of the degrees of the vertices in the graph (18) and compare it with the number of edges in the network, what do you notice?

The sum of the degrees of the vertices for this graph is eighteen and it has nine edges. Draw a few graphs of your own and confirm that the sum of degrees of the vertices is always twice the number of edges in the graph.

More formally , this is known as the **Handshaking theorem** and can be written:

Σ is the Greek letter sigma. It is used in mathematics to indicate 'the sum of'.

$$\Sigma \deg v = 2e$$

Can you explain why this is the case?

Since the sum of the degrees of the vertices is always even, it follows that there will always be an even number of odd vertices.

Isomorphic graphs

Two graphs are **isomorphic** if they have the same number of vertices and if the degrees of corresponding pairs of vertices are the same. As an example compare the following two graphs.

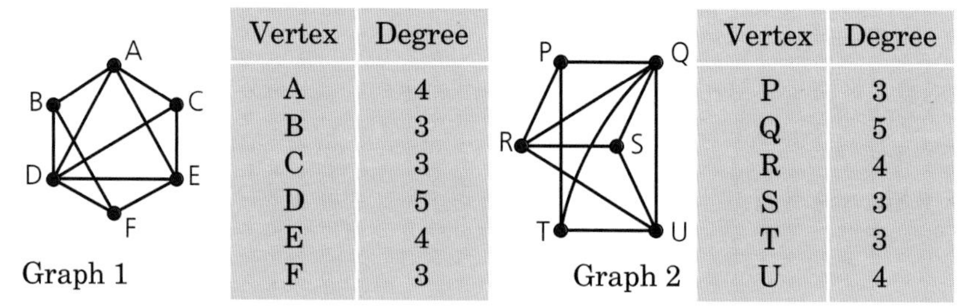

Vertex	Degree
A	4
B	3
C	3
D	5
E	4
F	3

Graph 1

Vertex	Degree
P	3
Q	5
R	4
S	3
T	3
U	4

Graph 2

These graphs are isomorphic. Both have six vertices and there is a vertex correspondence A – R, B – P, C – S, D – Q, E – U, F – T.

Do you remember the exploration at the beginning of Chapter 1?

Exploration 5.2

Konigsberg revisited

The town of Konigsberg in East Prussia was built on the banks of the River Pregel, with islands which were linked to each other and the river banks by seven bridges. The citizens of the town tried for many years to find a route for a walk which would cross each bridge only once and allow them to end their walk where they had started. Can you find a suitable route?

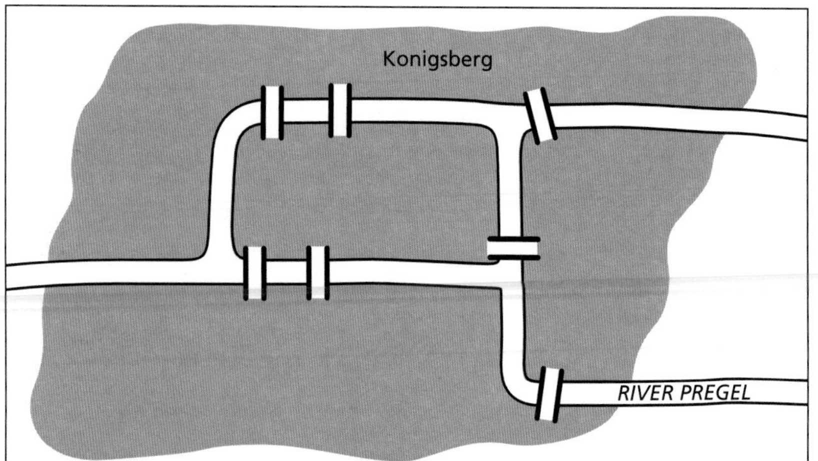

He published his proof in a paper entitled *Solution Problematis ad geometriam situs a pertinentis* (The solution of problem relating to the geometry of position) in 1736.

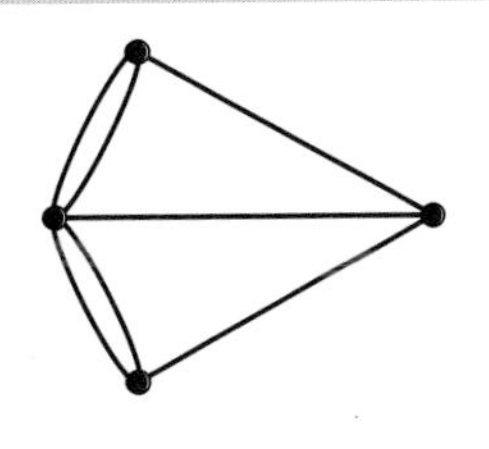

It is said that the inhabitants of Konigsberg tried unsuccessfully to find a route for many years, but it was not until Euler tackled the problem that it was proved to be impossible. Although his proof was not written in the language of modern graph theory, it contains many of the ideas that have formed the basis of the subject and can reasonably be described as the first paper on the subject.

We can now represent the town as a graph where the land areas are the vertices and the seven bridges are the edges.

The problem becomes that of finding an Euler cycle in this graph. We met the idea of an Euler cycle in Chapter 4, *Graphs and networks 3: The travelling salesman problem*. We need a path which passes along every edge of the graph exactly once and returns to its starting point. Euler not only showed that it is impossible to find such a path in Konigsberg, but extended his ideas to any graph and hence discovered a rule for finding when a graph contains a cycle which passes along every edge exactly once.

Exploration 5.3

Eulerian graphs

Consider the following graphs: are they Eulerian?
Can you suggest a method by which you might identify a graph which is Eulerian?

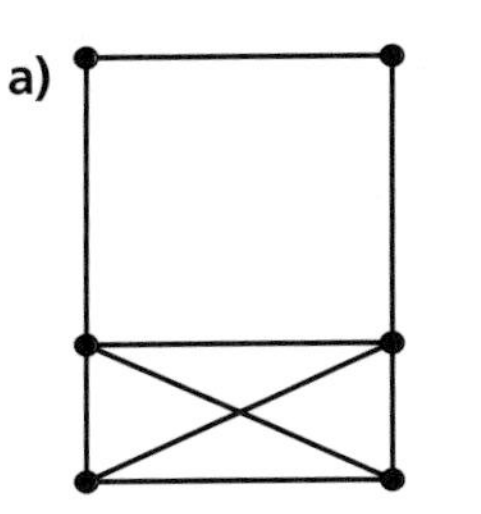

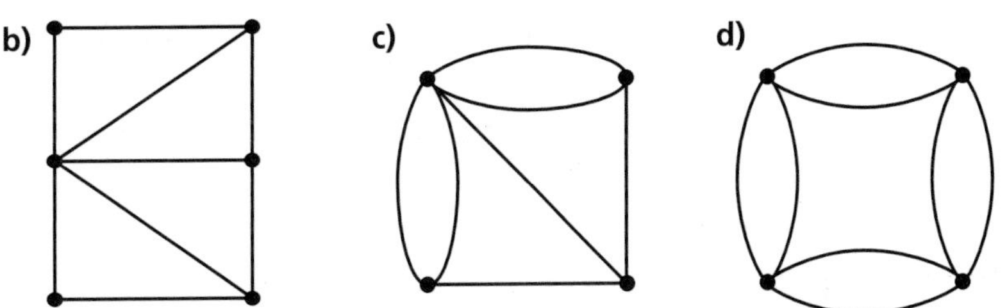

Solving the Konigsberg bridge problem

Euler discovered that for a connected graph to contain a cycle of this type, the degrees of all the vertices must be even numbers. The third graph (c) in Exploration 5.3 is isomorphic to the one in the Konigsberg problem and is therefore a representation of the Konigsberg bridges. We can see that this is not an Eulerian graph since all the vertices are odd. Thus the problem posed by the residents of Konigsberg has no solution, unless you repeat one or more of the bridges.

EXERCISES

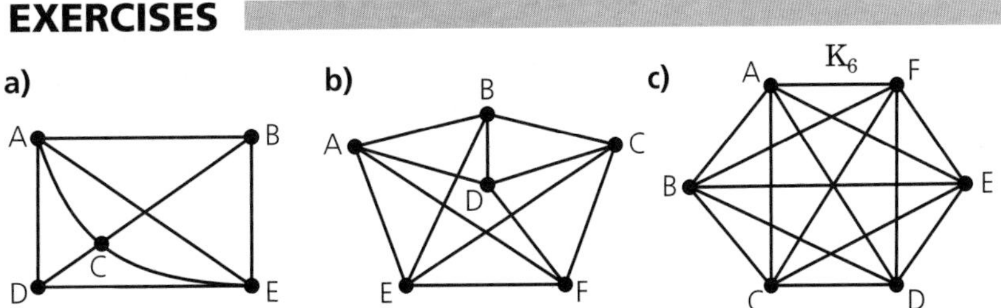

1 Identify which of the graphs in the figure above are planar, redrawing them where necessary. Check your results using Euler's formula.

2 For all the graphs in the figure above, state the degree of every vertex and confirm that the Handshaking theorem holds.

3 State which of the graphs in question 1 are Eulerian, giving an Euler cycle where possible.

4 Identify which of the following graphs are isomorphic.

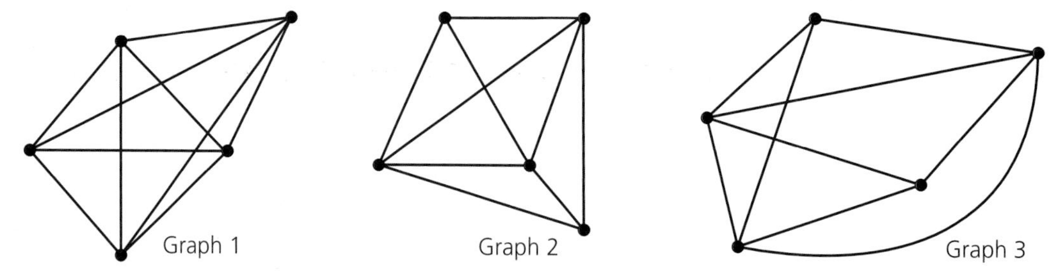

5 A traversable graph is one which you can draw without removing your pen from the paper and without going over the same edge twice.

This graph is traversable. This graph is not traversable.

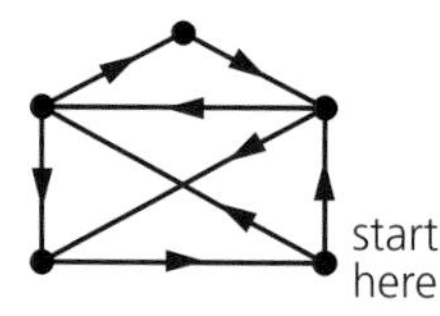

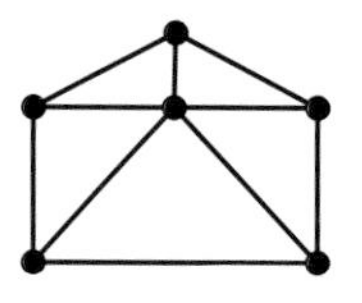

Are all traversable graphs also Eulerian? Explain your answer.

EXERCISES

5.1 HOMEWORK

1 Identify which of the following are planar graphs.

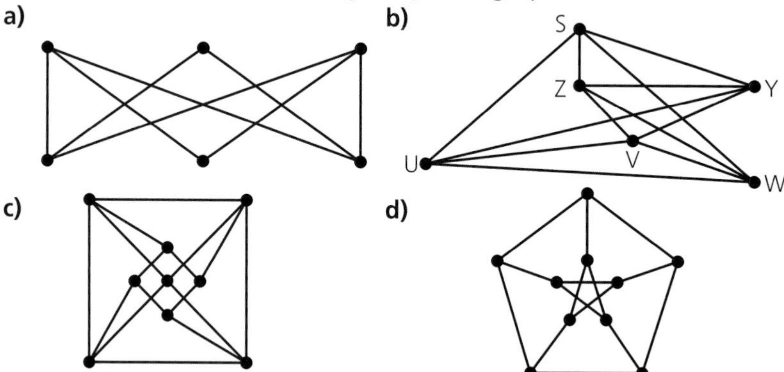

For those which can be redrawn with no edges crossing, check your result using Euler's formula.

For all the graphs above, determine the degree of each vertex and confirm that the Handshaking theorem holds. For each of them that is not planar, draw a graph that is isomorphic to it.

2 State which of the graphs in question 1 are Eulerian. For each that is, construct an Euler cycle.

3 Identify the edge-disjoint cycles in this Euler graph and use them to construct an Euler cycle.

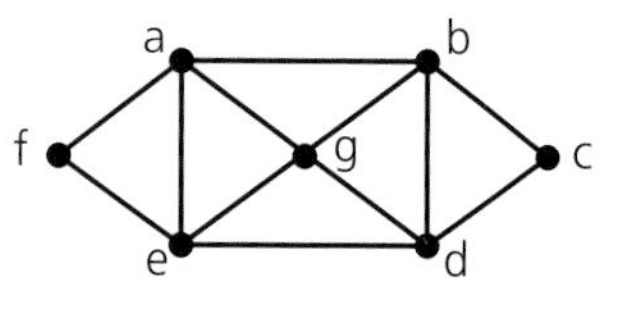

Note: *Edge-disjoint cycles are cycles, in the same graph, that have no edges in common.*

4 Use the two results $v - e + f = 2$ and $2e \geq 3f$ to deduce that $e \leq 3v - 6$. Hence confirm that K_5 is non-planar.

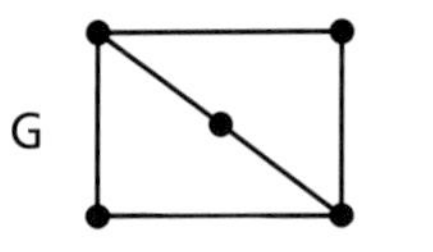

5 Consider this graph.

Show that the Handshaking theorem holds for the graph.
Draw a graph which has the same degree-sequence as G which
a) is isomorphic to G **b)** is not isomorphic to G.

ROUTE INSPECTION PROBLEMS

The 'Chinese postman'

The problem in Exploration 5.1 was to find a route through the network that traverses every edge exactly once and returns to its starting point. This type of problem, a route inspection problem, is often called the 'Chinese postman problem' because it was first published by the Chinese mathematician Mei-Ko Kwan in 1962.

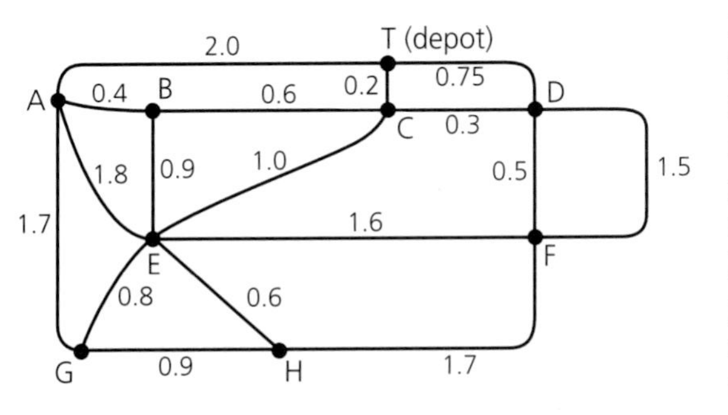

If the network contains an Euler cycle then there is no problem. The optimum length is simply the sum of the weights on all the edges. If the network does not contain an Euler cycle, then the postman will have to repeat some of the roads and the problem is then which should he repeat, to keep his journey to a minimum. To solve the problem we first need to draw a network to represent the area that the postman needs to cover.

The roads are the edges of the network and the intersections are represented by vertices.

The next step is to list the degrees of all the vertices.

A 4 B 3 C 4 D 4 E 6 F 4 G 3 H 3 T 3

We can see that there is no Euler cycle since the network contains four odd vertices: B, G, H and T, so we shall need to repeat some of the edges. The next question is which do we repeat?

Consider a vertex which has three edges joining it. If we move to vertex B along AB and leave along BC, we will need to re-enter B along BE and there is no way out, other than by repeating an edge, so a vertex of odd degree will need to have at least one edge at that vertex repeated. We choose to repeat the edges joining odd vertices which will add the minimum distance to the solution. The best way to do this is to consider the distances between all possible pairings of odd vertices like this.

Pairings of odd vertices	Distance	Route
TB and GH	0.8 + 0.9 = 1.7	(TCB + GH)
TG and BH	2 + 1.5 = 3.5	(TCEG + BEH)
TH and BG	1.8 + 1.7 = 3.5	(TCEH + BEG)

We choose to repeat the edges which add the smallest distance, in this case TB and GH. The length of the shortest route is then simply the sum of all the edge lengths plus the extra length due to the repeated edges.

17.25 + 1.7 = 18.95

The distance the postman will have to travel is 18.95 km and a possible route would be:

TABEAGHEGHFDFECDTCBCT

This is shown in the diagram.

Note: There may be several possible routes of this length.

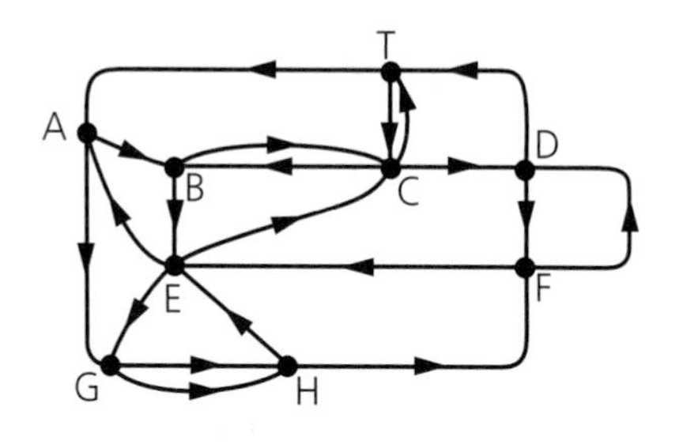

The algorithm can be stated like this:

Step 1: Identify all the odd vertices in the network.
Step 2: Consider all the routes joining pairs of odd vertices and choose the routes of shortest distance.
Step 3: Find the sum of the weights on all the edges in the network and add the distance found in step 2.
Step 4: Find a possible route round the network which repeats the edges identified in step 2 and is of the correct distance.

EXERCISES

5.2 CLASSWORK

1 During the school holidays, a group of children decide to plan a route which goes along all the paths in the local park, starting and ending at the main gate. Adrian says it is possible to do this without repeating any of the paths, but Mallika disagrees with him. Who is right? Devise a route they might take, saying how many paths, if any, they need to repeat.

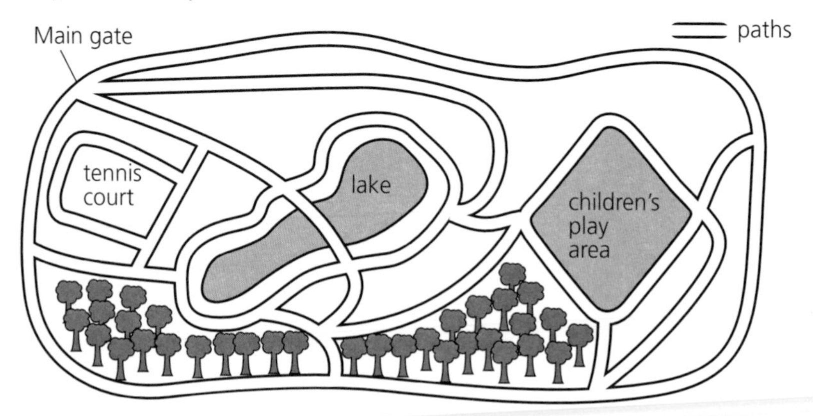

2 Mr and Mrs McCreedy do their shopping at Supasavers every week. The layout of the shop is shown in the plan below, the shaded areas are display shelves.

From the plan, draw a graph to represent Supasavers and plan a route which enables them to look at the goods on all the display shelves while covering the minimum distance.

3 Solve the route inspection problem for this network, starting and ending at S, stating clearly which edges you would need to repeat.

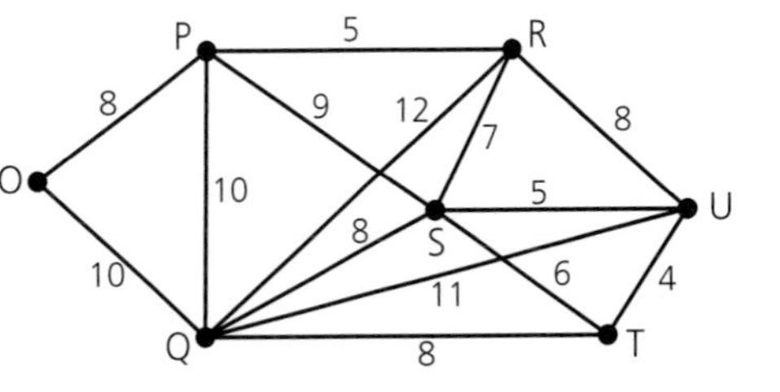

4 The morning after a severe gale, the Cambria highways department set off from their depot to inspect all the roads to make sure none of the towns are cut off by fallen trees. Find a route for them which is as short as possible.

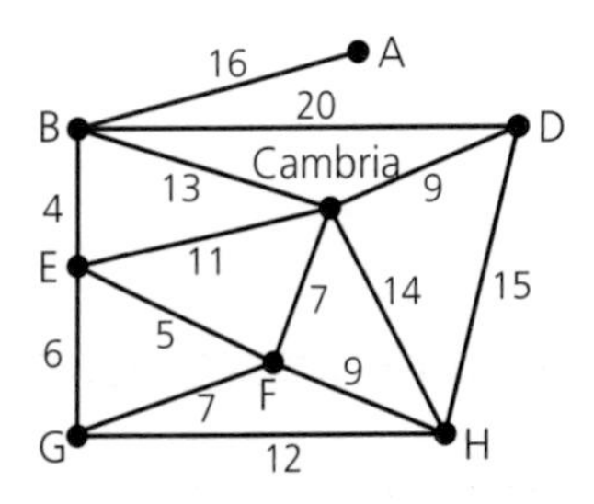

5 The distances between eight villages are given in the table below. Use the table to devise a route of minimum length, starting and ending at A, which travels along each route at least once.

	A	B	C	D	E	F	G
A	–	4	7	8	–	–	–
B	4	–	5	–	4	5	–
C	7	5	–	9	–	–	–
D	8	–	9	–	–	–	12
E	–	4	–	–	–	7	8
F	–	5	–	–	7	–	6
G	–	–	–	12	8	6	–

If the road between E and F is closed due to road works, how will this affect your method? What is the new solution?

EXERCISES

5.2 HOMEWORK

1 The ground floor plan of a house is given below.

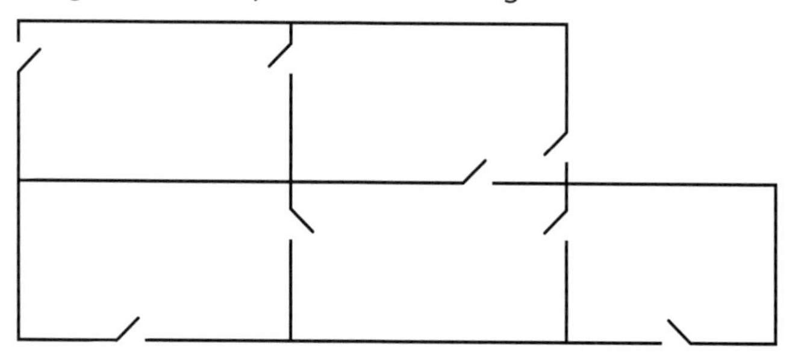

Draw a graph to represent the above. Discuss whether it is possible to walk in and out of the house so that each door of the house is used exactly once.
(**Hint:** Consider the cases whether you are going to return to the starting position or not.)

2 You are given a domino set with the doubles removed. We can represent the pieces as (0, 1), (0, 2), ... , (0, 6), (1, 2), ... , (1, 6), ... , (5, 6). Discuss the possibility of arranging the pieces in a connected series such that one number on a piece (as illustrated here) always touches the same number on its neighbour).

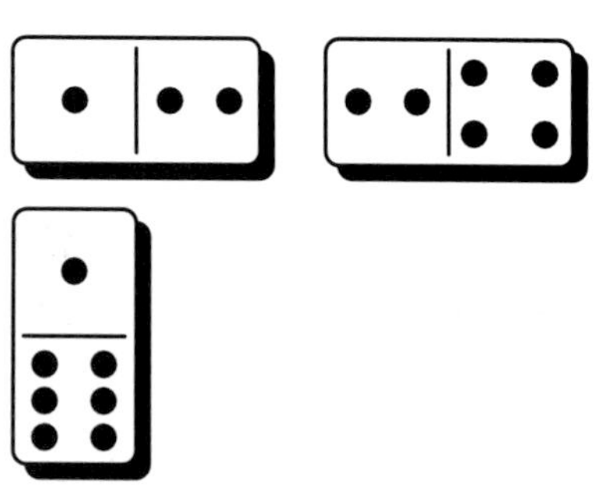

(**Hint:** Use a seven-vertex complete graph and check if it is Eulerian.) Would inclusion of the doubles alter your conclusions?

3 The distances between villages in the Welsh valleys are given in the table below. Devise a route of minimum length, starting and ending at Dowlais, which visits every village. Determine this minimum length if all distances are quoted in miles.

	Dowlais	Treharris	Aberdare	Abercynon	Nelson	Bargoed
Dowlais	–	10	6	–	–	8
Treharris	10	–	–	4	2	–
Aberdare	6	–	–	7	–	–
Abercynon	–	4	7	–	3	–
Nelson	–	2	–	3	–	8
Bargoed	8	–	–	–	8	–

4 A gas pipeline has to be laid alongside some of the existing roads in a region in order to link all the towns in the map. At each town there is a pumping station which ensures the gas can be pumped to or from adjacent towns.

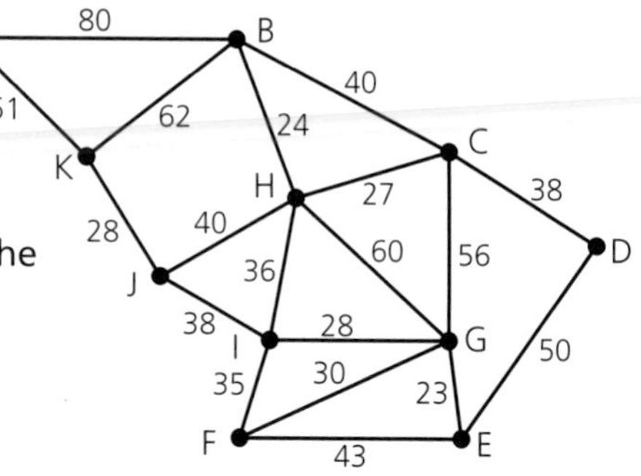

a) Before the project can go ahead the entire length of the road network has to be inspected. Determine the minimum distance for the inspection route.

b) The connectivity needs to be increased and one extra pipeline can be built between any pair of towns. The pipeline can go across country but it must not cross an existing pipeline. Draw in the new pipeline and give an estimate of the distance.

5 A clothing company, MH limited, is to make sweaters that will include their logo which is in the form of a network incorporating the company's initials. A diagram of the rectangular shaped logo is shown here, where the numbers indicate the lengths, in millimetres, between vertices.

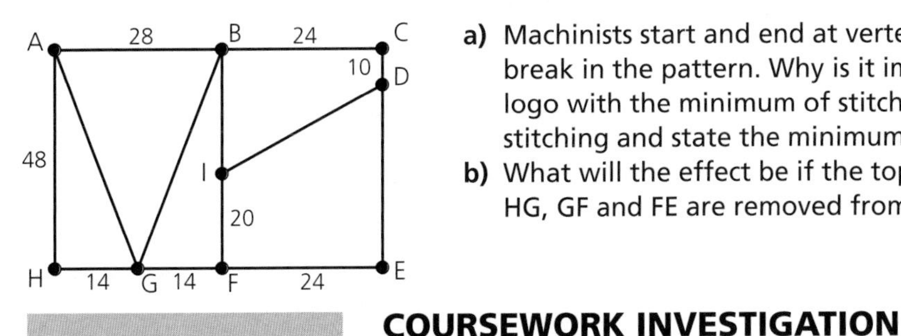

a) Machinists start and end at vertex A, stitching without a break in the pattern. Why is it important to produce the logo with the minimum of stitching? Determine the order of stitching and state the minimum length.

b) What will the effect be if the top and bottom edges AB, BC, HG, GF and FE are removed from the logo design?

COURSEWORK INVESTIGATION

Investigation

Museum visit

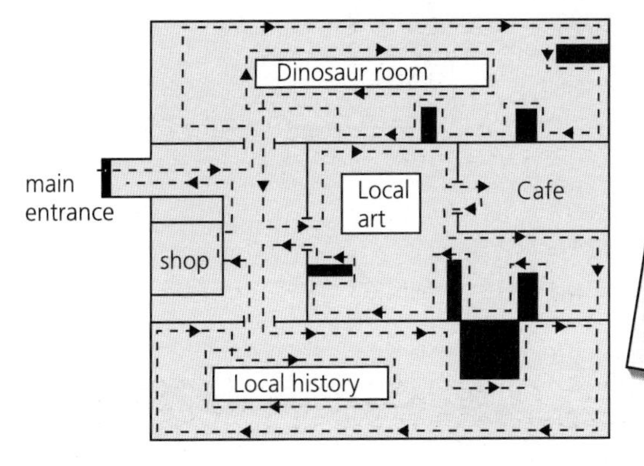

You are organising a trip to a local museum with a party of children. You want them to see all the exhibits, but do not want them to have to walk further than necessary. By modelling the museum as a network, plan a suitable route.

By considering sensible timings, draw up an itinerary for the visit.

CONSOLIDATION EXERCISES FOR CHAPTER 5

1 A connected graph G has six vertices and nine edges.

a) State the sum of the degrees of the vertices of G.

b) If G has two vertices with order four, draw G and give the degrees of the other vertices.

c) State, with reasons, whether G is (i) planar (ii) Eulerian.

2 A simple graph, G, has five vertices, and each of those vertices has the same degree, d.

a) State the possible values of d.

b) If G is connected, what are the possible values of d?

c) If it is Eulerian, what are the possible values of d?

d) If G is connected and planar, what is the only possible value of d? Give an illustration of such a graph.

(AEB Question 3, Discrete Mathematics Specimen Paper Pure 4, 1996)

3 **a)** How many continuous pen strokes are needed to draw the graphs in the diagrams below?

b) If a graph has k vertices of odd degree, what is the smallest number of continuous pen strokes needed to draw the graph?

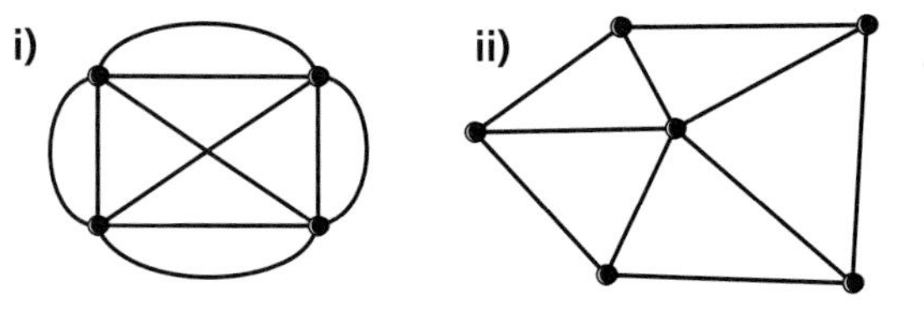

4 Solve the route inspection problem for this network.

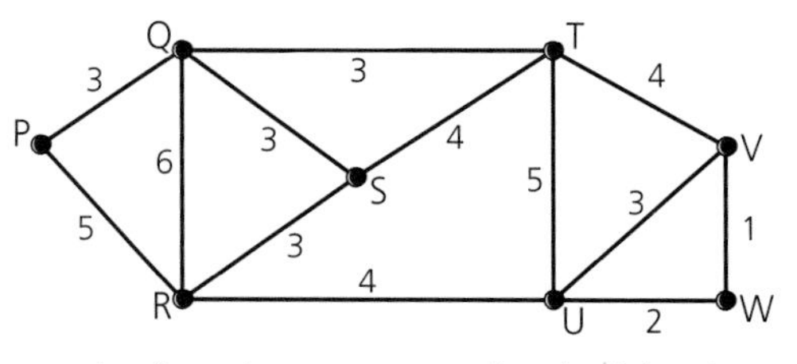

5 Sarah and Martin decide to spend a few days on a touring holiday in the region shown in this network. They wish to travel along all routes shown, using the local buses. The bus fares between towns are shown in the table. Assuming that they wish to start and finish at S, what route should they take in order to minimise the cost?

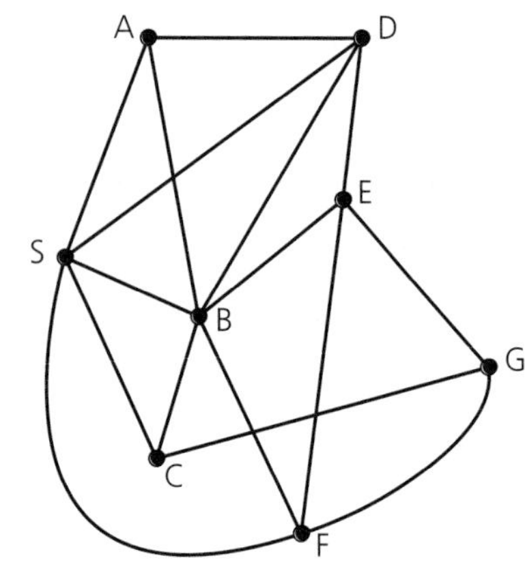

	S	A	B	C	D	E	F	G
S	–	83	95	100	150	–	195	–
A	83	–	75	–	88	–	–	–
B	95	75	–	60	115	90	120	–
C	100	–	60	–	–	–	–	135
D	150	88	115	–	–	73	–	–
E	–	–	90	–	73	–	110	65
F	195	–	120	–	–	110	–	76
G	–	–	–	135	–	68	76	–

6 The clothing company, New Xperience, has decided to break into the sportswear market. It has been decided to decorate all garments with a simple logo, in the form of a network incorporating the company's initials. A diagram of the logo is shown below. Vertices have been labelled A, B, C, D, E, F, G for convenience. The numbers indicate distances in cm between vertices.

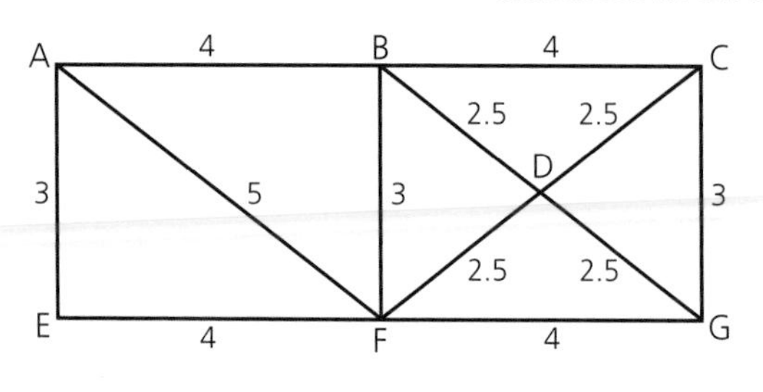

a) Plans involve the production of millions of garments, so it is important to produce the logo with the minimum of stitching. Machinists will start and end stitching at A, with no breaks. Explain why it will be necessary to oversew some edges. Use an appropriate algorithm to determine the order of stitching which minimises the length sewn. Explain each step carefully. Give your minimum length, and a corresponding order in which vertices should be stitched.

b) Amongst the items produced will be an ice-skating outfit. It is decided that, for this outfit, the standard logo will be decorated by sewing on a sequin at each of the seven vertices. The stitching for this additional feature is to be done without breaks, starting and ending at A.
Construct a minimum spanning tree for the network, explaining your criteria for including each of the arcs. Use it to calculate an upper bound for the length a machinist must sew in order to attach the sequins. Indicate how to reduce the upper bound by introducing circuits. The machinists start at A. They always stitch along arcs to the nearest vertex with no sequin, before returning to A. Give the three orders of vertex stitching that are possible under this policy, and the corresponding lengths of stitching.

(UODLE Question 12, Decision Mathematics AS, 1991)

7

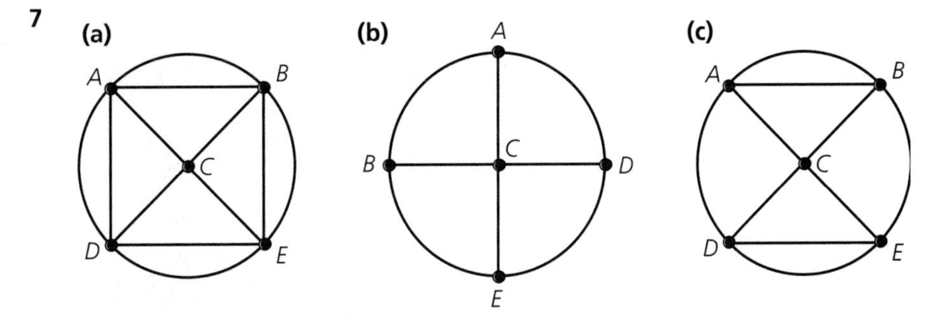

i) By considering the order of each node, classify each of the graphs **(a)**, **(b)** and **(c)** shown in the diagram as Eularian, semi-Eulerian or neither.

ii) Explain briefly how your classification in part **(i)** relates to the problem of finding a route through the graph that includes each arc exactly once. Are there any restrictions on where such a route can start and finish?

(OCR Discrete Mathematics 1 Specimen Paper 2000)

8 The map shows a number of roads in a housing estate. Road intersections are labelled with capital letters and the distances in metres between intersections are shown. The total length of all of the roads in the estate is 2300 m.

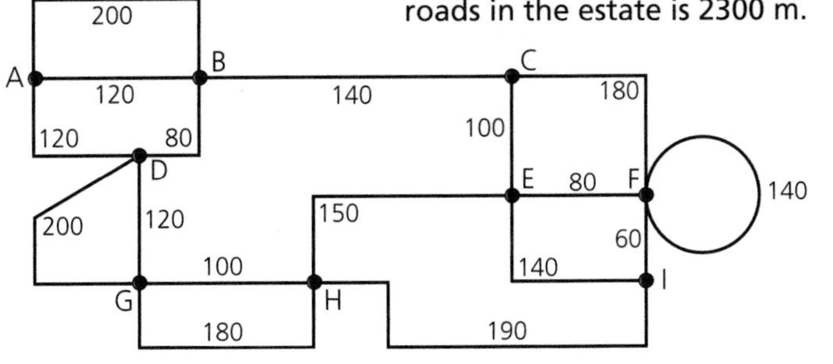

a) A newspaper deliverer has to walk along each road at least once, starting and ending at A. The shortest route to achieve this is required.

i) List those intersections which are of odd order and explain their significance to the problem.

ii) By investigating all possible pairings of odd intersections, find the minimum distance which the newspaper deliverer has to walk. (You are not required to apply a shortest distance algorithm to solve this. You are required to show the distance computations which lead to you choosing a particular pairing of odd intersections.)

iii) For each intersection other than A, give the number of times that the newspaper deliverer must pass through that intersection whilst following the shortest route.

Intersection	B	C	D	E	F	G	H	I
No. of visits								

b) The newspaper deliverer only calls at a proportion of houses. The postwoman has to call at most houses, and since the roads are too wide to cross continually back and forth, she finds it necessary to walk along each road twice, once along each side. She requires a route to achieve this in the minimum distance.

i) Describe how to produce a network to model this problem.

ii) Without drawing such a network, say why it will be traversable and calculate the minimum distance which the postwoman will have to walk.

c) The street-cleaner needs to drive his vehicle along both sides of each road. He has to drive on the correct side of the road at all times. He too would like a shortest route.
Explain how the street-cleaner's problem differs from the postwoman's, and say how the network would have to be modified to model this.

(O & C MEI Question 4, Decision & Discrete Mathematics Paper 19, 1996)

9 a) Prove that the graph G_1 is non-planar.

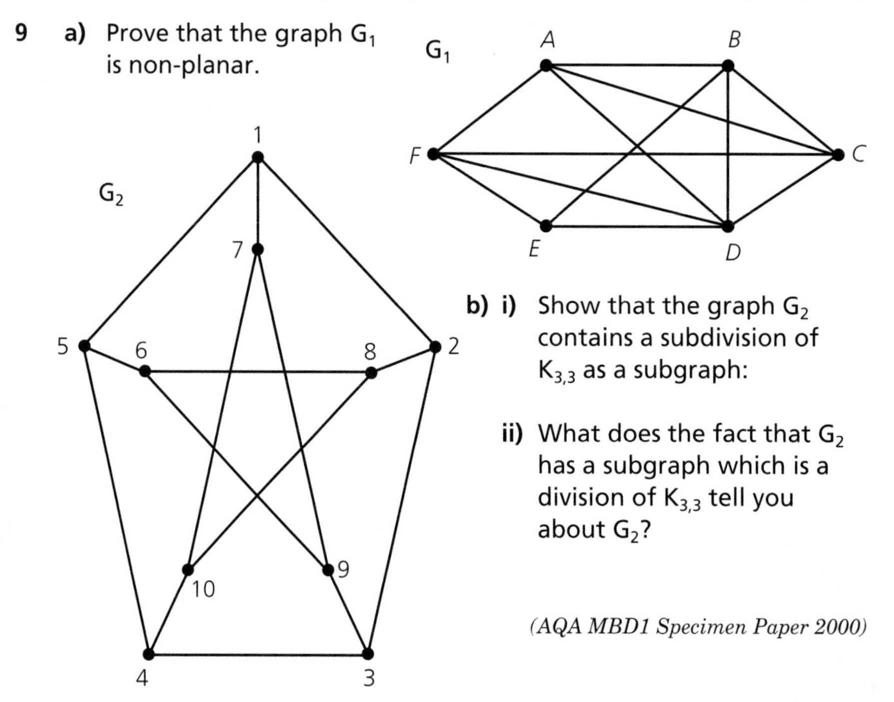

b) i) Show that the graph G_2 contains a subdivision of $K_{3,3}$ as a subgraph:

ii) What does the fact that G_2 has a subgraph which is a division of $K_{3,3}$ tell you about G_2?

(AQA MBD1 Specimen Paper 2000)

10 A commercial baker is to produce an iced and decorated cake in large numbers for retail in a supermarket chain. The decoration consists of a pattern of piped icing as shown here.

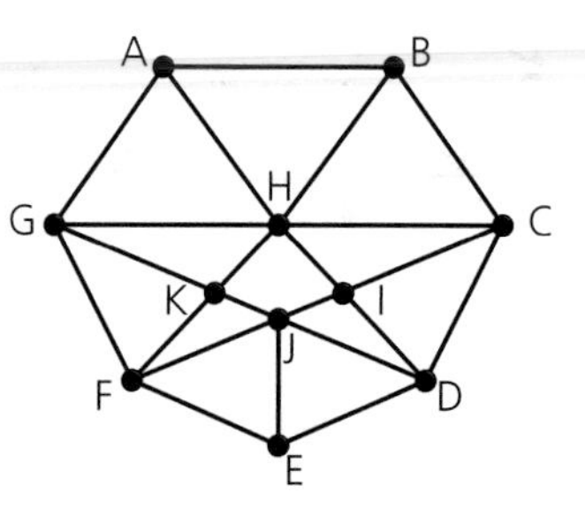

	A	B	C	D	E	F	G	H	I	J	K
A	–	120	–	–	–	–	100	100	–	–	–
B	120	–	100	–	–	–	–	100	–	–	–
C	–	100	–	90	–	–	–	110	90	–	–
D	–	–	90	–	110	–	–	–	80	100	–
E	–	–	–	110	–	110	–	–	–	80	–
F	–	–	–	–	110	–	90	–	–	100	80
G	100	–	–	–	–	90	–	110	–	–	90
H	100	100	110	–	–	–	110	–	70	–	70
I	–	–	90	80	–	–	–	70	–	60	–
J	–	–	–	100	80	100	–	–	60	–	60
K	–	–	–	–	–	80	90	70	–	60	–

The piping consists of straight lines of icing connecting the nodes A–K as shown above. The piping nozzle can be programmed to move from a node directly to any other node and to dispense or not to dispense icing *en route*. The approximate distances between nodes are shown in mm in the matrix.

a) The sum of lengths of the arcs of the network is 2020 mm. By referring to the orders of the vertices explain why it will be necessary for the nozzle to cover more than 2020 mm in order to complete the icing pattern.

b) Use Dijkstra's algorithm to find the shortest distances from A to J and from A to E. Show all of your working, including the order in which you assign permanent labels to vertices. Give routes associated with your shortest distances, explaining how you found them.

c) Use the method of solving the route inspection (Chinese postman) problem to find the shortest distances that the nozzle can travel to complete the pattern. (You are not required to use an algorithm to find the shortest distances between vertices other than those required in **b)**.)
Give the shortest distance and an appropriate route, indicating when then nozzle should not be dispensing icing.

d) An alternative design for the icing pattern is shown below. Count how many odd nodes there are in this pattern. In how many ways can they be paired together to construct possible solutions for the corresponding route inspection problem?

(UODLE Question 9, Applied Mathematics Paper 4, 1994)

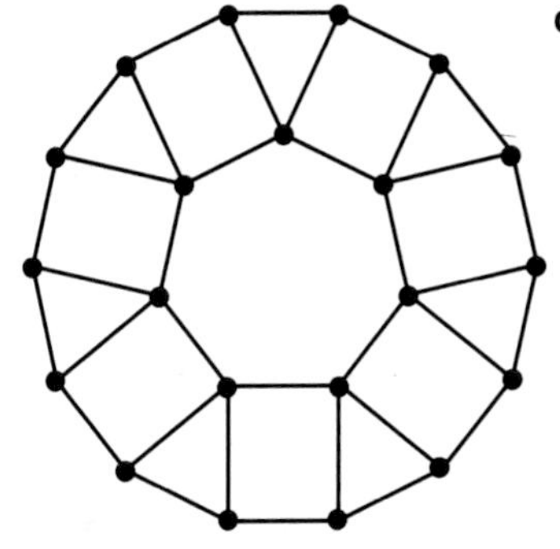

Summary

After working through this chapter you should:

- know that planar graphs are graphs that can be drawn in the plane with no edges crossing and that a simple graph is one in which there are no loops and any pair of vertices is connected by no more than one edge
- know that isomorphic graphs are graphs that have the same number of vertices and the degrees of corresponding pairs of vertices are the same
- know that the degree of a vertex is the number of edges incident on the vertex
- be able to say whether a graph is Eulerian and find an Euler cycle, which travels along every edge of a graph and finishes up at its starting point
- be able to apply these techniques to solve route inspection (Chinese postman) problems in networks.

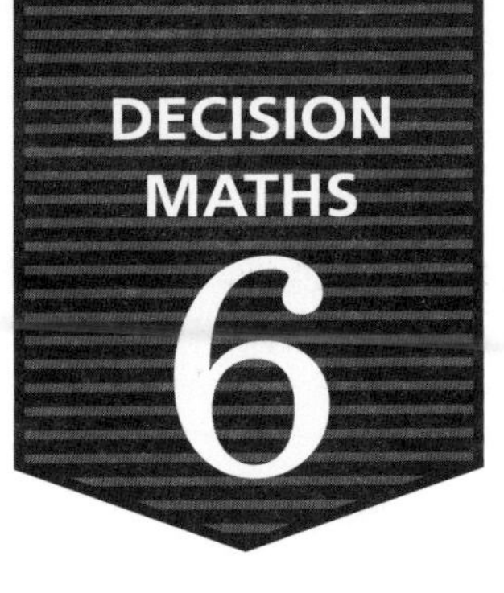

Critical path analysis

By the end of this chapter you should be able to:

- *draw precedence networks, which show the 'order' in which steps have to be carried out, using activity on vertices and activity on arcs procedures*
- *find the critical path, which is the minimum time needed to complete a project, following the longest path through the precedence network and including all the critical activities which are activities that must be started on time to avoid delaying the whole project*
- *draw cascade charts showing precedence relations, in the form of bars drawn against a time scale, that help to make best use of all resources*
- *use resource levelling to balance the amount of work to be done and the resources available for a project.*

Exploration 6.1

Kate's decorating – 1

Kate decides to redecorate her room. She buys some wallpaper and paint, and then sits down to plan the job. What tasks will she need to do and in what order? She is hoping to get it done over a weekend. Is she being realistic?

PRECEDENCE NETWORKS

Many companies use network analysis to help in the planning of large projects in order to ensure efficient use of time and resources. If, like Kate in the exploration, you are planning a project, the first thing to do is make a list of the steps involved in the project from start to completion, then estimate the length of time each step will take. Kate's decorating job might look like this.

	Activity	Duration (hours)
A	Tidy the room and move the furniture.	2
B	Remove the curtains and curtain rail.	0.25
C	Strip the old wallpaper.	1.5
D	Wash and sand the paintwork.	2
E	Apply a first coat of paint to ceiling and woodwork and let it dry.	6
F	Apply a second coat of paint and let it dry.	6
G	Hang the wallpaper.	5
H	Hang the curtains.	0.25
I	Reinstate the furniture.	1.5

Next, Kate needs to consider which activities need to be completed before others can be started so she can draw up a list of **precedence relations**, showing which activities must immediately precede others.

	Activity	Duration (hours)	Preceding activity
A	Tidy the room and move the furniture.	2	
B	Remove the curtains and curtain rail.	0.25	
C	Strip the old wallpaper.	1.5	A, B
D	Wash and sand the paintwork.	2	C
E	Apply a first coat of paint and let it dry.	6	D
F	Apply a second coat of paint and let it dry.	6	E
G	Hang the wallpaper.	5	C
H	Hang the curtains.	0.25	F, G
I	Reinstate the furniture.	1.5	H

*Precedence networks are sometimes called **activity networks.***

Having done all of this we can construct a **precedence network**, which is a way of showing this information on a diagram.

Constructing precedence networks

Some examination boards specify one particular method for constructing these networks. You should check your syllabus before proceeding.

There are two methods for constructing precedence networks. We shall consider each in turn.

Activity on vertex (node)

In this type of diagram, the activities are represented by the vertices of a network. The edges show the order of precedence.

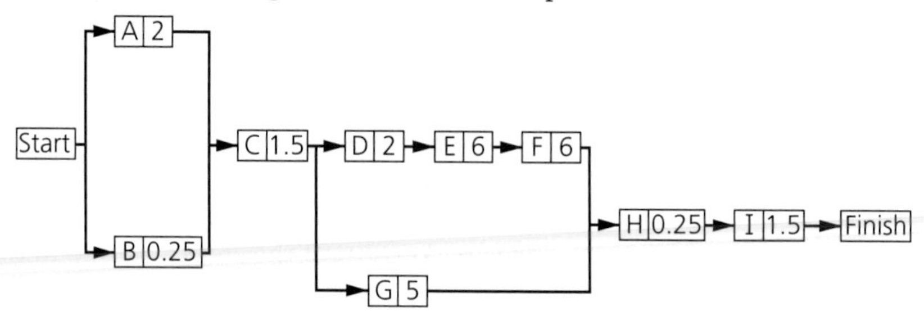

Here is the technique for drawing this type of network.

- You must have a start vertex and a finish vertex.
- Each activity is represented by a vertex which shows the activity and the duration.

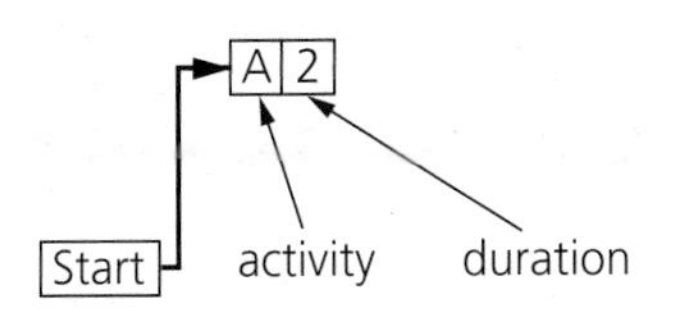

- Edges in the network show only the order in which activities must be done. They indicate where an activity cannot be started until another has been completed, so we can see that activity H (hanging the curtains) cannot be done until both the paint is dry (F) and the wallpaper has been hung (G).

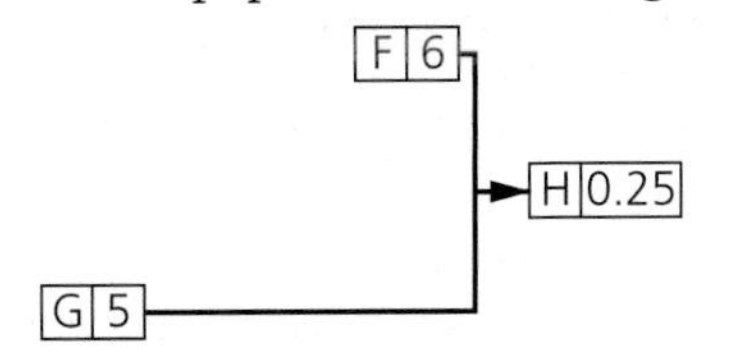

Activity on arc

In this type of network, the arcs represent the activities, while the vertices represent **events**, by which we mean the finish of one activity and the start of another.

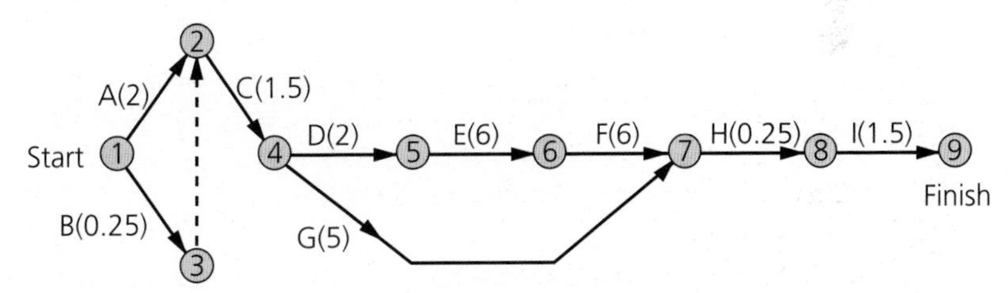

Here is the technique for drawing this type of network.

- You must have a start vertex and a finish vertex.
- Activities are represented by arcs, which show the activity and the duration.
- Vertices represent events which are numbered in such a way that each activity can be described uniquely by a pair of events (i, j) where i is the start event and j is the finish event. For example activity D is represented by (4, 5).
- **Dummy arcs** may need to be introduced to ensure that every activity is uniquely numbered or where an activity is directly preceded by more than one other, as in the case of C. Dummy arcs are marked as dotted lines and have zero duration.

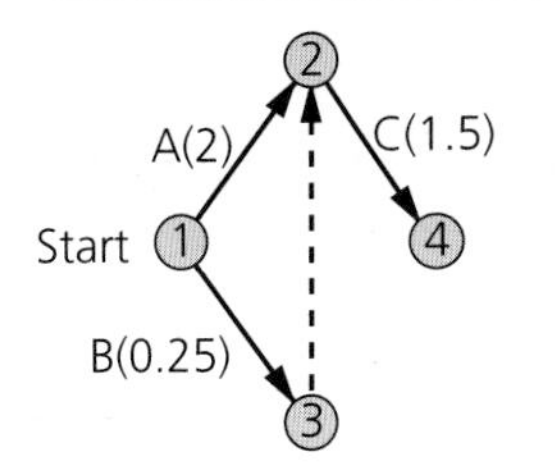

EXERCISES

In this exercise, you may use either type of network.

1 The two diagrams give alternative – but equivalent – ways of representing the precedence with which eight activities of a project must be performed.

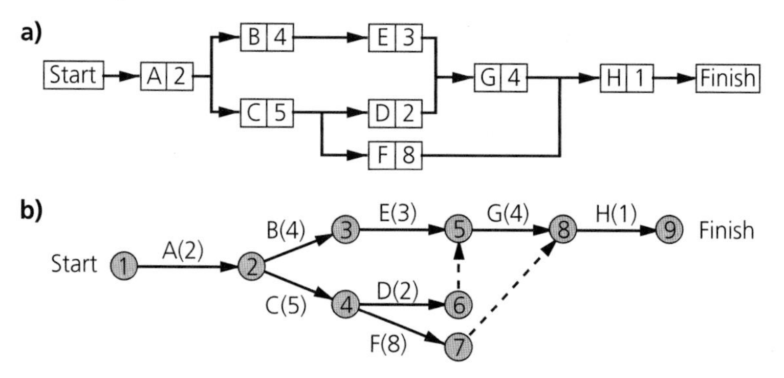

Use the networks to produce a table showing the duration of each activity and its immediate predecessor(s).

2 The table gives a series of tasks, the times taken to complete them and the immediately preceding activities.

Task	A	B	C	D	E	F	G
Duration (minutes)	3	5	5	4	5	8	3
Preceding activity			A	A	B	D, E	F, C

Show this information on a precedence network.

3 A conservatory is to be built on the back of a house. Details of the jobs to be done and timings are shown below.

	Activity	Duration (days)	Preceding activity
A	Lay the foundations.	3	
B	Build the walls.	5	C
C	Lay the drains.	2	A
D	Lay the floor.	1	B
E	Make the door and window frames.	2	
F	Erect the roof.	3	D
G	Fit the door frame and door.	0.5	E, F
H	Fit the windows.	2	G
I	Plaster the walls and ceiling.	8	H

Draw a precedence network to show this information.

4 When Marc returns home from work, he has to prepare and eat his evening meal. His meat pie and dessert are already prepared; the pie needs to be cooked and the dessert has to be removed from the refrigerator just before eating.

The table shows the activities which need to be done.

	Activity	Duration (minutes)	Preceding activity
A	Set the table.	5	
B	Heat the oven.	3	
C	Peel the potatoes.	5	
D	Cook the potatoes.	20	
E	Put the pie in the oven.	1	
F	Cook the pie.	30	
G	Cook the frozen peas.	10	
H	Eat the main course.	15	
I	Take the dessert from fridge.	1	
J	Eat the dessert.	7	
K	Wash up.	8	

Copy the table and fill in the 'preceding activity' column.

Show the information on a precedence network.

5 You are planning a dinner party for friends and wish to serve a three-course meal. Decide on a menu, then list the activities that will need to be completed. Make a table of times and precedence and draw a network for preparing your meal.

EXERCISES

6.1 HOMEWORK

1 Each of the following represents a part of a complete network. They are not necessarily complete in themselves. Construct networks to represent them, using the activity on arcs method.

a) Activities X and Y depend on activities A and B.

b) Activity X depends on activity A and activity Y depends on activities A and B.

c) Activity X depends on activities A and B, activity Y depends on activities B and C and activity Z depends on activity C only.

d) Activity X depends on activity A only, activity Y depends on activity B only and activity Z depends on activities A, B and C.

Repeat the above exercise using the activity on vertex method.

2 Redraw the following networks to eliminate any redundant or unnecessary dummy activities.

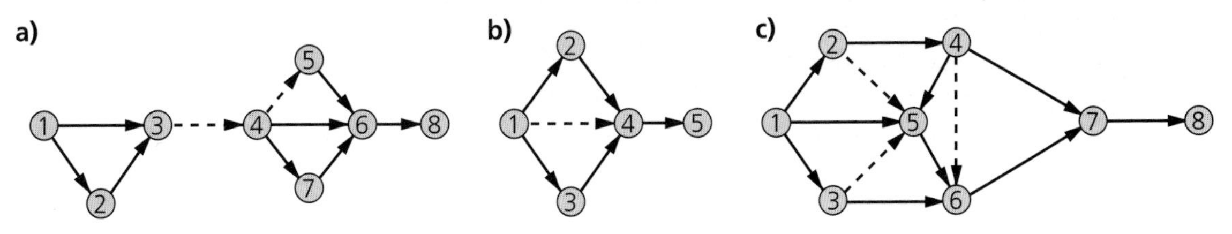

3 List the activities involved in each of the following situations. Construct the two types of precedence networks in each case. State your initial assumptions for each situation.

a) Coffee is made by boiling water then pouring it through a filter filled with coffee. Milk is boiled and added to the filtered coffee and the coffee is served.

b) Three taxis are to be cleaned by two cleaners. One works on the outside of a taxi and the other works on the inside. The inside must be completely cleaned before any work can start on the outside.

c) A picture is to be framed. Assume that the following activities need to be done.

A Obtain the frame materials.
B Make the frame.
C Obtain the glass.
D Fit the glass to the frame.
E Mount the picture.
F Fit the picture to the assembled frame and glass.
G Obtain adhesive tape for the dust excluder.
H Fix the adhesive tape.

4 The tables below give a series of activities, their completion times and immediate preceding activities (**predecessors**). Construct a precedence network.

a)

Task	A	B	C	D	E	F	G	H	I	J	K	L
Duration (weeks)	3	9	3	1	4	3	8	4	5	7	6	10
Preceding activities					A	B, E	C	C	D	D	H, I	J

b)

Task	A	B	C	D	E	F	G	H	I
Duration	5	6	10	6	12	4	11	5	5
Preceding activities		A	A	B	B, C	C	D, E	E, F	G, H

5 A company is to manufacture a piece of equipment made from two parts, A and B, which are assembled together. Each part requires the installation of a machine and the assembly then requires the installation of a third machine, C. The completed piece then needs testing for strength and durability on a specially-made testing rig. Assuming all materials are to hand, the activities and their duration times are as listed in this table.

Activity	Duration (weeks)
Obtain and install the machine for A.	5
Obtain and install the machine for B.	9
Obtain and install machine C.	16
Make part A.	4
Make part B.	7
Assemble the piece of equipment.	7
Make the test rig.	25
Test and dispatch.	8

CRITICAL PATHS

When we are planning a project we usually want to know the minimum time we will need to complete it. We can find this from the precedence network, by finding the longest path through the network from the start vertex to the finish vertex. This path is called the **critical path**. Any delay in the activities along this path would inevitably lead to a delay in completing the whole project.

The algorithm for finding the longest path is in two parts:

- the forward pass, in which we label each activity with its earliest start and finish times
- the backward pass, in which we find the latest time that we can risk starting each activity without delaying the whole project.

The **critical activities** can then be identified as being those that must start at a particular time or the whole project will be delayed. The critical path is the path through the network that includes all the critical activities.

We can demonstrate this on the two types of network developed in the previous section.

Activity on vertex

Consider the problem of decorating Kate's room introduced in Exploration 6.1.

Forward pass

We start by labelling the start vertex 0 (zero). We then label the vertices directly connected to the start with their earliest start time relative to the start time of the project, which is taken as zero. This is shown at the top left-hand corner of the box. On the top right-hand corner we show the earliest finish times which is calculated as:

earliest start time + duration of the activity

and is recorded as the total number of hours from the start of the project.

earliest start time for A → 0 2 ← earliest finish time for activity A

0 A 2

0 Start

0 0.25

B 0.25

Next we consider activity C, which must be preceded by both A and B; this means that the earliest start time for C is 2 hours after the project start time, since it cannot start until A is completed. The earliest finish time for C is 2 + 1.5 = 3.5 hours.

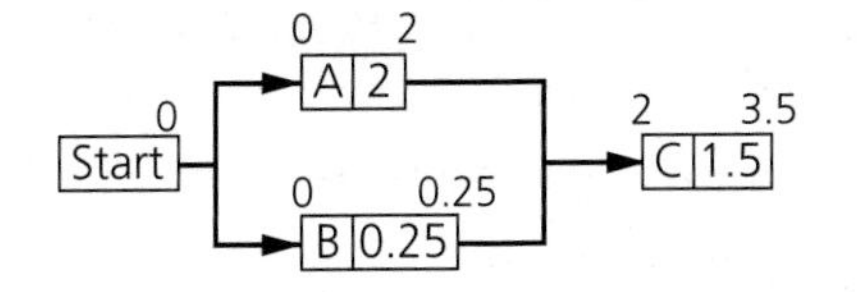

We continue in this way until we have labelled all the vertices with their earliest start and finish times. The network then looks like this.

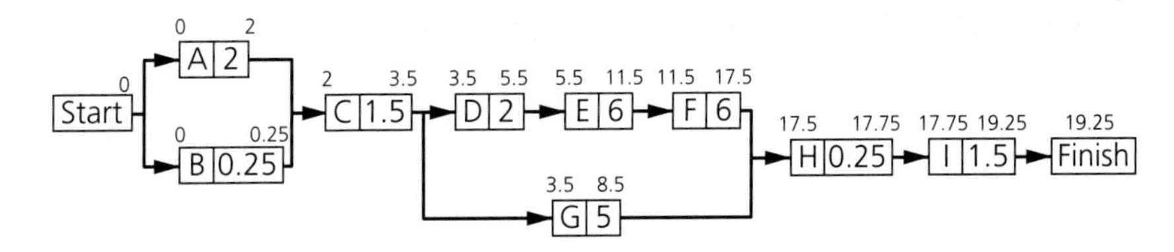

Backward pass

We now work backwards from the finish time of 19.25 hours, putting the labels below the vertices. The earliest finish time for activity I is 19.25 and the latest start time is 17.75, found by subtracting the duration of activity I from the latest finish time. In the same way the latest finish time and start times for activity H can be calculated and marked on the network.

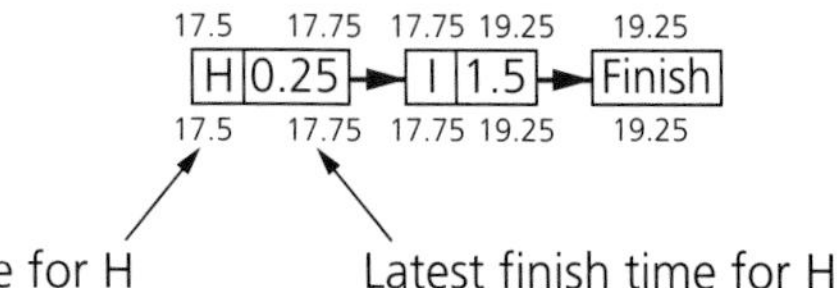

Activity H has two activities directly preceding it, F and G. The latest finish time for both of these will be 17.5 since they must both be completed before H begins. Subtracting the duration of these activities from their latest finish times will give us their latest start times.

activity F: latest start time = 17.5 – 6 = 11.5

activity G: latest start time = 17.5 – 5 = 12.5

These in turn can be marked on the network, which now looks like this.

We continue the backward pass until we reach vertex C, which precedes two other activities. Since we are considering latest start and finish times, we must give C a latest finish time of 3.5 since it must be completed before D is begun.

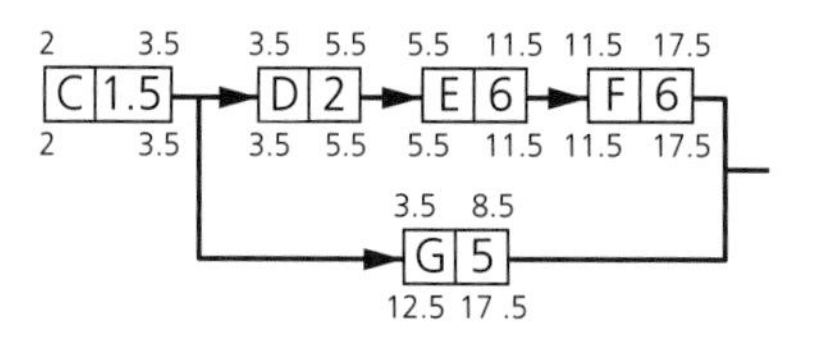

We continue in this way until we have labelled all the vertices with their latest start and finish times.
Our network now looks like this.

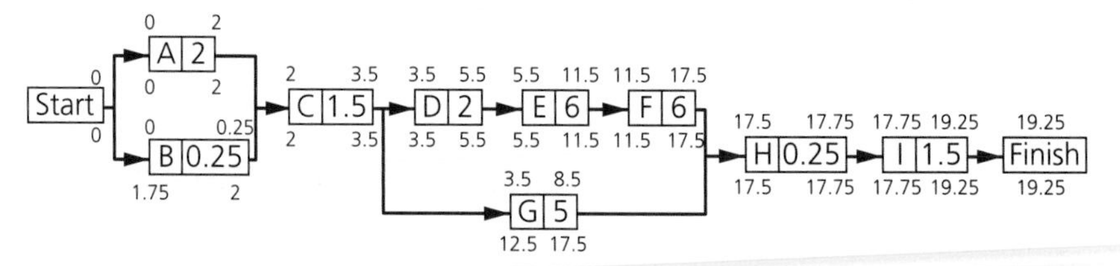

Notice that some of the activities have some flexibility in their start times, for example activity G can be started at any time from 3.5 hours after the start of the project to 12.5 hours after the start of the project without delaying the total time required. The maximum time by which an activity can be delayed without delaying the project is called the **float**. Activity G has a float of 9 hours. The activities with floats are shown below.

Activity	Latest start time	Earliest start time	Float
B	1.75	0	1.75 hours
G	12.5	3.5	9 hours

Activity on arc

Forward pass

Remember, an event is the finish of one activity and the start of another.

When the activities are represented on the arcs, the forward pass finds the earliest event times and the backward pass finds the latest event times. We annotate the network using a double box like this.

earliest event time — latest event time

The forward pass gives these values for the earliest event times.

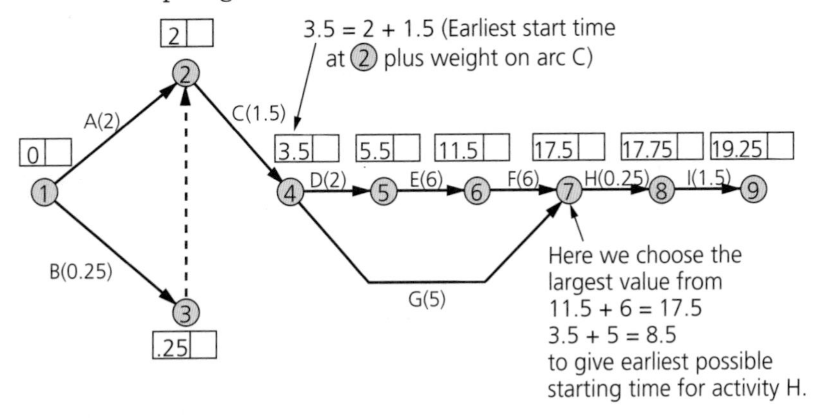

Backward pass

The backward pass gives these values for the latest event times.

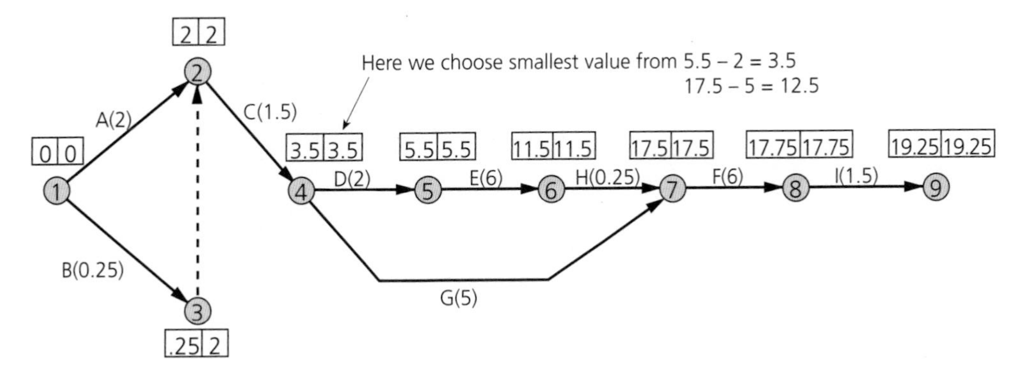

Once again we can identify the activities with float from the network by considering the event times. For activity (i, j):

float = latest event time for j − earliest event time for i − duration of activity

For example, in the case of activities B(1, 3) and G(4, 6):

for B: float = 2 – 0 – 0.25 = 1.75 hours

for G: float = 17.5 – 3.5 – 5 = 9 hours

Showing the critical path on a network diagram

As we stated earlier, the critical activities *must* start at a particular time or the whole project will be delayed. The activities that have no flexibility in their start times are those with a zero float. In our example these are A, C, D, E, F, H and I, so the critical path must be the path through these activities.

The critical path can be shown on the network diagram like this.

Activity on vertex

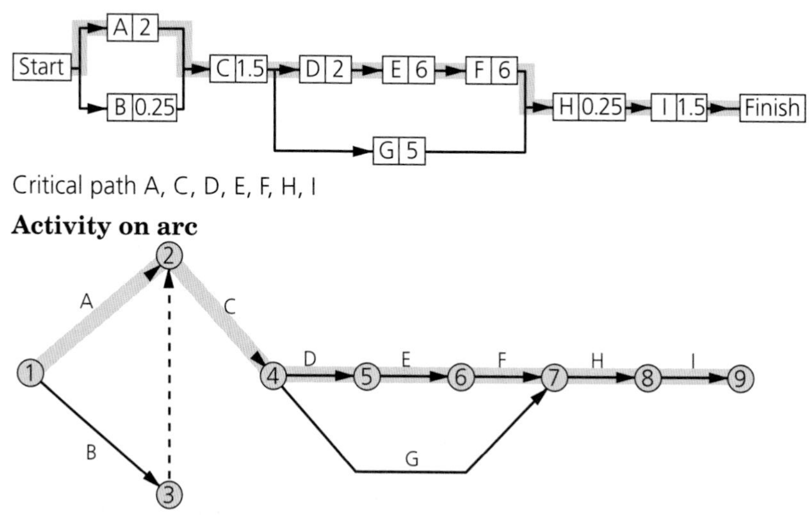

Critical path A, C, D, E, F, H, I

In the previous example, the activities which had float were independent of each other. This is called an **independent float**. Now consider this example.

Example 6.1

Kieran decides to make a wooden chair for his niece's third birthday. The activities necessary to complete the project are shown in this table.

	Activity	Duration (hours)	Preceding activities
A	Make the wooden base.	2	
B	Make the back.	2.5	
C	Make the legs.	5	
D	Cut the foam for the cushions.	1	
E	Make covers for the cushions.	3	D
F	Assemble the chair.	2	A, B, C
G	Assemble the cushions.	1	E
H	Paint the chair.	4	F
I	Fix the cushions to the chair.	0.5	G, H

If we construct the precedence network and find the critical activities and the activities with float, we can see that some of them are linked by precedence, so that if one activity with a float, say D, does not start at its earliest start time, it affects the amount of float available for activities E and G.

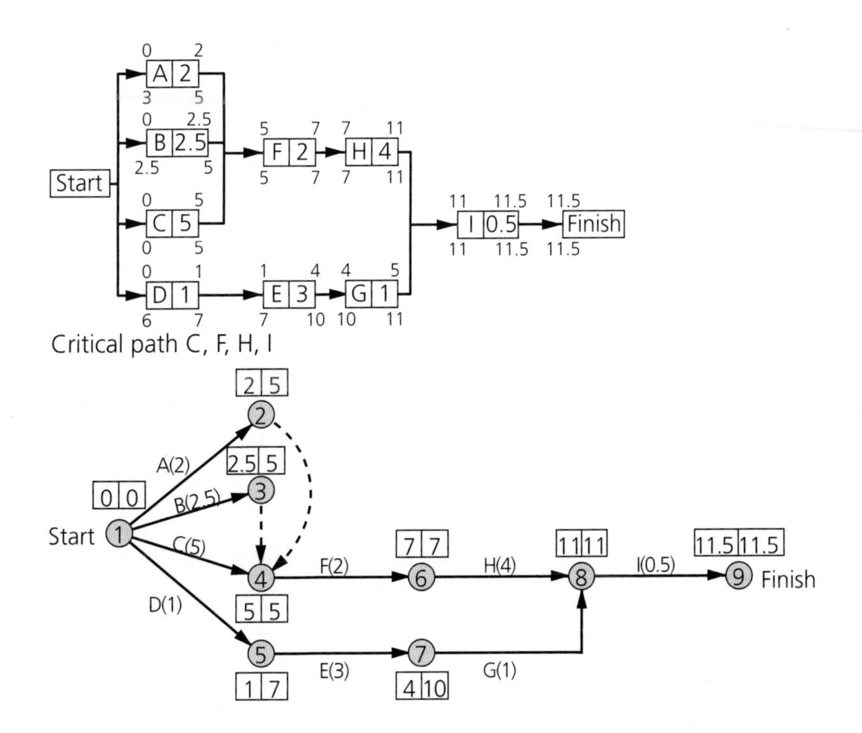

*The type of float in the example above is called an **interfering float**.*

EXERCISES

6.2 CLASSWORK

Perform a forward and backward pass through the network, for questions 1–5 in Exercises 6.1 CLASSWORK. Identify the critical activities. For each network give the critical path.

EXERCISES

6.2 HOMEWORK

1 Find the critical path and minimum project time for this network.

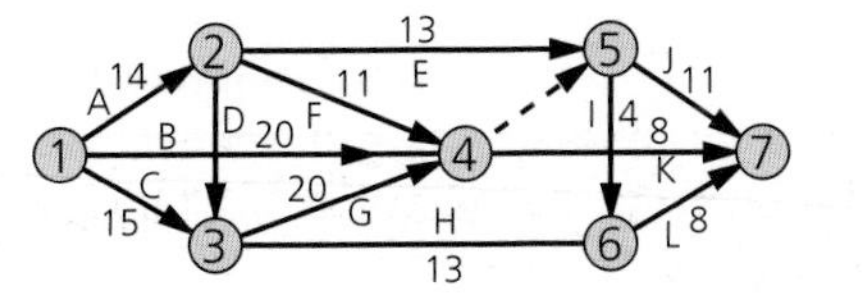

2 A hospital intends to computerise its medical records. A consultancy firm has identified the activities and their duration, in weeks, that are needed to obtain the system and make it ready for operating.

The network representing the activities is given below.

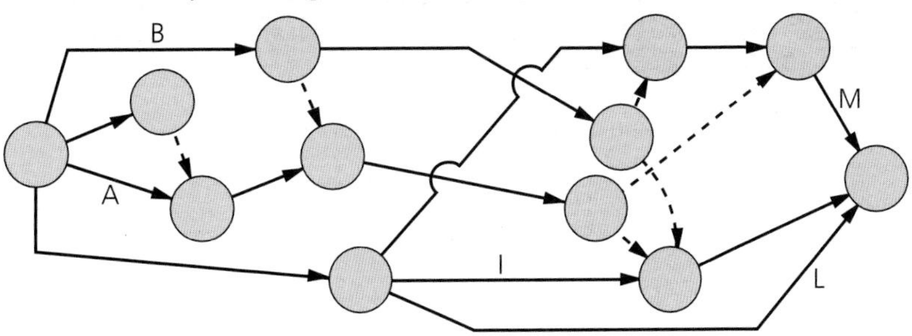

	Activity	Duration	Preceding activities
A	Order/delivery of hardware	6	–
B	Order/delivery of software	4	–
C	Preparation of computer site	5	–
D	Installation of hardware	1	A, C
E	Loading software	1	B, D
F	Preparing data	10	K
G	Loading data	2	E, F
H	Training operators	3	D
I	Initial user training	7	K
J	Hands-on user training	3	E, I, H
K	Agree final specification	2	–
L	Inform managers	1	K
M	System text	2	G, H

Number the events and complete the labelling of the activities. Determine the critical path and the minimum completion time.

3 Perform a forward and backward pass through the network in each of questions 3–5 in Exercises 6.1 HOMEWORK. Identify the critical activities. Give the critical path and minimum project time.

CASCADE CHARTS

Exploration 6.2

Kate's decorating – 2

In an unaccustomed fit of generosity, Kate's brother Rowan offers to help her to decorate her room. Rowan wants to know when he will be needed so he can plan his own activities. When should she ask him to be available to help her?

In the previous section, we saw how to find the minimum time needed to complete a project, providing that the non-critical activities could be carried out at the same time as the critical activities. In other words, there was no restriction on the number of workers. Kate's decorating project could be completed in 19 hours 15 minutes, provided someone else can remove the curtains for her while she moves the furniture and then come back and help with the painting or wallpapering. If she has to do the job by herself it will take her up to 24.5 hours, the total time for all the activities (unless she can save a bit of time by starting the wallpapering while the paint is drying).

If a project is to be planned efficiently, we need to take account of the resources (both people and equipment) needed at different stages. **Cascade charts** show the same information as a precedence network but in the form of bars drawn against a time scale to show when the activities will need to take place so as to make best use of resources as well as time. Before we can draw a cascade chart we need to allocate each activity a **cascade activity number** (CAN) to ensure a logical sequence. There is a method for this, and we can relate it to our example.

Activity	CAN		
			Step 1: Set CAN at 1; set time at 0. Begin at the start vertex and number the first activity connected to it 1.
A	1	These are connected directly to the start vertex.	**Step 2:** Allocate CANs to other activities connected directly to the start vertex.
B	2		
C	3	This is preceded by A and B, but step 4 is satisfied.	**Step 3:** Consider activities that immediately follow those already numbered. allocate CANs to them in turn.
D	4		
G	5	G, like D, is directly connected to C.	
E	6		**Step 4:** Where an activity is directly preceded by more than one other, it cannot be allocated a CAN until all directly preceding activities are numbered.
F	7		
H	8	Preceded by F and G – step 4 is satisfied.	
I	9	Stop.	**Step 5:** Stop when all activities have been numbered.

Having applied the CAN method to our example we have a chart that looks like this.

CAN		Activity	Duration (hours)	Preceding activities
1	A	Tidy the room and move the furniture.	2	
2	B	Remove the curtains and curtain rail.	0.25	
3	C	Strip the old wallpaper.	1.5	A, B
4	D	Wash and sand the paintwork.	2	C
5	G	Hang the wallpaper.	5	C
6	E	Apply a first coat of paint and let it dry.	6	D
7	F	Apply a second coat of paint and let it dry.	6	E
8	H	Replace the curtain rail and hang the curtains.	0.25	F, G
9	I	Reinstate the furniture.	1.5	H

Now we can display this information on a cascade chart. Activities are represented by bars with length proportional to the duration of the activity. The activities are assumed to start at the earliest start time. The dotted lines show precedence relations between activities, and float times for non-critical activities.

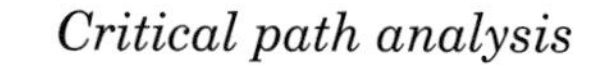

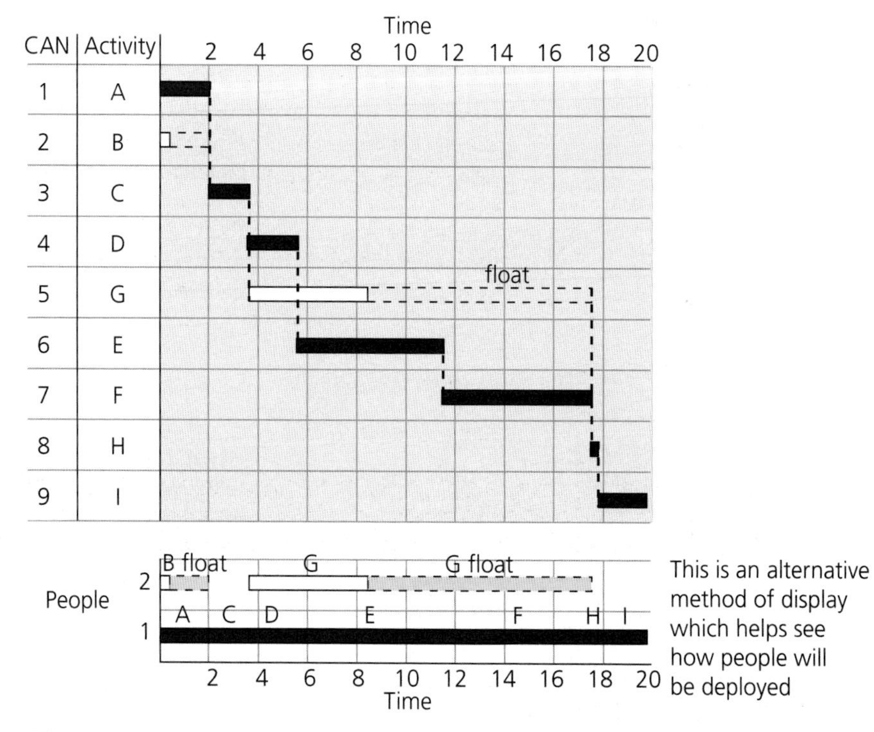

This is an alternative method of display which helps see how people will be deployed

The critical activities have been shaded; we know that these cannot be moved. Other activities could be delayed, but not beyond the dotted lines showing precedence otherwise they will move all subsequent activities along, making the duration of the project greater.

Resource levelling

Once we have drawn a cascade chart, it is possible to draw a **resource histogram** to show the number of people needed at any stage in the project.

For the relatively simple example of the decorating job, the resource histogram would look like this.

Resource histogram

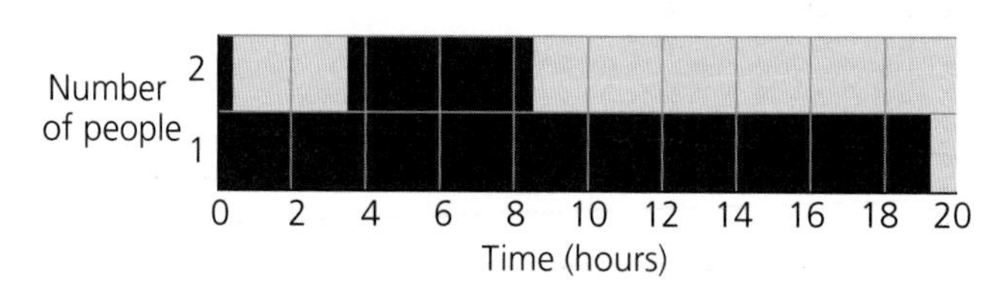

We can see from this that Rowan needs to be available to help Kate for 15 minutes at the start and for 5 hours some time between 3 and 17 hours into the project. If he helps her out she can probably get the job done in a weekend, but it will be hard work.

Suppose Kate starts work at 6 p.m. on Friday. Draw up a timetable for her so that she finishes on Sunday (and manages to eat and sleep).

With a more complicated project a resource histogram can be used as a planning tool, to ensure that there is a balance between the amount of work to be done and the resources available. This process is know as **resource levelling**.

Example 6.2

A youth club has been offered a long lease on a run-down building. The committee decide to have the building done up and a local builder offers to do the work at a reasonable price. They decide to have a party on the opening day and invite a local celebrity to perform the opening ceremony.

The only date that the celebrity is available is in three weeks' time, so the builder makes a list of jobs to be completed, along with timings and the minimum number of workers required to complete each job in the stated times. Will it be possible to complete the renovations in three weeks?

Here is the list the builder prepared.

	Activity	Duration	Preceding activity	Number of workers required
A	Agree the details of the lease.	4		1
B	Remove fireplaces and fittings.	2	A	2
C	Rewire.	2	A	2
D	Strip the wallpaper.	1	A	1
E	Burn off the old paint.	2	A	2
F	Install central heating.	3	B, C	2
G	Replaster.	3	D, F	2
H	Sand the floor and woodwork.	3	G, E	1
I	Varnish the floor.	2	H	1
J	Paint the ceiling.	1	I	1
K	Paint the woodwork.	2	I	1
L	Repaint the walls.	2	J	1
M	Replace fittings.	1	L, K	1
N	Clean up.	1	M	3

Solution

We start by drawing a precedence network and finding the critical path.

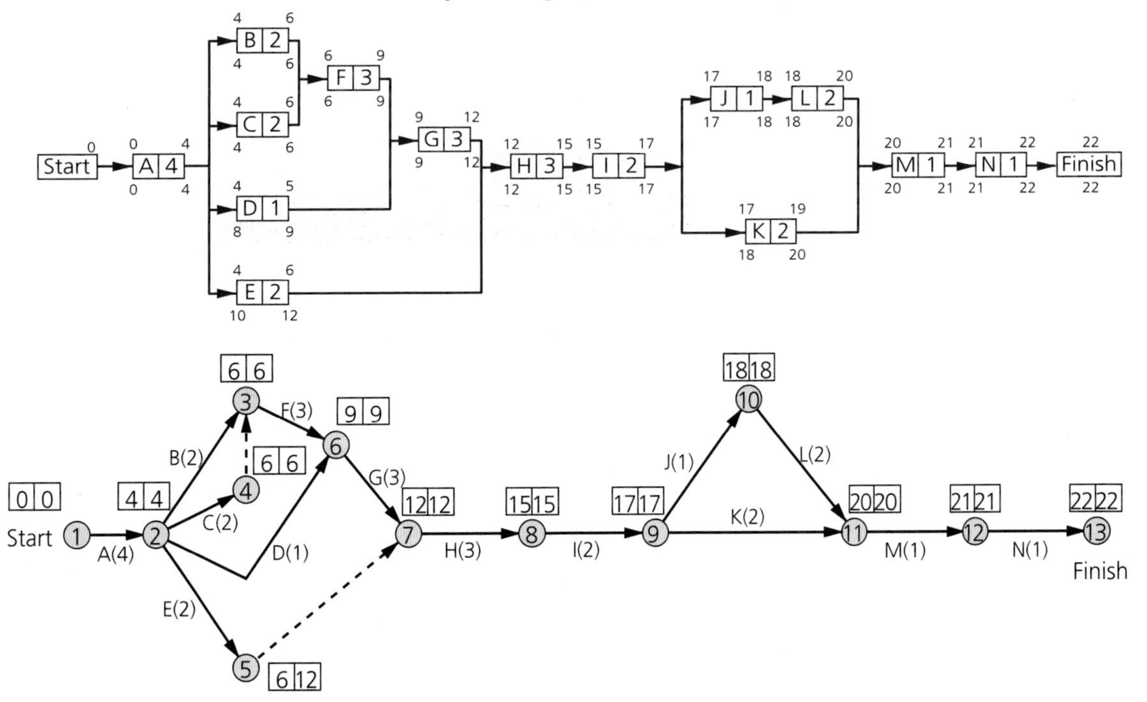

The critical activities are A, B, C, F, G, H, I, J, L, M, N. So there are two critical paths:

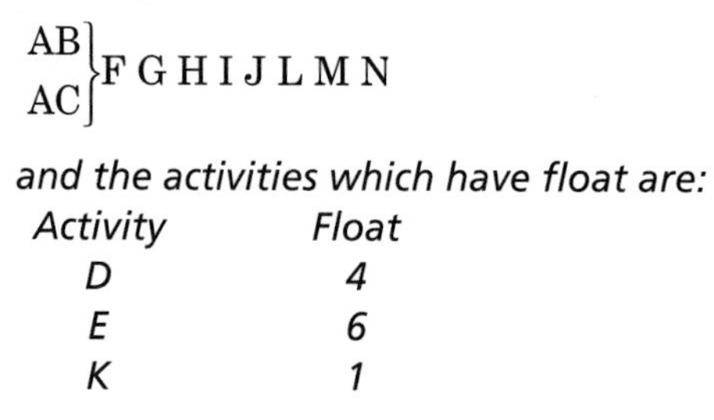

AB } F G H I J L M N
AC }

and the activities which have float are:

Activity	*Float*
D	*4*
E	*6*
K	*1*

Now allocate CAN numbers, using the method in the previous section.

CAN		Activity	Duration	Preceding activity	Number of workers required
1	A	Agree the details of the lease.	4		1
2	B	Remove fireplaces and fittings.	2	A	2
3	C	Rewire.	2	A	2
4	D	Strip the wallpaper.	1	A	1
5	E	Burn off the old paint.	2	A	2
6	F	Install central heating.	3	B, C	2
7	G	Replaster.	3	D, F	2
8	H	Sand the floor and woodwork.	3	G, E	1
9	I	Varnish the floor.	2	H	1
10	J	Paint the ceiling.	1	I	1
11	K	Paint the woodwork.	2	I	1
12	L	Repaint the walls.	2	J	1
13	M	Replace fittings.	1	L, K	1
14	N	Clean up.	1	M	3

Now we can display the information on a cascade chart.

Cascade chart

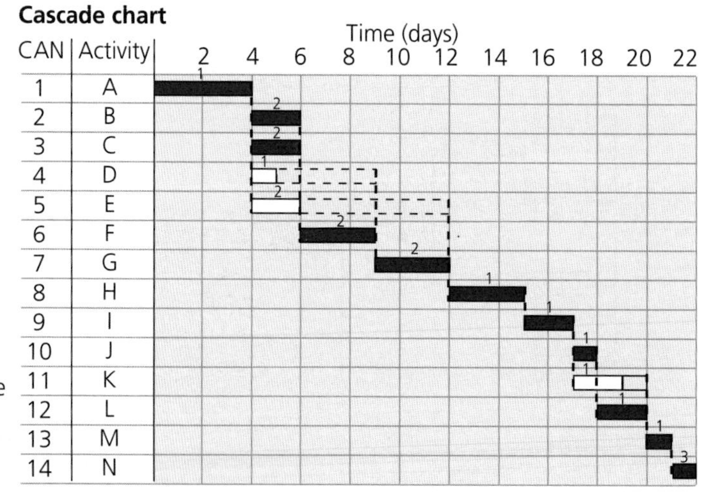

Note: The numbers above the bars give the number of workers required for each job

Then we can use the cascade chart to draw a resource histogram.

Resource histogram

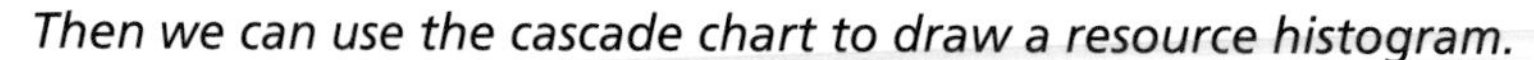

It seems that the work cannot be done in three weeks, though it is only a day over schedule. Looking at the histogram, we might wonder whether the best use is being made of the workers, for example, is it absolutely necessary to have seven workers on day 5?

Applying resource levelling

There are three ways in which resource levelling can be applied to a problem in order to increase the efficiency of a project.

1 Make better use of resources (in this case workers) by using the float to alter the timing of non-critical activities.

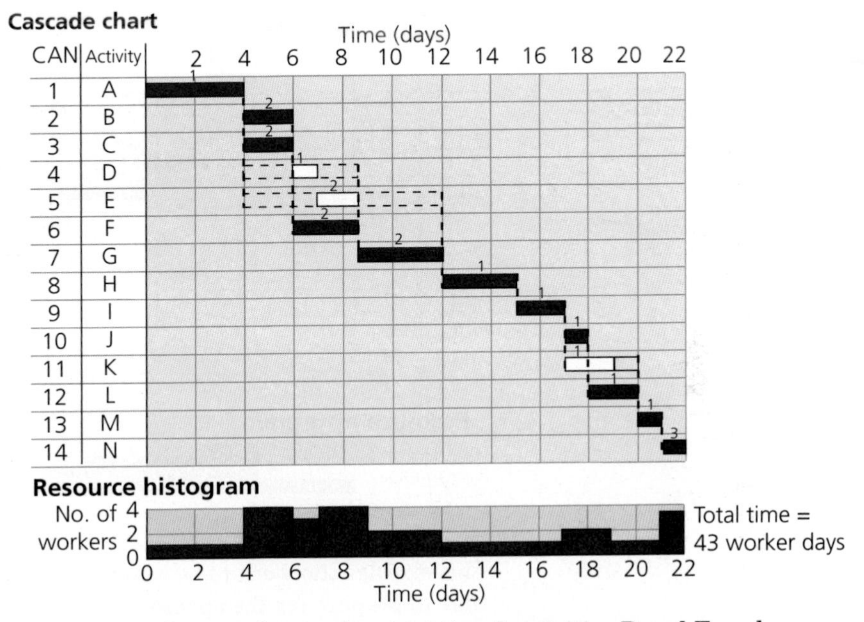

By using the float to change the start timed activities D and E we have *evened out* the number of workers, getting rid of the peak value of 7.

2 Use the float time on activity E to reduce the number of workers from two to one and increase the duration of the activity from two days to four.

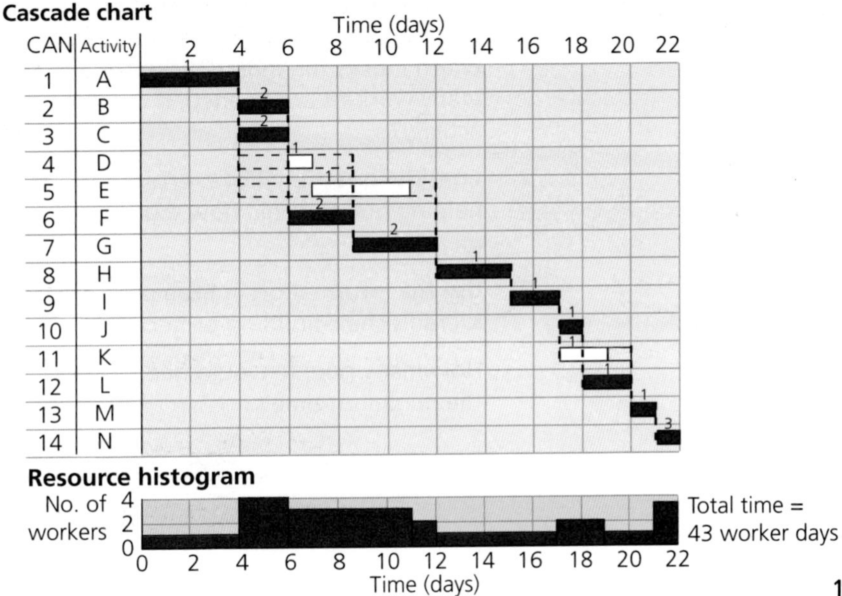

This gives a more even use of workers which is more cost-effective but still does not solve the problem of running over time.

3 If it is important to stay within a time limit, then we can choose to increase the number of workers on some activities, thus reducing the time. If the builder were to put two workers on activities H and I, we could reduce the time for the two from five days to two and a half days.

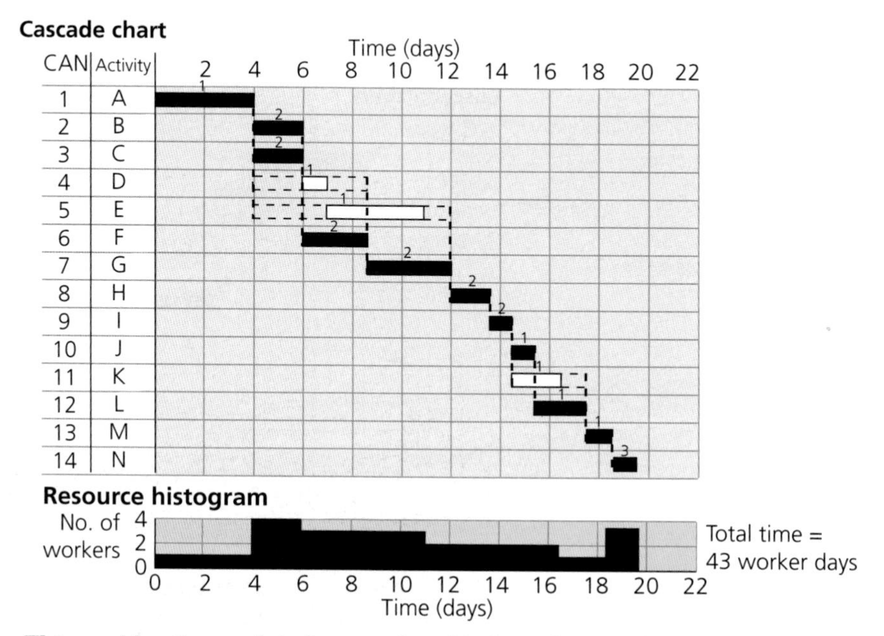

This enables the work to be completed in less than 20 days, leaving time to prepare for the opening ceremony. It also has the advantage of using only the same number of worker-days as the other plans, so should not, in theory, cost more!

EXERCISES

6.3 CLASSWORK

1 Draw a cascade chart for the information given in question 2 of Exercises 6.1 CLASSWORK. Use resource levelling to obtain a solution which makes the most efficient use of the workers available. How many workers are needed, if only one person is needed for each activity?

2 Consider question 4 of Exercises 6.1 CLASSWORK. If Marc's girlfriend is helping him to cook, how much time could be saved in preparing the meal?

3 Use the information in Example 6.1 to draw a cascade chart for Kieran's chair-building project. Be careful with the interfering floats.

How many people would need to help in order to finish the project in the minimum time?

4 Study this chart.

Activity	Duration	Preceding activity	Number of workers required
A	4		1
B	7		1
C	5		1
D	2	B, C	2
E	3	A, B	1
F	5	E	1
G	2	E, H	1
H	6	D	2
I	2	G, F	1
J	3	F	1
K	3	H	1

a) Draw a precedence network and find the critical path.
b) Use a cascade chart and resource histogram to schedule the activities as efficiently as you can.

5 The table below shows a simplified schedule for building a house.

	Activity	Duration	Preceding activity
A	Dig the foundations.	12	
B	Pour the foundations.	8	A
C	Build the walls.	10	B
D	Erect the roof.	6	C
E	Make the windows and doors.	7	
F	Install the windows and doors.	2	C, E
G	Install the plumbing.	4	C
H	Install the electrics.	3	D, F
I	Plaster the internal walls.	4	G, H
J	Decorate.	5	I
K	Landscape the gardens.	7	G

a) Produce a network for the project and identify the critical activities.
b) Display the information on a cascade chart.
c) For activity E find:
i) the independent float **ii)** the interfering float.
d) The cost associated with each task, when completed in the normal duration, is shown in the table overleaf, together with the extra cost per day which would be incurred if the task is to be completed in less time (sometimes called the **crash** cost).

Activity	Cost normal duration (£)	Extra cost for early completion per day(£)
A	12 000	2000
B	9000	2000
C	18 000	4000
D	15 000	3000
E	3000	850
F	2500	800
G	5000	1250
H	4500	1000
I	6000	1500
J	6000	1600
K	8000	2000

If the project had to be completed in one less day than the completion time found in **a)**, which activity (or activities) could the builder choose to do in one day less. Why?

e) What is the percentage increase in cost?

EXERCISES

6.3 HOMEWORK

1 Determine the critical path for the network below. Draw a cascade chart and use resource levelling to obtain a solution that makes the most efficient use of the workers available. How many workers are needed, if only one person is needed for each activity?

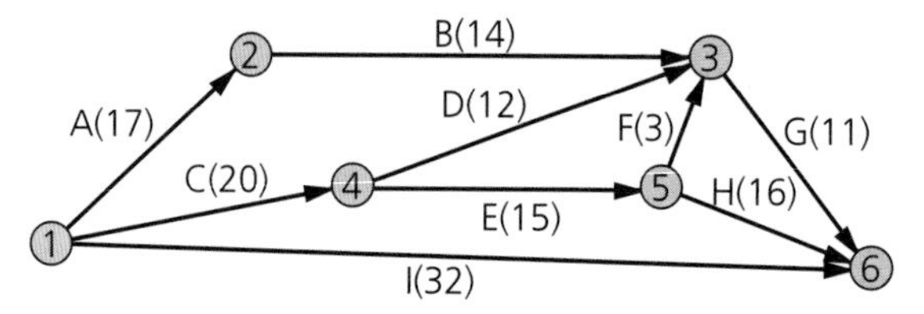

2 Analyse the following network and draw a cascade chart. Use resource levelling to obtain a solution which makes the most efficient use of the workers available.

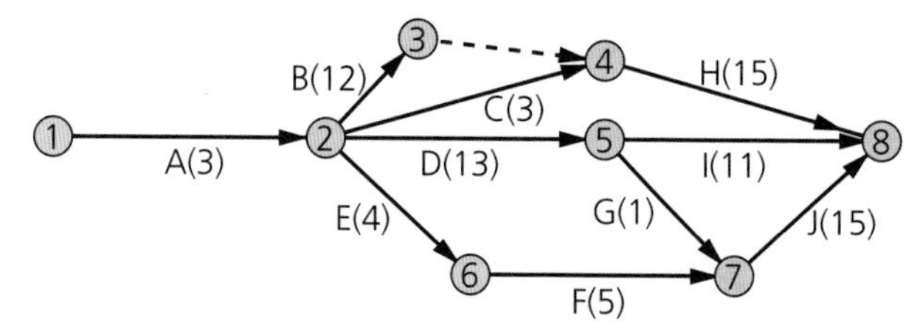

3 Use a cascade chart for the information given in each of the questions in Exercises 6.2 HOMEWORK. Use resource levelling to obtain a solution that makes the most efficient use of the workers available. How many works are needed if only one person is needed for each activity?

4 The table below gives the activities to prepare, cook and serve a plate of fish and chips.

	Activity	Duration (minutes)	Preceding activity
A	Peel the potatoes.	6	–
B	Prepare the oil.	6	–
C	Fillet the fish.	10	–
D	Prepare the batter.	4	–
E	Mix the batter.	4	D
F	Chip the potatoes.	4	A
G	Fry the chips.	14	B, F
H	Coat the fish with batter.	5	C, E
I	Fry the fish.	12	B, H
J	Serve the fish with the chips.	3	G, I

a) Produce a network for the above and identify the critical activities. State the minimum time for completion.

b) Display the information on a cascade chart. Use resource levelling to obtain a solution which makes most efficient use of manpower.

MATHEMATICAL MODELLING ACTIVITY

Going to your school or college

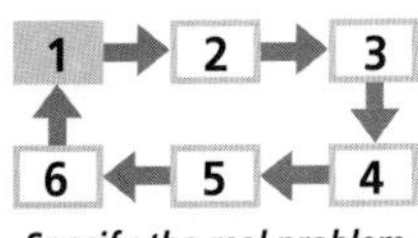

Specify the real problem

Problem statement

Identify the activities you perform from the moment of rising from your bed to leaving for school or college on a typical morning. Identify the minimum time this takes and the critical activities.

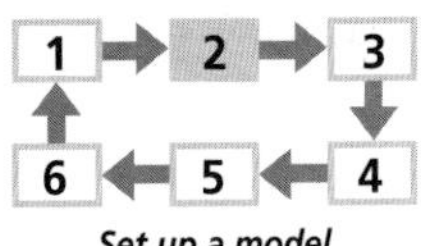

Set up a model

Set up a model

Initially you should take seven or eight activities you perform before leaving home in the morning. Make a record of the time it takes to carry out each activity. You may need to do this on several occasions and take the average time.

Assume initially that you are the only person at home, so you will need to prepare your own breakfast.

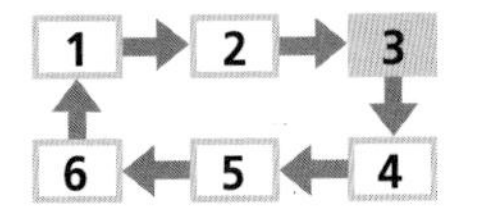

Formulate the mathematical problem

Mathematical model

Certain activities should run concurrently, for instance you may complete your assignments or homework whilst eating your breakfast. State clearly all assumptions you make about the activities you selected.

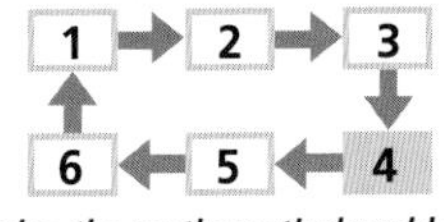

Solve the mathematical problem

Mathematical solution

Construct a precedence network using the activity on vertex method. Determine the critical path and the latest time that you need to get up, to leave by a set time. Draw up a cascade chart and a resource histogram, so that if, at any stage, there is an added resource (for instance someone to prepare breakfast for you), then your activities may be planned more efficiently.

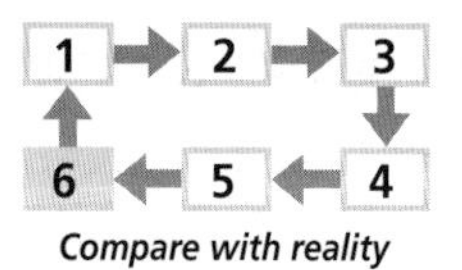

Compare with reality

Refinement to the model

Now re-inspect each activity and consider whether it may be broken down into components, particularly if these components may be performed concurrently. Rework the solution using your revisions.

CONSOLIDATION EXERCISES FOR CHAPTER 6

1 The two diagrams give alternative but equivalent ways of representing the precedences with which the seven tasks of a project must be performed. In this question you may use whichever one of these two ways you prefer.

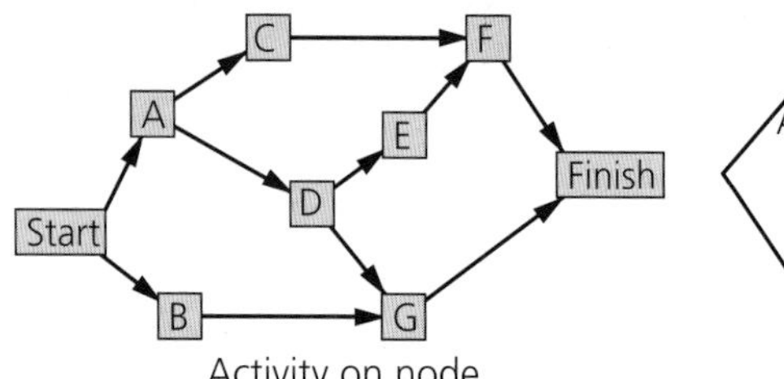

Activity on node

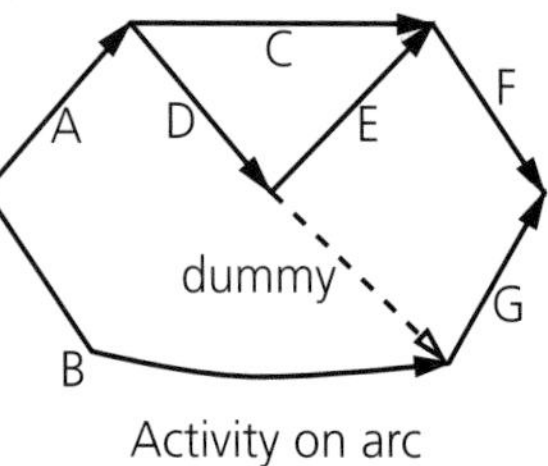

Activity on arc

The times required to complete each task are:

Task	A	B	C	D	E	F	G
Time required (minutes)	2	4	4	3	4	6	3

a) Produce a table showing the immediate predecessor(s) for each activity.

b) Perform a critical path analysis to obtain the critical path and the completion time for the project.

c) What flexibility is there in the scheduling of activity C?

d) Show how to modify your network to incorporate an activity H which cannot start until activities A and B are completed, and which must itself be completed before activity D can start.

e) i) Produce a modified precedence table incorporating activity H and any consequential changes.

ii) If the time needed for activity H is two minutes, what effect will your modification have on the completion time and on the critical path?

(UODLE Question 19, Paper 4, 1995)

2 The project of assembling a piece of furniture delivered as a 'flat-pack' was broken down into the activities shown in the following table. The table also shows each activity with its preceding activities and the duration of each activity (in minutes).

a) Draw the weighted network which models this project, numbering each event.

b) Calculate the earliest time after the start of the project at which each activity may start.

c) Hence determine the length of the critical path.

d) Write down the activities which lie on the critical path.

	Activity	Duration (minutes)	Preceding activity
A	Open the box.	10	
B	Check the sections against the parts list.	30	A
C	Check the fixings against the parts list.	20	A
D	Assemble the shell.	40	B
E	Cut back to size.	40	B
F	Screw the shell together.	20	C, D
G	Assemble the drawers.	20	B
H	Glue the drawers together.	10	G
I	Put the whole together.	30	E, F, H

(ULEAC Question 8, Specimen Paper D1, 1996)

3 **a)** The activity network shows the precedences between seven activities involved in a project. The activities are represented by arcs and are labelled by letters. Their durations are also shown.

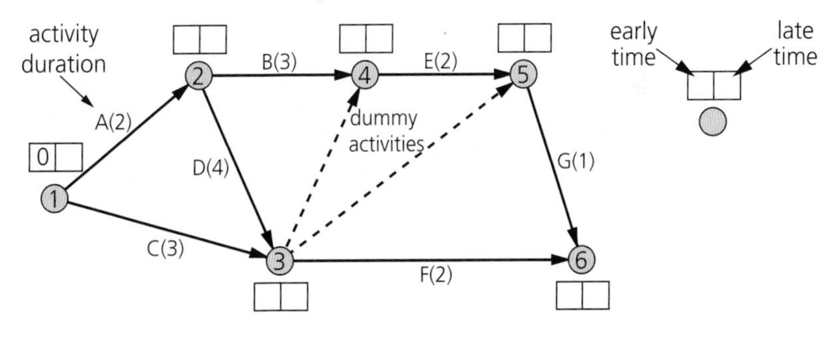

i) Explain why the dummy activity shown from event number 3 to event number 5 is not needed.

ii) Complete this table showing the immediate predecessor(s) for each activity.

Activity	A	B	C	D	E	F	G
Immediate predecessor(s)	none		none				

iii) Perform a forward pass and a backward pass on the activity network to determine the earliest and latest event times.

iv) Complete the table (overleaf) showing the float for each activity.

Activity	A	B	C	D	E	F	G
Float							

v) State the minimum time for completion and the activities forming the critical path.

b) A second activity network with six activities is as follows.

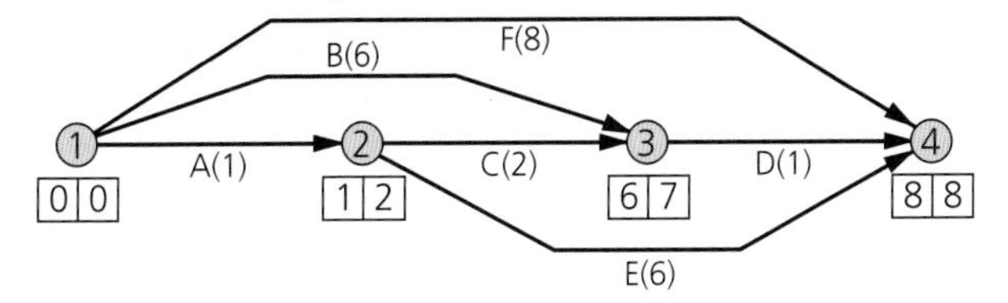

Activity C has interfering float of 4 (because 7 – 1 – 2 = late time of 3 – early time of 2 – duration of C = 4). It has independent float of 2 (because 6 – 2 – 2 = early time of 3 – late time of 2 – duration of C = 2). Explain the relevance of these measures when scheduling activity C.

c) The activity network below is the same as that in part **b)**, except that activities B and E have reduced durations, and activity D has increased duration. This has affected some early and late times.

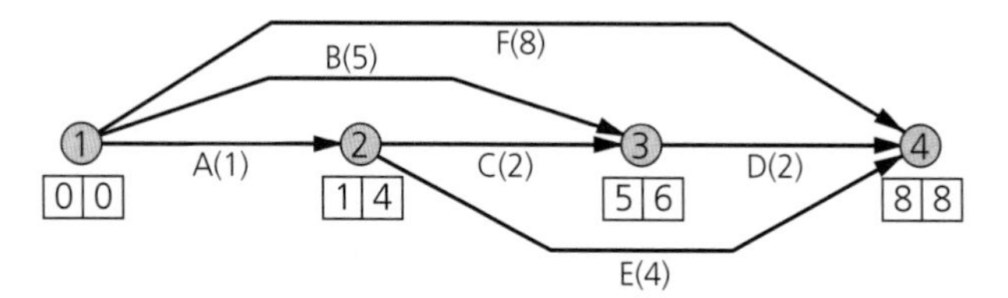

Give the interfering float and the independent float for activity C in this network.

(O & C MEI Question 1, Decision & Discrete Mathematics Paper 19, January 1996)

4 **a)** Find the early and late times for each activity in the precedence diagram in the figure.

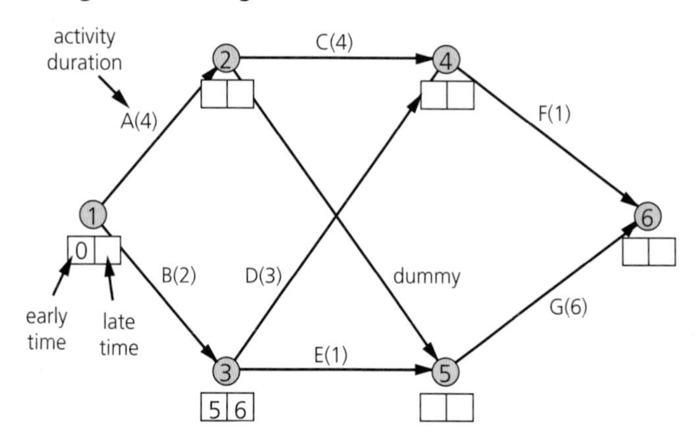

b) List the critical activities.

c) Give the total float for activity C.

(OCR MEI D&D1 Specimen2000)

5

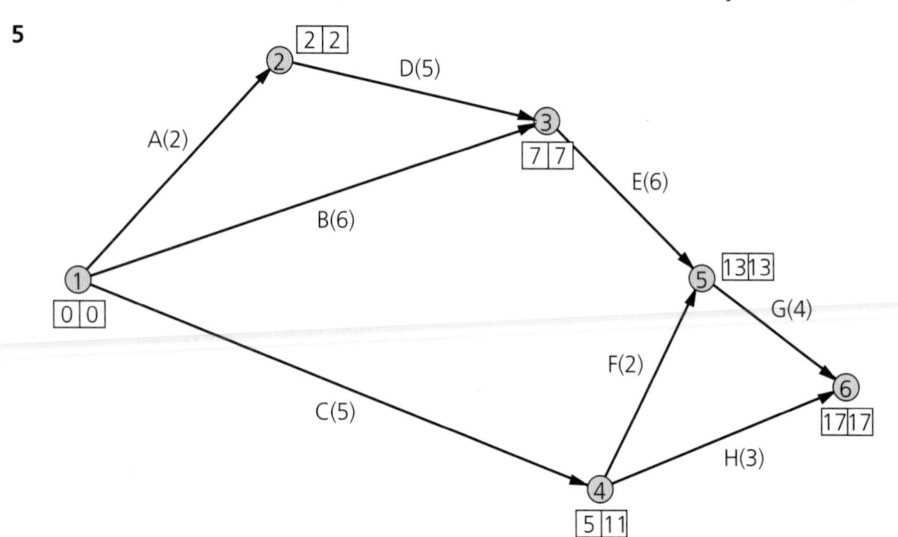

A project is modelled by the activity network in the figure. The activities are represented by the arcs. The number in brackets on each arc gives the time, in hours, taken to complete the activity. The left box entry at each vertex is the earliest event time and the right box entry is the latest event time.

a) Determine the critical activities and the length of the critical path.
b) Obtain the total floats for the non-critical activities.
c) On a grid, draw a cascade (Gantt) chart showing the information found in parts (a) and (b).

Given that each activity requires one worker:

d) draw up a schedule to determine the minimum number of workers required to complete the project in the critical time. State the minimum number of workers.

(Edexcel D1 Specimen 2000)

6 The table gives information about a construction project.

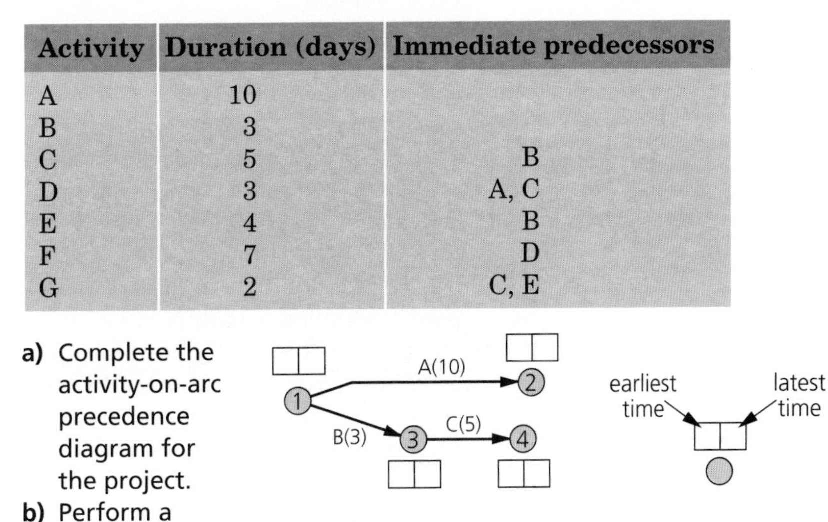

Activity	Duration (days)	Immediate predecessors
A	10	
B	3	
C	5	B
D	3	A, C
E	4	B
F	7	D
G	2	C, E

a) Complete the activity-on-arc precedence diagram for the project.
b) Perform a forward and backward pass on your precedence network to determine the earliest and latest event time.
c) State the minimum time for completion and the critical path.
d) Use an appropriate method to produce an ordering of the activities, and hence draw a cascade chart for the project on the axes given.

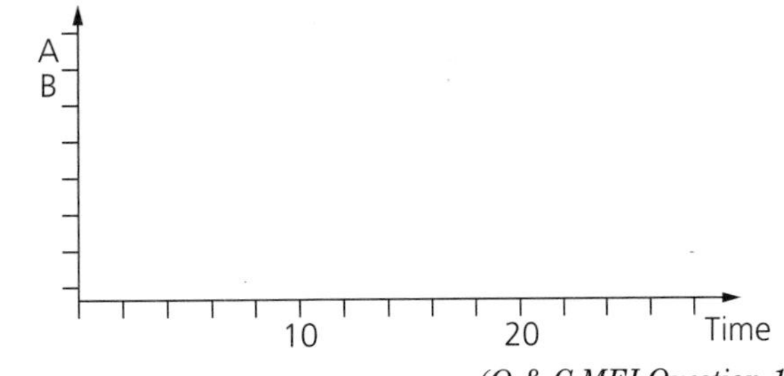

(O & C MEI Question 1, Specimen Paper, 1994)

7 The diagram shows an activity-on-arc precedence diagram. Activity durations are in days, and are shown in brackets against the arc representing the activity. Activity X connects event 6 to event 8.

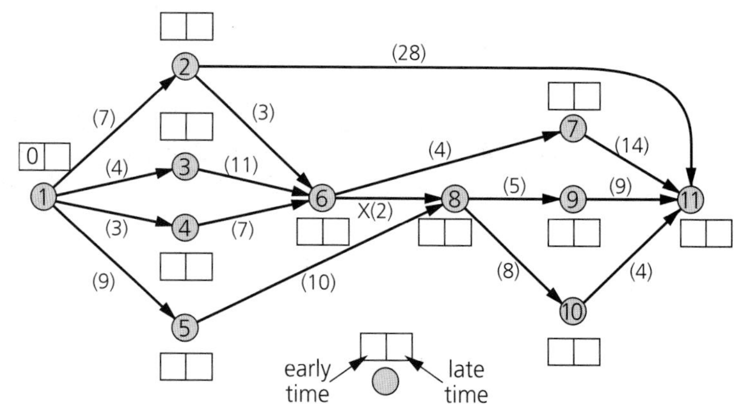

a) Perform a forward pass on this network. Give the early time for event number:
i) 6 **ii)** 8.

b) Perform a backward pass on the network. Give the late time for event number:
i) 8 **ii)** 6.

c) The independent float for activity X is defined as:
early time for event 8 – late time for event 6 – duration of activity X (provided that it is non-negative).
Give the independent float for activity X.
The total float for activity X is defined as:
late time for event 8 – early time for event 6 – duration of activity X
Give the total float for activity X.
Explain the significance of each of independent float and total float scheduling activity X.

d) The network below is part of a larger network. It shows all of the activities which are connected to activity X, *and only those activities*, together with some early and late times.

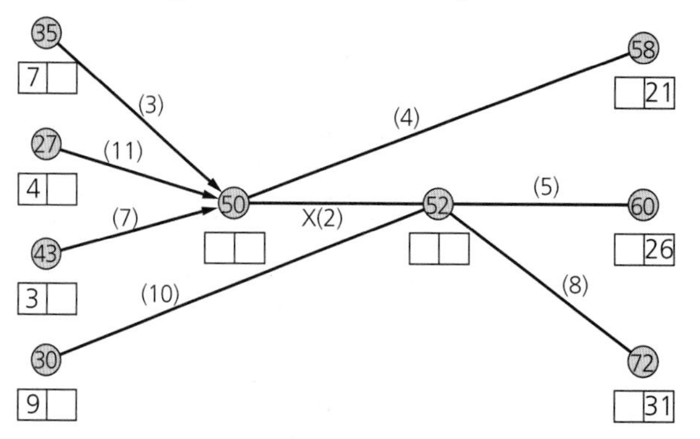

Say which of the following it is possible to compute from the information given on this network. In each case do the computation if it is possible, and say why it is not possible if not.
the early time for event 52
the early time for event 58
the late time for event 50
the late time for event 35

(O & C MEI Question 2, Decision & Discrete Mathematics Paper 19, June 1995)

8 The table shows the activities involved in a project, their durations, and their immediate predecessors.

Activity	A	B	C	D	E	F	G	H
Duration (days)	3	1	1	4	5	2	2	2
Immediate predecessors	–	–	B	A	A, C	B	F	D, E

a) Draw an activity network for the project.

b) Perform a forward pass, and a backward pass on your activity network. Give the minimum completion time and the activities forming the critical path.

c) In a cascade chart each activity is represented by a bar showing when it is scheduled. In the chart below, activities A and B are shown, both starting as early as possible.

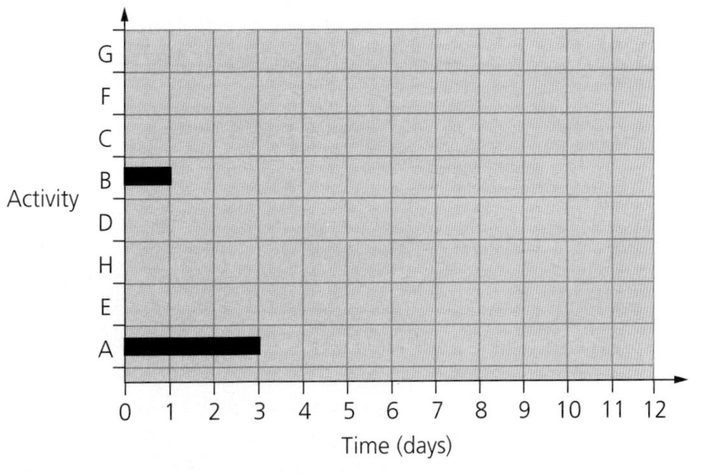

Copy and complete the cascade chart, with the activities ordered as shown, and with each activity starting as early as possible.

d) The number of people needed for each activity is as follows.

Activity	A	B	C	D	E	F	G	H
People	3	2	2	2	3	1	1	4

Activities are to be scheduled, some later than their earliest possible start time, so that at most five people are needed at any one time, whilst the project is still completed in the minimum time. What are the new scheduled start times for activities D, F and G?

(AQA MBD1 Specimen 2000)

9 The table shows the activities involved in a project, their durations, and their immediate predecessors.

Activity	A	B	C	D	E	F	G	H
Duration (days)	1	2	1	1	4	2	3	2
Immediate predecessors			A	B	B	C, D	E, F	E

a) Complete the activity-on-arc precedence diagram for the project.

activity duration
A(1)
B(2)
1 2 3
0
early time
late time

b) Perform a forward pass and a backward pass on your precedence diagram to determine the earliest and latest event times.
State the minimum time for completion and the activities forming the critical path.

c) Complete a cascade chart for the project on the given axes, with the activities ordered as shown.

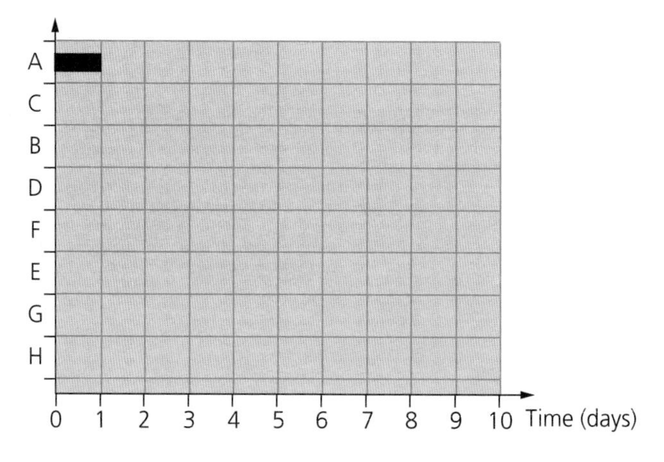

d) The numbers of people needed for each activity are as follows.

Activity	A	B	C	D	E	F	G	H
People	1	4	2	3	1	2	3	2

Activities C and F are to be scheduled to start later than their earliest start times so that only five people are needed at any one time, whilst the project is still completed in the minimum time. Specify the scheduled start times for activities C and F.

(O & C MEI, Question 1, Decision & Discrete Mathematics Paper 19, January 1995)

10 The following table gives details of a set of eight tasks which have to be completed to finish a project. The 'immediate predecessors' are those tasks which must be completed before a task may be started.

a) Produce either an activity-on-arc or an activity-on-node network for the project.

b) Find the minimum time to completion and the critical activities.

Task	Duration (days)	Immediate predecessor(s)
A	3	
B	8	
C	7	
D	4	A
E	2	D, B
F	6	C
G	3	E
H	2	E

The cost associated with each task, when completed in the normal duration, is given in the following table, together with the extra cost that would be incurred in using extra resources to complete the task in one day less than the normal duration.

Task	Cost (£) Normal duration	Extra cost (£) Completion 1 day sooner
A	3000	1000
B	6000	800
C	5000	700
D	5000	1200
E	1200	600
F	5500	1000
G	2800	1100
H	1700	500

c) Find the minimum cost that would be entailed in completing the project in one day less time than the minimum time to completion which you found in **b)**.
Give the percentage increase in cost that this involves.

(UODLE Question 7, Paper 4, 1994)

Summary

After working through this chapter you should be able to:

- draw precedence networks to show the necessary steps in a project, using activity-on-vertices and activity-on-arcs procedures
- find the critical path, which is the minimum time needed to complete a project, following the longest path through the precedence network and including all the critical activities which are activities that must be started on time to avoid delaying the whole project
- draw cascade charts showing precedence relations, to make best use of all resources
- use resource levelling to balance the amount of work to be done and the resources available for a project.

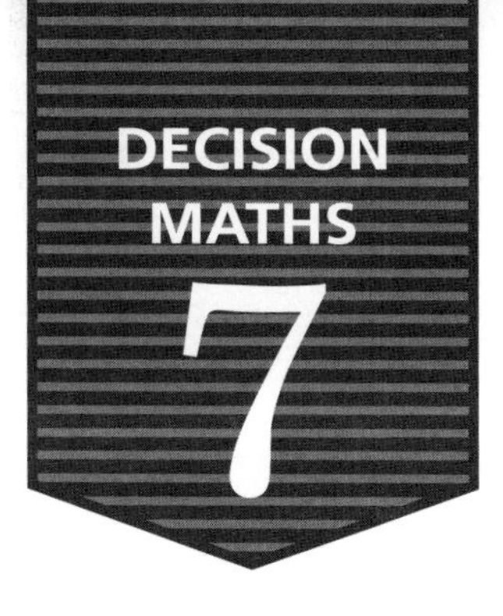

Matchings

By the end of this chapter you should be able to:

- *understand how a bipartite graph, which is one that can be divided into two subsets in such a way that each edge of the graph joins a vertex from one subset to a vertex in the other, can be applied in modelling*
- *subdivide a bipartite graph by inserting vertices of degree 2*
- *apply a matching to a bipartite graph by linking some of the vertices in one subset to some of the vertices in the other*
- *use the algorithm for maximum matching of the two subsets in a bipartite graph.*

Exploration 7.1

Building a timetable

Mrs Frawt, deputy head of Everbrite School, is trying to complete the timetable. She must fit Environmental Studies, Geography, Graphics, Mathematics and Science into the last timetable slot and fortunately she has five suitable teachers available. The table shows which subjects each teacher can cover.

Teacher	Subjects
Miss Abraham	Environmental Studies, Geography
Mr Bowen	Environmental Studies, Mathematics, Science
Mrs Carter	Geography, Graphics
Ms Delgardo	Geography, Mathematics, Science
Mr Eggham	Graphics, Mathematics, Science

How should Mrs Frawt allocate her staff? Try solving the problem using your own methods, before reading on. Then you can compare the method you use with the techniques developed in this chapter.

GRAPH THEORY – BIPARTITE GRAPHS

A **bipartite graph** is a graph in which the vertices can be divided into two subsets, A and B, in such a way that each edge of the graph joins a vertex in A to a vertex in B. We distinguish the vertices in subset A from those in subset B by using a small, solid circle for those in A and an open circle for those in B.

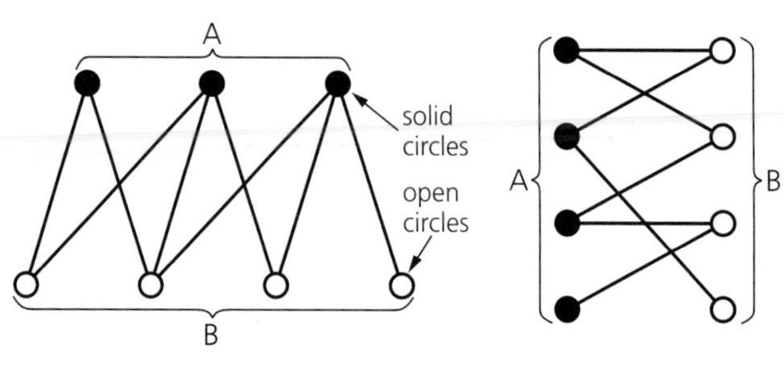

A **complete bipartite graph** is one in which each vertex in A is joined to every vertex in B by an edge. Complete bipartite graphs are denoted by $\mathbf{K}_{r,\,s}$ where r is the number of vertices in A and s is the number of vertices in B.

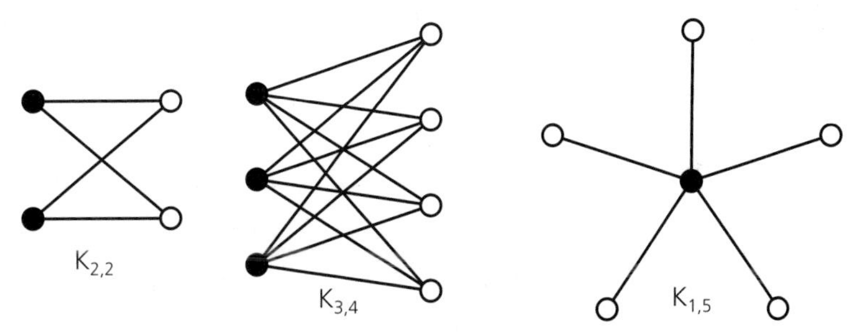

Unlike most of the graphs we have met so far, not all bipartite graphs are planar. For example, is $K_{3,\,3}$ planar?

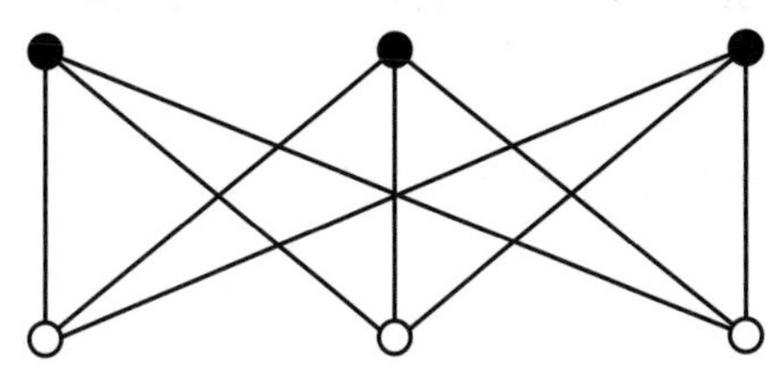

If $K_{3,\,3}$ is planar we should be able to draw it with no edges crossing. If you try it, you will soon see that it is not possible.

A **subdivision** of a graph is formed by inserting vertices of degree 2.

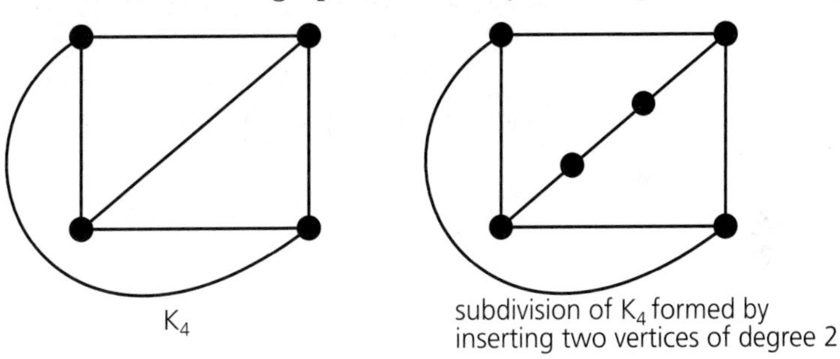

If a graph is non-planar, every subdivision of that graph must also be non-planar. The Polish mathematician Kuratowski proved that all non-planar graphs contain subgraphs which are subdivisions of either $K_{3,\,3}$ or K_5 (we proved that K_5 is non-planar in Chapter 4, *Networks 3: Route inspection problems*), thus we can show that a graph is non-planar by finding a subdivision of $K_{3,\,3}$ or K_5 in it.

Example 7.1

Show which of these graphs are non-planar because they contain subgraphs which are subdivisions of either $K_{3,\,3}$ or K_5.

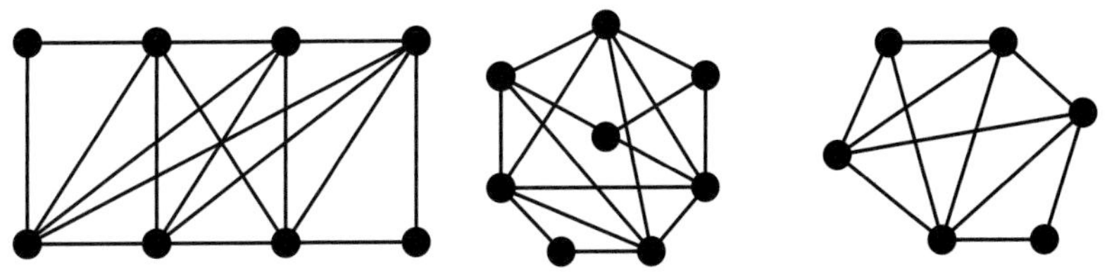

Solution

a) *This graph contains $K_{3,3}$ and is non-planar.*

b) *This graph contains a subdivision of K_5 and is non-planar.*

c) *This graph does not contain subdivisions of either $K_{3,3}$ or K_5 and thus is planar.*
$v - e + f = 6 - 11 + 7 = 2$
Euler's formula is satisfied.

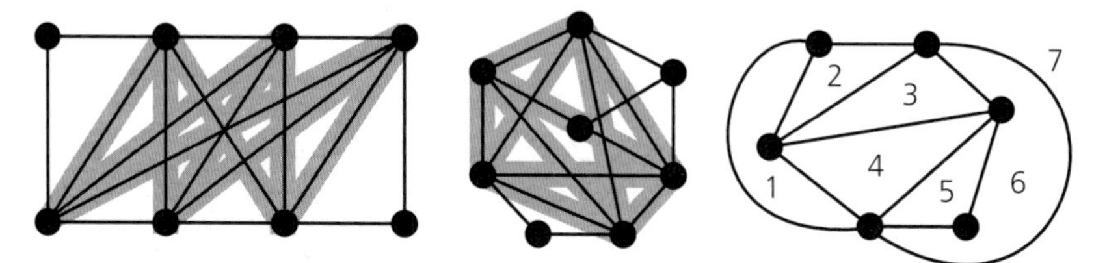

Modelling with bipartite graphs

A situation such as the one in Exploration 7.1 can be modelled using a bipartite graph. We represent the teachers by the vertices in subset A and the subjects by the vertices in subset B. To simplify the graph we will abbreviate the teachers' names to A, B, C, D and E and denote the subjects by numbers.

Teacher	Subjects
Miss Abraham (A)	Environmental Studies (1), Geography (2)
Mr Bowen (B)	Environmental Studies (1), Mathematics (4), Science (5)
Mrs Carter (C)	Geography (2), Graphics (3)
Ms Delgardo (D)	Geography (2), Mathematics (4), Science (5)
Mr Eggham (E)	Graphics (3), Mathematics (4), Science (5)

Now the edges of the graph represent 'can teach' and they join each teacher to the subjects which they are able to teach.

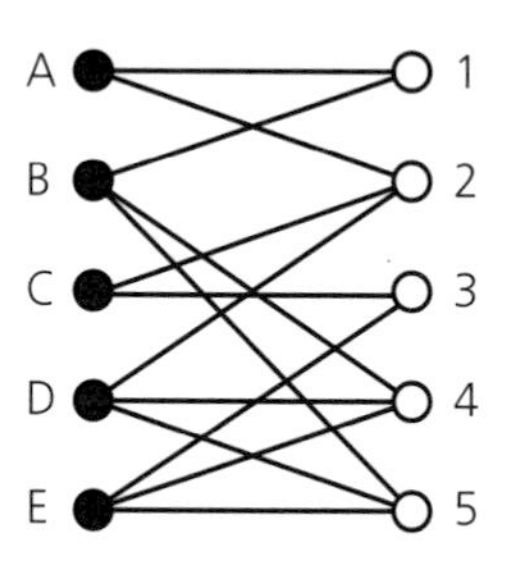

EXERCISES

1 a) Draw the following complete bipartite graphs.
i) $K_{2,3}$ ii) $K_{4,4}$ iii) $K_{1,6}$

b) How many vertices and edges does the complete bipartite graph $K_{r,s}$ have?
What are the degrees of the vertices?

2 Show which of the following graphs are non-planar because they contain subdivisions of either $K_{3,3}$ or K_5.

*The graph in part **a)** is called a **Petersen graph** and is named after Danish mathematician Julius Petersen (1839–1910) who published a paper on its properties in 1898.*

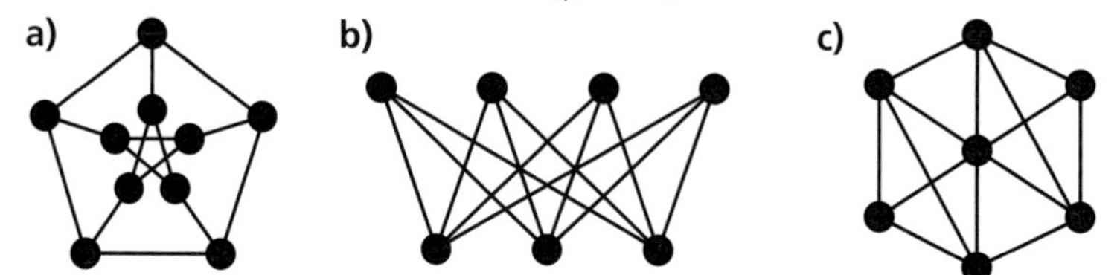

3 This bipartite graph shows five people (A, B, C, D and E) and five tasks which they are qualified to do (1–5). Use the graph to state which tasks each person is qualified to do.

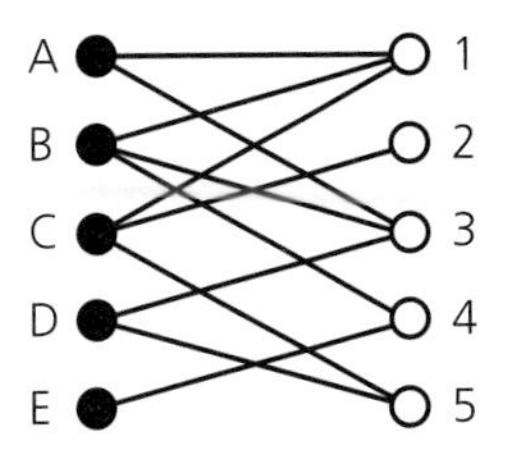

4 Everbrite School has entered a team in the local schools' swimming gala. They make a list of their best female and male swimmers and the events in which they can compete.

	Student	Event
Girls	Meena	breaststroke, backstroke
	Lauren	freestyle, breaststroke
	Tara	backstroke, breaststroke, freestyle
	Kerry	butterfly, freestyle
Boys	Gary	butterfly, freestyle, breaststroke
	Paul	backstroke, freestyle
	Eddie	butterfly, backstroke
	James	freestyle, backstroke, breaststroke

Show the two sets of information on bipartite graphs.

5 The manager on a building site has to assign five tasks to five members of the workforce. The workers who are qualified to do the jobs are shown in this table.

Worker	Jobs for which they are qualified
Anne	plastering, joinery, glazing
Brian	bricklaying, roofing
Carl	bricklaying, plastering, joinery
Daljit	joinery, roofing, glazing
Errol	bricklaying, plastering, roofing

Show the information on a bipartite graph.

EXERCISES

7.1 HOMEWORK

1 Determine which of the following are non-planar graphs by trying to find subdivisions of $K_{3,3}$ or K_5.

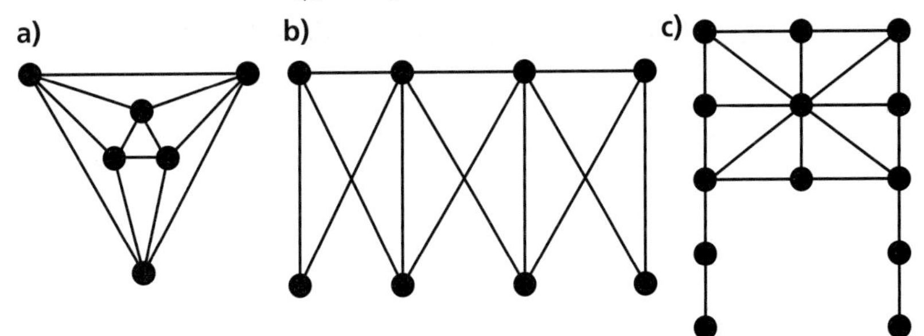

2 **a)** Draw the following complete bipartite graphs.
i) $K_{4,2}$ **ii)** $K_{2,4}$
Comment on the resulting graphs.
b) If one edge is removed from K_5 show that the resulting graph is planar.

c) If one edge is removed from $K_{3,3}$ show that the resulting graph is planar.

3 a) How many vertices and edges does the graph K_{2n} have? For what values of n is K_{2n} planar?
b) Show that $K_{2,4}$ is a planar graph by giving a planar drawing of it and verify that it satisfies Euler's theorem.

4 In an oral examination, five students (A, B, C, D and E) are to be examined by four lecturers.
Lecturer 1 examines A, D, and E. Lecturer 2 examines B, C and E.
Lecturer 3 examines A, C and D. Lecturer 4 examines B, C and D.
Show the information on a bipartite graph.

5 Give examples of:
a) a subgraph of a non-planar graph which is planar
b) a non-planar graph for which Euler's formula is true.

AN ALGORITHM FOR MAXIMUM MATCHING

Let us return to Mrs Frawt, in Exploration 7.1. Having represented her timetabling problem using a bipartite graph, Mrs Frawt can use it to help find a solution. Although this example can easily be solved by inspection, we shall use it to demonstrate an algorithm that will always provide an optimal solution. A graph that links some of the vertices in A with some of the vertices in B is called a **matching**. The optimal solution is called the **maximum matching**.

The first part of this algorithm involves using a labelling procedure. We start with any matching, **M**. An example follows.

Step 1:

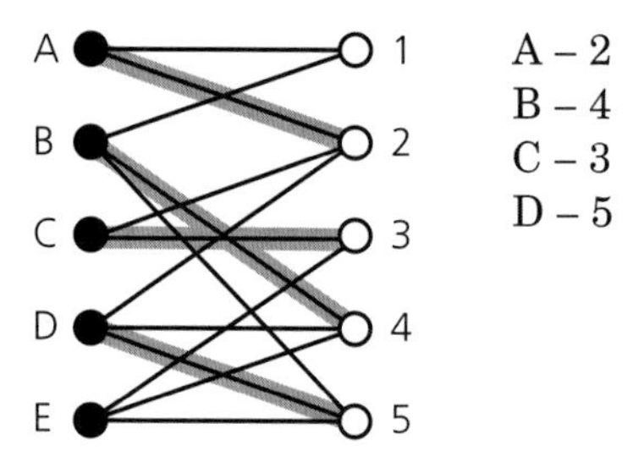

A – 2
B – 4
C – 3
D – 5

Note: It is easy to find an initial matching. It need only consist of one edge, though it speeds up the algorithm if the matching is close to optimal.

This is not a maximum matching, since Mr Eggham has not been allocated a subject and there is no one to teach Environmental Studies, so we proceed as follows.

Step 2:

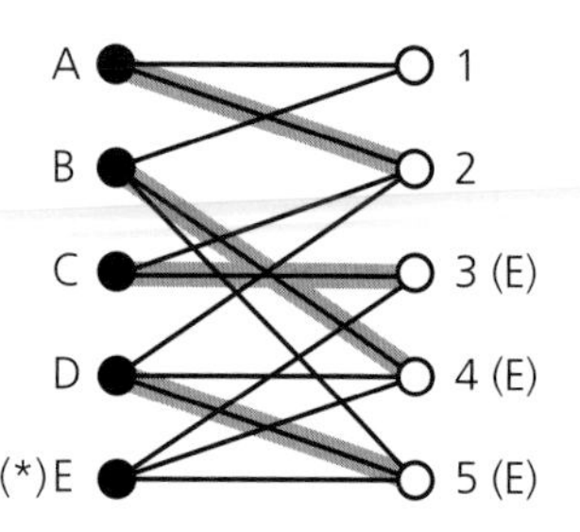

Label all the vertices on the left (subset A) which are not in the initial matching with a star (*). In this example the only vertex in subset A *not* in the matching is E.

Then we find all the vertices on the right (subset B) that are directly connected to E by edges *not* in the initial matching. We label these (E) since they are connected to vertex E.

If there were more than one right-vertex not in our initial matching, we should repeat this process until they were all labelled in this way.

Step 3:

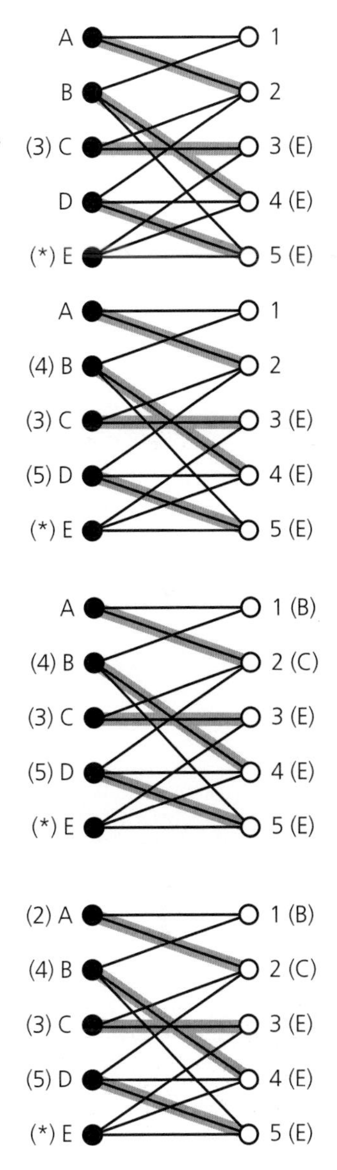

Next we choose one of the newly-labelled right-vertices, say 3, and label any left-vertices which are connected directly to it by an edge which is in the initial matching. So we should label vertex C with 3.

We then repeat this process with other newly-labelled right-vertices until all the labelling at this stage is complete.

Step 4:

We now look in turn at the newly-labelled left-vertices, B, C and D, and label the right-vertices connected to them but not in the initial matching which are not already labelled, giving this situation.

Step 5:

We repeat steps 3 and 4 until all possible vertices have been labelled.

When no further labelling is possible, we move on to the next stage of the algorithm.

Finding an alternating path

An **alternating path** in a bipartite graph is one which satisfies the following conditions:

- it joins vertices from subset A to those in subset B
- alternate edges of the path are in **M**, and the other edges are not in **M**
- the initial and final vertices are not incident with an edge of **M**.

Step 6:

We now look for a labelled vertex in subset B which is not in our initial matching, in this case vertex 1.

Step 7:

Starting here, we go to the vertex indicated by its label (B), and from B we go to the vertex indicated by its label and so on until we reach the vertex labelled (*):

The alternating path is 1–B–4–E.

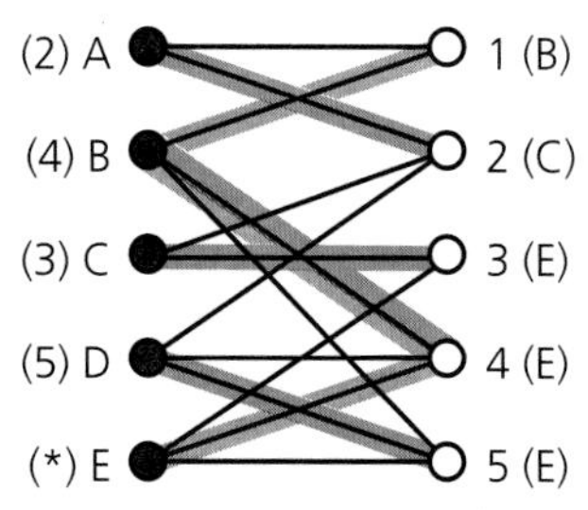

Our new matching consists of:

- the edges of the initial matching that are not in the alternating path: A – 2, C – 3, D – 5
- the edges in the alternating path that are not in the initial matching: B – 1, E – 4
  giving this solution.

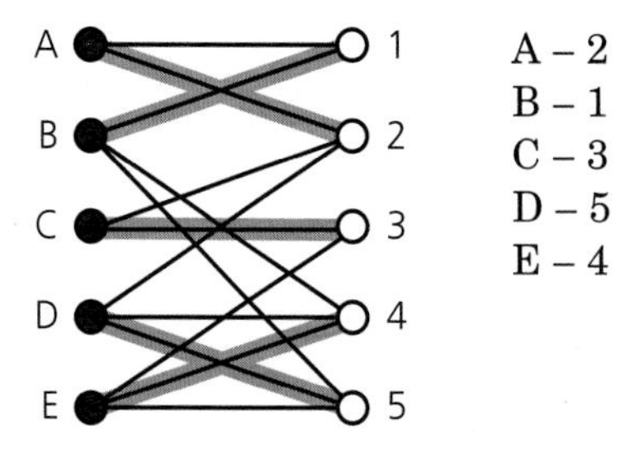

A – 2
B – 1
C – 3
D – 5
E – 4

This is a maximum matching, since all five vertices in A have been matched to vertices in B.

Note: Don't forget to interpret your solution in the context of the original problem.

So Mrs Frawt should timetable Miss Abraham to teach Geography, Mr Bowen to teach Environmental studies, Mrs Carter to teach Graphics, Ms Delgardo to teach Science and Mr Eggham to teach Mathematics.

It is important to note that this solution is not unique. If we had started with a different initial matching we may well have obtained a different solution.

The algorithm for finding a maximum matching is listed below.

Step 1: Find an initial matching (**M**).

Step 2: Label with (*) all the vertices in A which are not in **M**.

Step 3: Choose a newly-labelled vertex in A, say a, and label with (a) all the unlabelled vertices in B joined to a by an edge not in **M**. Repeat for all newly-labelled vertices in A before going on to step 4.

Step 4: Choose a newly-labelled vertex in B, say b, and label with (b) all the unlabelled vertices in A joined to b by an edge in **M**. Repeat for all newly-labelled vertices in B before going on to steps 5.

Step 5: Repeat steps 3 and 4 until no further labelling is possible.

Step 6: Look for a labelled vertex in B which is not in **M**. If such a vertex is found, go to step 7. If there is no such vertex, the matching **M** is optimal. STOP!

Step 7: Find an alternating path, starting at the labelled vertex in B and ending at (*).

Step 8: The new matching consists of:

- the edges of **M** which are not in the alternating path
- the edges in the alternating path but not in **M**.

The previous section was set out with a diagram at each step so that it would be easy to see what was happening. In setting out questions of this type it is not necessary to draw each step, though coloured pens are useful to indicate the initial matching and the alternating path, as shown below.

Example 7.2

A youth club wishes to enter a team at the regional athletics meeting. They have six good runners, A, B, C, D, E and F, who can take part and there are six track events. Runner A can run the 100 m and 200 m, runner B can run the 800 m and 1500 m, C the 200 m and 100 m hurdles, D the 400 m, 800 m and 1500 m, E the 400 m and 800 m and F the 100 m and 100 m hurdles. How should the captain pick the team?

Solution

Initial matching:

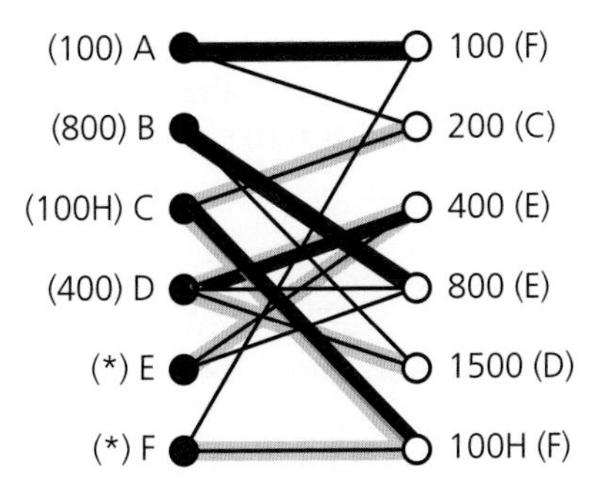

Initial matching **M**
A – 100
B – 800
C – 100H
D – 400

Note: This leaves out E *and* F. Thus we get two alternating paths – one starting at 200 (pale highlighter) and one starting at 1500 (dark highlighter).

The alternating paths are:

- 200 – C – 100H – F
- 1500 – D – 400 – E

Edges of **M** that are not in either alternating path are:

- A – 100
- B – 800

Edges in alternating paths but not in **M** are:

- C – 200
- F – 100H
- D – 1500
- E – 400

So a solution is:
A – 100 m
B – 800 m
C – 200 m
D – 1500 m
E – 400 m
F – 100 m hurdles

EXERCISES

7.2 CLASSWORK

1. Repeat Mrs Frawt's timetabling problem using the initial matching: A – 1, C – 2, D – 4, E – 5.

2. Use the maximum matching algorithm to solve question 3 of Exercises 7.1 CLASSWORK.

3. Use the maximum matching algorithm to choose the teams which Everbrite School will enter for the swimming gala in question 4 of Exercises 7.1 CLASSWORK.

4. Using the information in question 5 of Exercises 7.1 CLASSWORK, allocate the jobs on the building site.

5. The managing director of a store is seeking to appoint five supervisors for five of his departments: Electrical goods, Furnishings, Hardware, Men's wear and Ladies' fashions. The short-listed applicants have a variety of skills.

   Annabel has experience of retailing in both ladies' and men's wear.
   Bettina has never worked in clothing, but could work in any of the other departments.
   Chris has spent his working life in electrical and hardware shops.
   Doreen's last post was as supervisor in a small hardware shop and before that she spent several years in a ladies' fashion store.
   Imran has worked in men's tailoring and in furnishings.
   Who should be appointed to supervise each department? Is your solution unique?

EXERCISES

7.2 HOMEWORK

1. In this bipartite graph, the thick lines denote an initial matching. Using this initial matching, find a maximum matching.

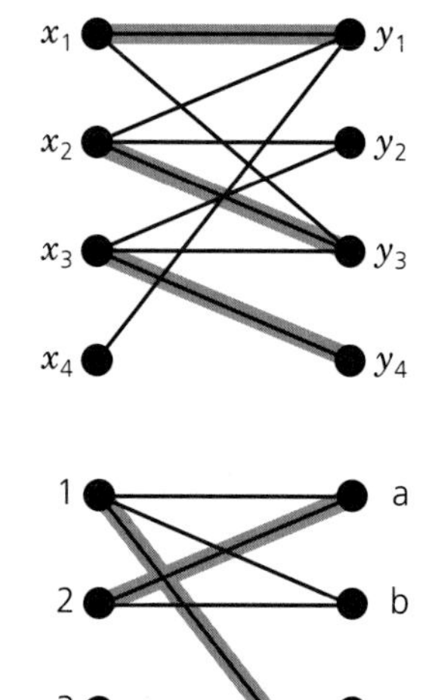

2. In this bipartite graph, the thick lines denote an initial matching. Write down an alternating path with respect to this matching and hence find a maximum matching in the graph.

3. A university has seven vacancies (a, b, c, d, e, f and g) left on their courses. There are seven applicants (A, B, C, D, E, F and G) left in clearing, and their qualifications for the different vacancies are shown in the following table.

Course	Suitable applicant
a	A, D, E
b	B, E, G
c	D
d	B, C, E, G
e	D, F
f	B, D, F
g	A, B

Draw a bipartite graph to represent the problem of allocating courses to applicants. Write down an initial matching and hence find a maximum matching. Is this solution unique?

4 Angela, Beth, Caroline, Denise and Erica are going clubbing on St Valentine's Day and the table below gives the males they would like to accompany them. Determine a suitable scheme for partners for the night out (if one exists).

Angela	Dan, Mike, Gavin
Beth	Simon, John
Caroline	Simon, Mike
Denise	Dan, Gavin, John
Erica	Mike, John, Dan

5 There are five clues left, to complete a crossword.
1 across begins with D.
16 across ends in the letter G.
20 across needs a word with letter A in the middle.
8 down requires an R in the second position and a T in the last position.
12 down is the clue 'student monetary award'.
The five words that come to mind are DOING, DRAFT, FRAME, GOING and GRANT. Can these words be uniquely fitted into the crossword? If so, in which positions should they be placed?

MATHEMATICAL MODELLING ACTIVITY

Supply and demand

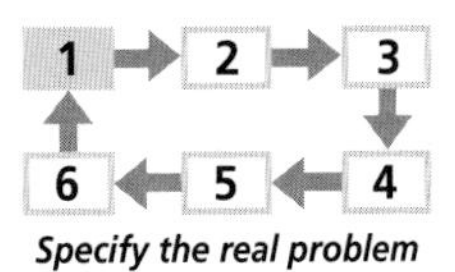

Specify the real problem

Problem statement

Five builders' merchants, A, B, C, D and E, supply building sites P, Q, R, S and T. The cost of supplying bricks from each builders' merchant to each building site is shown in the table (in £s per tonne).

	P	Q	R	S	T
A	15	14	16	15	10
B	12	10	13	14	10
C	13	10	12	13	12
D	10	16	11	13	13
E	14	12	14	13	15

The sites all require the same amount of bricks. Allocate a builders' merchant to supply each site so that the total cost is minimised.

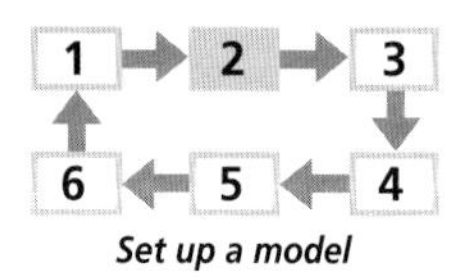

Set up a model

Set up a model

A good starting point is to reduce the table to a 'relative cost' matrix by subtracting the minimum cost from all entries. List any assumptions which have been made.

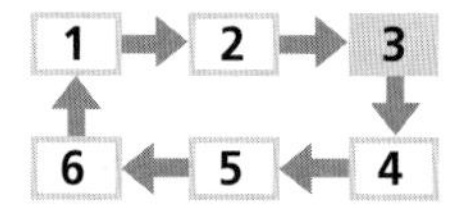

Formulate the mathematical problem

Mathematical model

We will try to allocate as many as possible using the routes with zero relative cost, so we can use the relative cost matrix to produce a bipartite graph showing zero cost edges.

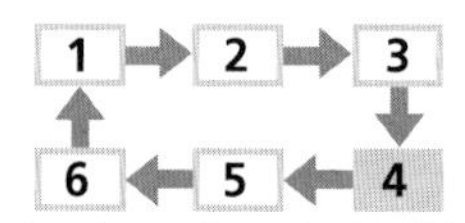

Solve the mathematical problem

Mathematical solution

Obtain an optimal matching using the bipartite graph and consider how you would deal with any builders' merchants or building sites not in this solution.

Inspect your possible solutions to consider which gives a minimum delivery cost.

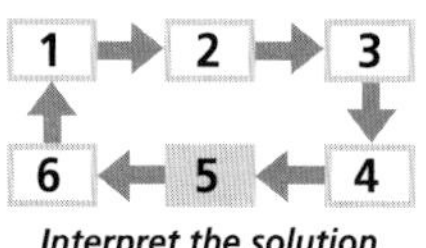

Interpret the solution

Refinements to the model

Instead of simply subtracting the minimum cost from all entries, try firstly subtracting the minimum value in each row from all entries in the row, then subtracting the minimum value from each column of your new matrix from all the values in that column. Use this new matrix to draw a bipartite graph using maximum matching algorithm. Check your solution against your solution from the first attempt.

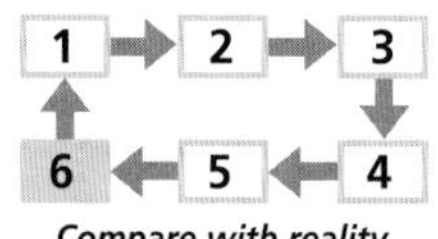

Compare with reality

Interpret the solution

Interpret your final solution in the context of the original problem.

CONSOLIDATION EXERCISES FOR CHAPTER 7

1 Draw the complete bipartite graphs for **a)** $K_{4,3}$ **b)** $K_{5,2}$.

2 Which of the following graphs are planar?

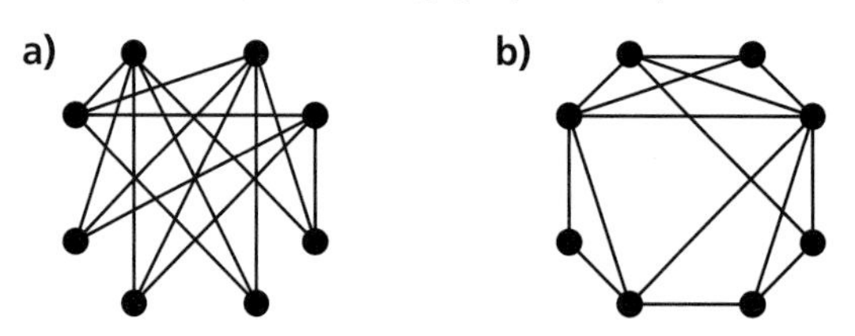

3 Five packs of sandwiches have been prepared for lunch, one each of egg, cheese, ham, tuna and salmon. Five people have been invited for lunch and the sandwiches which they like are given in the table.

Mr Large	Egg, Cheese
Mrs Moore	Egg, Tuna, Salmon
Ms Nice	Cheese, Ham
Mr Oliver	Cheese, Tuna and Salmon
Miss Patel	Ham, Tuna, Salmon

a) Draw a bipartite graph to model this situation.

The host allocates the egg sandwich to Mr Large, the cheese to Ms Nice, the tuna to Mr Oliver and the salmon to Miss Patel.

b) Indicate this initial matching in a distinctive way on a bipartite graph.

c) Starting from this matching use the maximum matching algorithm to find a complete matching. Indicate clearly how the algorithm has been applied.

(Edexcel Decision Mathematics 1 Specimen Paper 2000)

4 Granny has bought Christmas presents for her five grandchildren.

The teddy bear is suitable for Cathy, Daniel or Elvis.
The book is suitable for Annie or Ben.
The football is suitable for Daniel or Elvis.
The money box is suitable for Annie or Daniel.
The drum is suitable for Cathy or Elvis.

Draw a bipartite graph, G, to show which present is suitable for which grandchild.

Granny decides to give Annie the book, Cathy the teddy bear, Daniel the money box and Elvis the drum. This leaves Ben without a present, since the football is not suitable for him.

i) Show the incomplete matching, M, that describes which present Granny had decided to give to each child.

ii) Use a matching algorithm to construct an alternating path for M in G, and hence find a maximal matching between the presents and the grandchildren.

(OCR Discrete Mathematics 2 Specimen Paper 2000)

5 Alex likes ready-salted and salt-and-vinegar crisps, Bill likes prawn-cocktail and salt-and vinegar crisps, while Charles prefers ready-salted and salt-and-vinegar crisps.

a) Show this information on a bipartite graph.

b) Just one of each type of these crisps is available. Alex chooses the ready salted bag and Bill chooses the salt-and-vinegar bag. Charles is then disappointed. Demonstrate, by using an algorithm to find an alternating path from this initial matching, how these selections could be made so each has a bag of crisps of their own liking.

(AQA Mathematics A Discrete 1 Specimen Paper 2000)

6 A small firm gets a rush order. To finish the job on time, the manager asks each of the five employees to work overtime to watch the machines on one evening during the week. Due to other commitments, the employees are only available on the following evenings.

A can work late on Monday or Wednesday.
B can work late on Monday or Thursday.
C can work late on Tuesday, Wednesday or Friday.
D can only work late on Thursday.
E can work late on Tuesday or Friday.

In how many ways can the manager organise the overtime rota?

7 A manager wishes to appoint supervisors to four of his departments: Electrical, Food, Stationery and Watches. He has four candidates for these posts. The table shows which departments the candidates are qualified to supervise.

Ahmed	Stationery, Watches
Bridget	Electrical, Food, Stationery
Chris	Food, Stationery, Watches
Diane	Electrical, Stationery

a) Model this situation with a bipartite graph.
The manager allocates Ahmed, Chris and Diane to the first department in their individual lists.
b) Starting from this matching show clearly how a matching in which each candidate gets a department can be obtained.

(ULEAC Question 4, Specimen Paper D1, 1996)

8 Everbrite School is mounting its annual play and several staff have volunteered to take charge of various aspects of the production. Mrs Frawt and Miss Delgardo have both said they will do either publicity or ticket sales, Miss Perry can do make-up and costumes, Mr Quince offers help with lighting or scenery, Ms Rouse would like to be the producer but wouldn't mind doing make-up, Mrs Standish enjoys dressmaking but can also paint scenery, Mr Taylor has produced shows before and can also do sound or lighting and Mr Unwin says he'll have a go at anything except sewing! Who should do which task?

9 One Saturday six potential customers visit the local travel agent looking for 'last minute' bargain holidays. The agent has holidays available with the following accommodation.

1 hotel with a twin room	1 hotel with a family room (for four)
1 apartment sleeping four	1 flat sleeping up to four
1 apartment sleeping six	1 flat sleeping up to eight

The customers are: one married couple, a group of three friends, a family of four, a family of five, two families (seven people in all) who wish to stay together and a group of five friends. Can the agent arrange holidays for all the customers? If not, who will be disappointed?

10 A medley relay swimming team consists of Mike, Neal, Oliver and Pete. The legs which they can swim are:
Mike: backstroke or butterfly
Neal: breaststroke, backstroke or freestyle
Oliver: breaststroke, freestyle or butterfly
Pete: freestyle or backstroke.

a) In how many ways can the swimming order be decided?
b) The table below gives each swimmer's times for the different legs. Which solution gives the minimum time?

	Breaststroke	Backstroke	Butterfly	Freestyle
Mike		66	64	
Neal	66	64		60
Oliver	63		65	61
Pete		65		63

Summary

After working through this chapter you should:

- understand that a bipartite graph is a graph in which the vertices can be divided into two subsets such that each edge of the graph joins a vertex in one subset to a vertex in the other subset
- understand the term 'subdivision' to mean the insertion into a bipartite graph of vertices of degree 2
- be able to show when a graph is non-planar by finding a subdivision of $K_{3,3}$
- know how to use bipartite graphs to model real problems such as matching workers with particular skills to a job that needs to be done
- use the maximum matching algorithm to find the best solution to the matching problem.

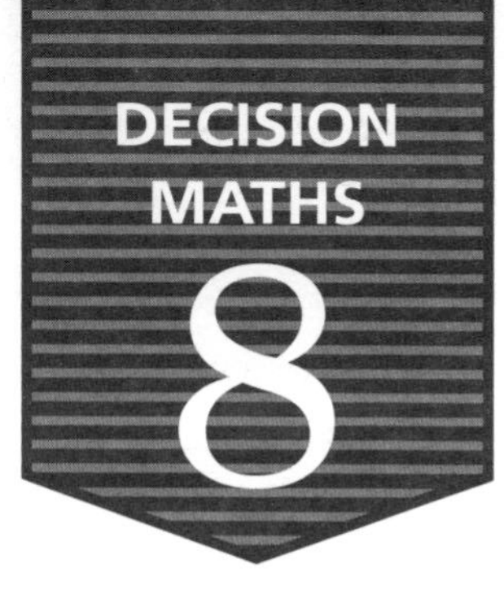

Linear programming 1: Modelling and graphical solution

By the end of this chapter you should be able to:

- *recognise problems that involve linear programming and know how to formulate them as mathematical problems*
- *define the variables in a problem and formulate the objective function which relates these variables*
- *establish the constraints which are limits on the values and ranges of the variables*
- *use a graphical method for solving linear programming problems in two variables by identifying the feasible region which includes all the possible combinations of values of the variables.*

LINEAR PROGRAMMING PROBLEMS

Exploration 8.1

A balanced diet

The parents of a two-year-old child want to ensure that they include at least the minimum satisfactory amount of protein, energy and vitamin C. The daily amounts recommended by good health guides are 32 g of protein, 5500 kilojoules (kJ) of energy and 30 mg of vitamin C. Suppose that the parents decide to achieve the daily amounts through milk and orange juice. The following table shows the amount of protein, energy and vitamin C available in these drinks.

Item	Measure of drink	Protein (g)	Energy (kJ)	Vitamin C (mg)
milk	100 ml	3.0	300	1
juice	100 ml	0.2	40	10

Milk costs 38p per litre and orange juice costs 48p per litre. The parents want to find the minimum costs to achieve the recommended daily intake.

If x and y are the number of measures of milk and juice, write down:

- the total costs of the drinks
- three inequalities for the amount of protein, energy and vitamin C so that the minimum satisfactory amounts are achieved.

Cost function

The **cost function** is the relationship between the cost and the variable components included in the cost. One measure of milk or orange juice is 0.1 litre so the cost of a measure of milk is 3.8p and the cost of a measure of juice is 4.8p.

x measures of milk will cost $3.8x$ and y measures of orange juice will cost $4.8y$.

The total cost function is given by $C = 3.8x + 4.8y$.

Constraints

For protein, the number of grams in x measures of milk is $3x$ and in y measures of orange juice it is $0.2y$. The minimum amount of protein required is 32 g. So:

$$3x + 0.2y \geq 32$$

Similarly, the amount of energy in x measures of milk and y measures of orange juice is $300x + 40y$. Since the minimum requirement is 5500 kJ:

$$300x + 40y \geq 5500$$

For vitamin C:

$$x + 10y \geq 30$$

Finally, since it is not possible for the child to drink negative amounts of milk and orange juice:

$$x \geq 0 \text{ and } y \geq 0$$

The mathematical problem that we need to solve can be summarised like this.

Find a minimum value for	$C = 3.8x + 4.8y$	**1**
where	$3x + 0.2y \geq 32$	**2**
	$300x + 40y \geq 5500$	**3**
	$x + 10y \geq 30$	**4**
	$x \geq 0, y \geq 0$	**5**

This is an example of a linear programming problem. It is made up of two parts:

- line 1: the **objective function**
- lines 2–5: the **constraints**.

The objective function is a **linear function** that has to be minimised or maximised depending on the nature of the problem.

The constraints are linear equalities or linear inequalities that restrict the values of the variables x and y. The constraints given by line 5 are called the **trivial constraints**.

The above mathematical problem will be solved graphically and algebraically later in this chapter.

Formulating linear programming problems

Example 8.1

Ben has two evening and weekend jobs, mowing lawns and cleaning cars. He can spend a maximum of 15 hours per week on these jobs. He has several regular clients for whom he spends six hours per week mowing lawns and three hours per week cleaning cars. Ben is paid £2.50 per hour for mowing lawns and £3.50 per hour for cleaning cars.

Formulate a linear programming problem for Ben to maximise his earnings.

Solution

Start by defining the variables.

- *Let x be the number of hours per week spent mowing lawns.*
- *Let y be the number of hours per week spent on cleaning cars.*
- *Let I be the income per week.*

Now set up the objective function.
The amount Ben earns by mowing lawns is $2.5x$ and from cleaning cars it is $3.5y$. The objective function is:

$I = 2.5x + 3.5y$

Establish the constraints.
Since the maximum number of hours available is 15:

$x + y \leq 15$

Ben is committed to 6 hours of mowing lawns and 3 hours of cleaning cars:

$x \geq 6, y \geq 3$

The linear programming problem is:

Find a maximum value of	$I = 2.5x + 3.5y$	**1**
where	$x + y \leq 15$	**2**
	$x \geq 6, y \geq 3$	**3**

Notice that in this example we do not need the trivial constraints because they are satisfied by line 3.

EXERCISES

8.1 CLASSWORK

1 Chris is making coffee cakes and chocolate cakes for a cake stall at the village fete. Each chocolate cake needs 2 eggs and 250 g of margarine. Each coffee cake needs 3 eggs and 150 g of margarine. Chris buys three dozen eggs and a 3 kg tub of margarine. Formulate a linear programming problem so that Chris can make as many cakes as possible.

2 A factory produces two types of toys: trucks and bicycles. In the manufacturing process of these toys three machines are used. These are a moulder, a lathe and an assembler. The table shows the length of time needed for each toy.

	Moulder	Lathe	Assembler
bicycle	1 hour	3 hours	1 hour
truck	0 hours	1 hour	1 hour

The moulder can be operated for 3 hours per day, the lathe for 12 hours per day and the assembler for 7 hours per day. Each bicycle made gives a profit of £15 and each truck made gives a profit of £11. Formulate a linear programming problem so that the factory maximises its profit.

3 A physical fitness enthusiast, Karen, decided to exercise by a combination of jogging and cycling. She jogs at 6 mph and cycles at 18 mph. A measure of the benefit of these activities is given through aerobic points. An hour of jogging gains 4 aerobic points and an hour of cycling gains 3 aerobic points. Each week she would like to earn at least 12 aerobic points, cover at least 54 miles and cycle at least twice as much as she jogs.

Formulate a linear programming problem so that Karen minimises the total time spent exercising while achieving her goals.

4 A house builder builds two types of home.

The first type requires one plot of land, £40 000 capital, 150 worker days of labour and is sold for £8000 profit.
The second type requires two plots of land, £105 000 capital, 200 worker days of labour to build and is sold for £11 500 profit.
The house builder owns 150 plots of land, has available £9 600 000 and 24 000 worker days of labour.
Formulate a linear programming problem to advise the builder on a strategy to maximise the profit.

5 A food manufacturer makes Alma margarine from vegetable and non-vegetable oils. The raw oils are refined by the manufacturer and blended to form the margarine. One objective of the manufacturer is to make as large a profit as possible; however, there are various constraints. It is important that the raw oils are refined separately to avoid contamination and the final blend of oils must be soft enough to spread, but not too runny. To achieve this quality each ingredient and the final product has a hardness factor. The following table gives the cost and hardness of the raw materials used to make Alma margarine.

	Oil	Cost (£ per kg)	Hardness factor
Vegetable	Groundnut	0.7	1.2
	Soya bean	1.05	3.4
	Palm	1.05	8.0
Non-vegetable	Lard	1.30	10.8
	Fish	1.10	8.3

Alma margarine sells for £1.55 per kg and to ensure good spreading, its hardness must be between 5.6 and 7.4. The manufacturers can refine up to 40 000 kg of vegetable oil and up to 32 000 kg of non-vegetable oil per day. Formulate the problem of maximising the profit as a linear programming problem.

EXERCISES

8.1 HOMEWORK

1 As part of a charity stunt a student wishes to travel around a circular route in 80 minutes or less. The first five miles are to be travelled on top of a horse-drawn tram and the final three miles in a pink hearse. If the speeds of tram and hearse are x mph and y mph respectively, write down the constraints for the above.

2 For a dog to remain healthy its daily diet must provide it with at least 2 megajoules (MJ) of energy and at least 0.3 kg of protein, but not more than 0.04 kg of salt. There are two items in the diet of most dogs, biscuits and meat.
1 kg of meat costs 90p and provides 5 MJ of energy, 0.5 kg of protein and 0.04 kg of salt.
1 kg of biscuits costs 25p and provides 20 MJ of energy and 0.2 kg of protein and 0.08 kg of salt.
Formulate the linear programming problem to enable you to find the cheapest combination of biscuits and meat in the dog's daily diet, that will keep it healthy.

3 A soap company manufactures two powders: XTRA for extra cleaning power and YTER for a whiter wash. These powders contain ingredients A and B in the following proportions.

	A	B
XTRA	90%	10%
YTER	50%	50%

The company can store up to 20 000 kg of ingredient A and up to 7000 kg of B. To obtain favourable price reductions their supplier requires a purchase of at least 5000 kg of A and at least 2000 kg of B per week. The maximum amounts of XTRA and YTER in total that can be produced by the plant is 25 000 kg per week. The profit on XTRA is £1.50 per kg and on YTER it is £2.10 per kg.
Devise a set of inequalities to represent all of the above constraints in terms of the number of thousand kg of each of XTRA and YTER produced each week. Write down the algebraic expression representing the weekly profit.

4 A fruit farmer has two orchards A and B, both of which produce apples of class 1, class 2 and class 3.

In orchard A, the labour and equipment costs for picking are £20 per hour and picking for one hour produces six bushels of class 1 apples, two bushels of class 2 apples and four bushels of class 3 apples.
In orchard B, the labour and equipment costs for picking are £16 per hour and picking for one hour produces two bushels of class 1 apples, two bushels of class 2 apples and twelve bushels of class 3 apples.

The farmer is under contract to supply twelve bushels of class 1 apples, eight bushels of class 2 apples and fourteen bushels of class 3 apples to a market each day.
Formulate a linear programming problem to determine how many hours each day the labour force should be consigned to each orchard if the total costs are to be minimised, given that the labour force will not work more than eight hours per day.

The farmer may offset his costs by selling picked apples, which are surplus to those under contract, to a cider manufacturer at £1 per bushel (regardless of class). Determine the new cost function in this case.

5 New World Wines Ltd produce a wine by blending 'Chateau Australia House' (CAH) with an inferior 'Bondi Beach Muddie' (BBM). The blend becomes unstable if the proportion of BBM exceeds 80 per cent or falls below 25 per cent of the mixture, so this is to be avoided. The maximum wine-handling capacity per day is 5000 litres and the plant is so constrained that certain pipes and mixing vessels must always be kept full, so that the total amount of BBM processed and half the amount of CAH processed is at least 2000 units. The cost to the company of BBM is 22 cents per litre and that of CAH is 30 cents per litre and the blended wine is sold at 200 cents per litre.

Formulate a linear programming problem that will:

- minimise the costs for the company,
- maximise the profit for the company.

SOLVING LINEAR PROGRAMMING PROBLEMS

We can find an optimal solution for linear programming problems with two variables by drawing straight-line graphs for the constraints. This defines an acceptable region called the **feasible region**. Then an optimal solution can be found at one of the points of the feasible region.

Exploration 8.2

A graphical solution

A linear programming problem is given by the following summary.

Maximise the function	$f = 80x + 70y$	**1**
where	$2x + y \leq 32$	**2**
	$x + y \leq 18$	**3**
	$x \geq 0, y \geq 0$	**4**

- Draw the straight-line graphs $2x + y = 32$ and $x + y = 18$.
- Shade in the region of the x–y plane in which inequalities 2–4 hold.
- Investigate at which point the objective function f takes its largest value.

Solving graphically

Referring to Exploration 8.2, consider the constraint 2 in the form:

$y \leq 32 - 2x$

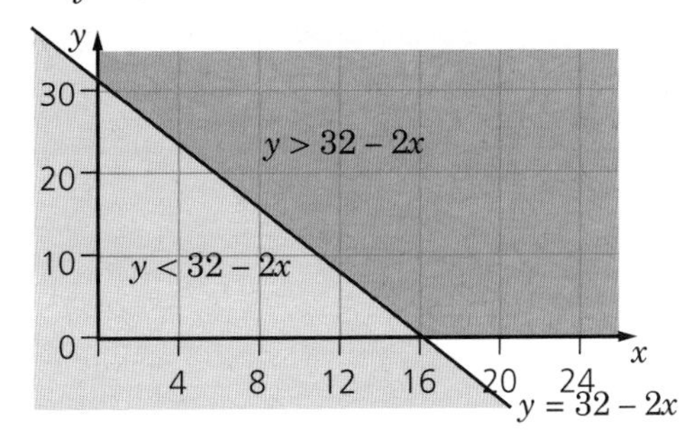

For all points on the line, $y = 32 - 2x$. Therefore the points satisfying $y < 32 - 2x$ must lie below the line i.e. for y values that are smaller than $32 - 2x$.

Repeating the same argument for inequalities 3 and 4, the allowable region that satisfies the constraints is shown in the diagram. This is the feasible region and in this case it includes the edges of the region.

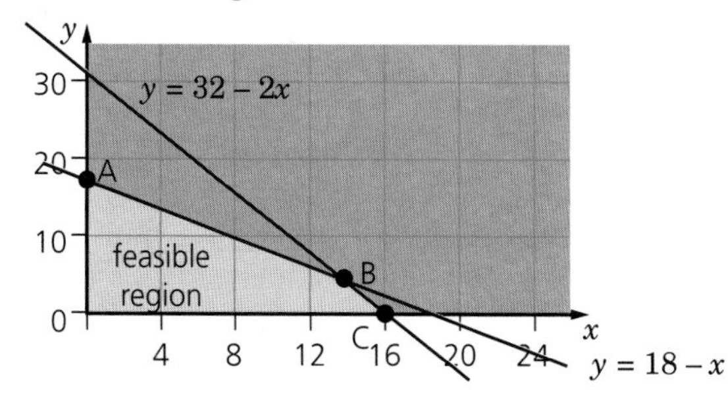

The objective function is $f = 80x + 70y$. By trial and error you may have found that this function takes its largest value at point B(14, 4). At this point:

$$f = 80 \times 14 + 70 \times 4 = 1400$$

The result that the optimal value occurred at one of the vertices of the feasible region is no mere chance, as the following theorem shows.

The fundamental theorem of linear programming

The maximum (or minimum) value of the objective function occurs at one of the vertices of the feasible region.

We can verify this theorem using the feasible region and objective function for Exploration 8.2. Consider the objective function:

$$f = 80x + 70y$$

Rearranging this gives:

$$y = \tfrac{1}{70}f - \tfrac{8}{7}x$$

For given values of f this is an equation of a straight line with slope $-\frac{8}{7}$.

The diagram shows the feasible region and the graph of $y = \frac{1}{10}f - \frac{8}{7}x$ for $f = 700$, $f = 1400$ and $f = 2100$.

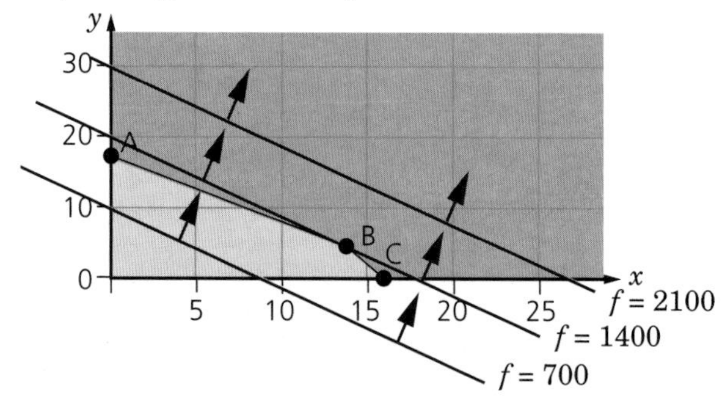

The three lines are all parallel. As f increases in value the line $y = \frac{1}{10}f - \frac{8}{7}x$ moves through the feasible region until it reaches B when it is about to leave the feasible region. Beyond B the constraints are no longer satisfied. So the maximum value of f for points inside the feasible region occurs at vertex B.

Exploration 8.3

Investigating the fundamental theorem

A linear programming problem is given by the following summary.

Maximise the function	$f = x + y$	**1**
where	$y - 2x \leq 0$	**2**
	$x + 2y \leq 100$	**3**
	$4x + 3y \leq 300$	**4**
	$x \geq 0, y \geq 0$	**5**

- Draw the feasible region.
- Verify the fundamental theorem of linear programming for $f = x + y$.

Investigating the theorem

The diagram shows the feasible region and the line $y = f - x$ for $f = 100$.

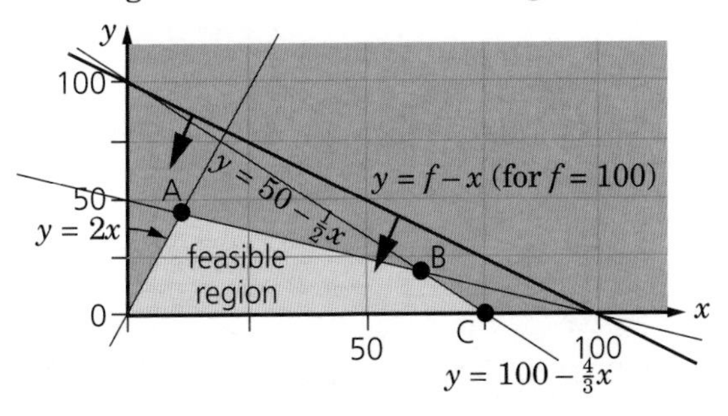

As the line $y = f - x$ moves towards the feasible region the value of f decreases until the line reaches the vertex B. At this point $x = 60$ and $y = 20$ and $f = x + y = 80$. If the line moves into the feasible region then f decreases further.

Hence the maximum value of f occurs at point B and has value 80.

Example 8.2

Solve this linear programming problem.

Minimise the function	$f = 2x + 3y$
where	$x + y \geq 4$
	$3x + 5y \geq 16$
	$x \geq 0, y \geq 0$

Solution

The feasible region is the region bounded by the straight lines $x + y = 4$, $3x + 5y = 16$ *and* $y = 0$.

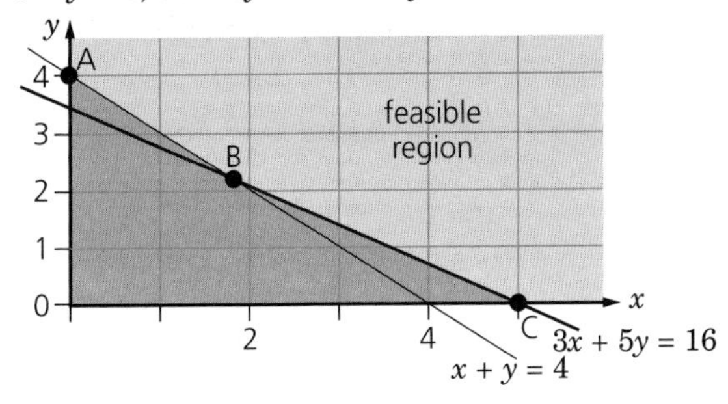

This table shows the coordinates of the vertices and the value of the objective function.

Vertex	x	y	$f = 2x + 3y$
A	0	4	12
B	2	2	10
C	$\frac{16}{3}$	0	$\frac{32}{3}$

The minimum value of the objective function is $f = 10$ *and occurs at* $x = 2$, $y = 2$.

Example 8.3

Solve this linear programming problem.

Minimise the function $f = 7x + 4y$

where

$$2x + y \geq 11$$
$$x + y \leq 10$$
$$x + 3y \leq 18$$
$$x + 4y \geq 16$$

Solution

The feasible region is the region bounded by the straight lines $2x + y = 11$, $x + y = 10$, $x + 3y = 18$ *and* $x + 4y = 16$.

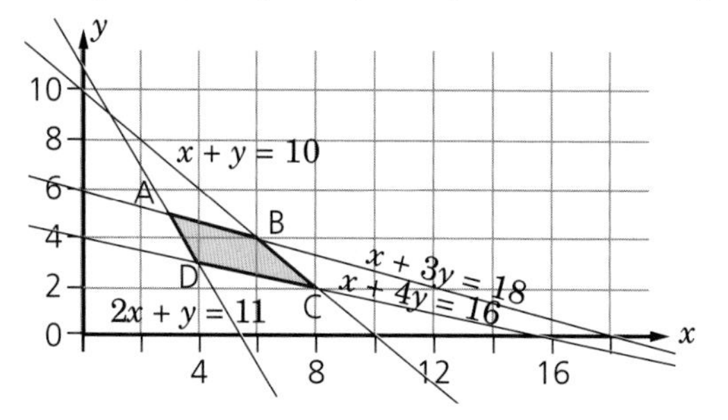

The following table shows the coordinates of the vertices and the value of the objective function f.

Vertex	x	y	$f = 7x + 4y$
A	3	5	41
B	6	4	58
C	8	2	64
D	4	3	40

The minimum value of f is 40 and occurs at $x = 4$, $y = 3$.

Example 8.4

Integer Solutions

Consider the following linear programming problem where x *and* y *must take integer values.*

Minimise the function $f = 5x + y$

subject to constraints

$$30x + 9y \leq 3700$$
$$30x + 4y \leq 2000$$
$$y \geq 75$$

Solution

The graphical solution looks like this

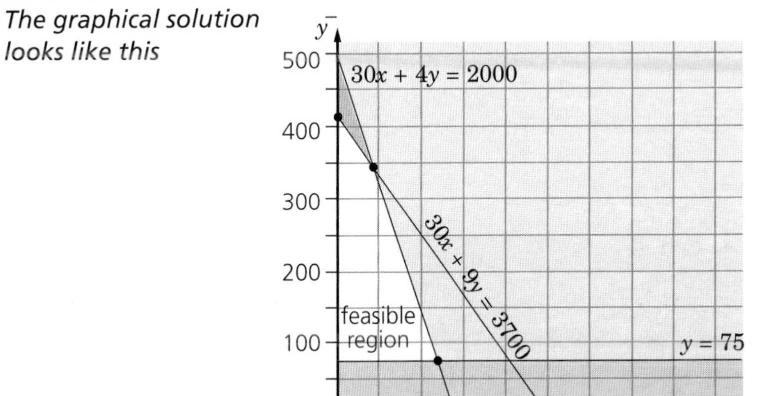

The maximum value of f is $446\frac{2}{3}$, when $x = 21\frac{1}{3}$ and $y = 340$

However, the problem states that x and y must take integer values, so this is not a valid solution. The obvious point to check is (21,340) which gives f = 445 BUT TAKE CARE.

The point (21, 341) is also in the feasible region which gives f = 446, so the solution is actually x = 21, y = 341 giving f = 446.

Exploration 8.4

A non-unique solution

■ Investigate the fundamental theorem applied to this problem.

Maximise the function $f = x + y$

where
$$y - 4x \leq 0$$
$$x + y \leq 10$$
$$4x + y \leq 20$$
$$x \geq 0, y \geq 0$$

More than one solution

In this problem the slope of the objective function straight line $y = f - x$ is equal to the slope of the boundary of the feasible region. The maximum value of f is 10 and occurs at any point of the boundary between A and B.

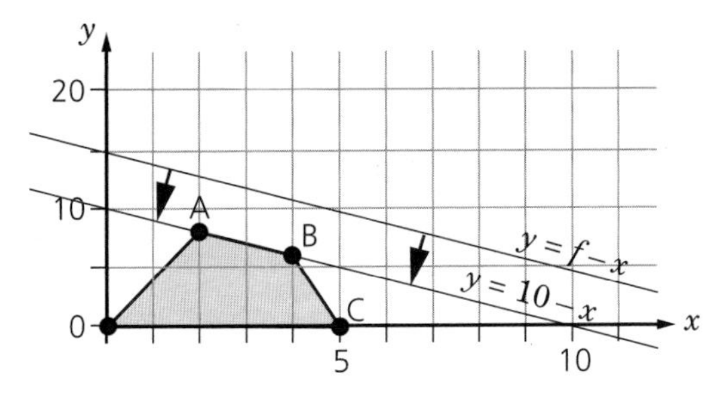

EXERCISES

8.2 CLASSWORK

1 Solve the following linear programming problems using the graphical method of solution.

a) Maximise the function $f = 2x + 3y$
where
$x + 3y \le 120$
$7x + 2y \le 156$
$x \le 20$
$x \ge 0, y \ge 0$

b) Maximise the function $f = x + 2y$
where
$x + y \le 100$
$y - 3x \ge 60$
$x - y \ge 0$

c) Minimise the function $f = x + y$
where
$2x + y \le 6$
$2x + 3y \le 12$
$4x + y \le 10$
$x \ge 0, y \ge 0$

d) Minimise the function $f = 5x + 2y$
where
$x + y \le 20$
$2x + y \ge 20$
$4x - 3y \ge 10$
$y \ge 4$

e) Maximise the function $f = \frac{2}{3}x + y$
where
$x + y \ge 15$
$x + y \le 45$
$x \le 30$
$y \le 25$
$x \ge 0, y \ge 0$

2 **a)** Maximise the function $c = 10x + 17y - 1500$
where
$3x + 2y \le 3600$
$x + 4y \le 1800$
$x \ge 0, y \ge 0$

b) Minimise the function $f = 2x + y$
where
$x + y \le 27$
$x + y \ge 15$
$y \le 2x$
$x \ge 3, y \ge 3$

c) Maximise the function $f = x + \frac{1}{2}y$
where
$2x + y \le 36$
$x + y \ge 15$
$y \le 2x$
$x \ge 3, y \ge 3$

3 The constraints of a linear programming problem are given by:

$2y - x < 40$
$3x + 4y \ge 180$
$x < 90$
$y \ge 0$

a) Find integer values of x and y which maximise the objective function $f = 10x + 10y$.

b) Find integer values of x and y which minimise the objective function $c = x + 5y$.

4 Solve questions 1–4 from Exercises 8.1 CLASSWORK. Why can you not solve question 5?

EXERCISES

8.2 HOMEWORK

1 Shade in the feasible region determined by the simultaneous inequalities:
$3y + 2x \leq 12, x - 5y \leq 5, y - 3x \leq 3$.
Which of the points (0, 0), (–2, 3), (2, –3), (0, –1), (2, 2), (3, 5) lie in the feasible region?

2 A feasible region is defined by the area inside and on the square with the four vertices:
A(1, 1), B(–1, 1), C(–1, –1), D(1, –1).

At which corner does the maximum of the objective function $z = -2x - 3y$ occur? State the maximum value of z.

3 Draw the line segments joining the following points.
A(0, 0) to B(5, 9) B(5, 9) to C(10, 7) C(10, 7) to D(15, 0)
D(15, 0) to A(0, 0)
A feasible region is defined as that region on or inside these line segments. Where is the minimum of the objective function $f = y - 2x$ attained? What is the minimum value of f?

4 The feasible region M is defined by $y \geq 2, x \leq 1, y \leq 2x + 4$. Where does the objective function $f = -6x + 2y$ attain:
a) a minimum **b)** a maximum value in M?

5 Find solutions to questions 2–5 of Exercises 8.1 HOMEWORK.

MATHEMATICAL MODELLING ACTIVITY

Curve fitting

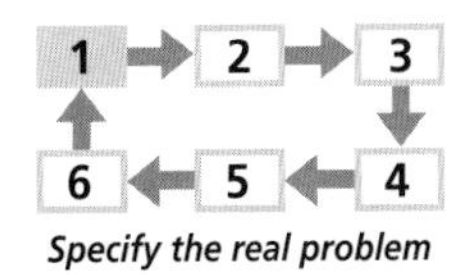

Specify the real problem

Problem statement

A pump's performance may be represented by three curves:
- production head* against flow rate
- absorbed power against flow rate
- efficiency against flow rate.

*Head refers to the height to which the pump can lift a column of water and is the most common way of describing the device's pressure-generating capability.

At a fixed speed of shaft rotation, these parameters are linked by:

$$\eta = 0.001\frac{\rho g h q}{P}$$

where η is the efficiency (%), ρ is the density (assumed as 1000 kg/m^3), g is the gravitational acceleration (taken as 9.81 m/s^2), h is the head generated in metres (m), q is the flow rate in litres per second (l/s) and P is the power in kilowatts (kW).

When a pump is being mathematically tested during the design process, the 'best efficiency point' (BEP) is first fixed in order to specify the optimum flow, head, efficiency and consequent power absorption.

For a particular pump, the BEP head = 20 m, BEP flow = 50 l/s, BEP efficiency = 80% and so BEP power = 12.26 kW.

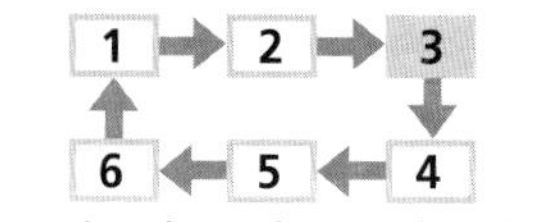

Formulate the mathematical problem

Mathematical problem

- In addition to the BEP, the head may be specified at two additional points, one at zero flow and another at, for example, 140% of BEP flow. A curve can be defined to join these three specified points of head against flow. Assume that at zero flow h = 25 m and at 140% of BEP flow h = 15 m. Based on these figures, deduce a method of joining the three points with a smooth curve and determine its equation.
- The efficiency will also be represented by a curve that is fixed by three conditions:
  a) its value will be zero at zero flow
  b) its value at BEP is known
  c) it must peak at BEP.
  Deduce a method of meeting these three efficiency conditions with another smooth curve and determine the equation relating to q.
- Given that head–flow and efficiency–flow curves are now defined in equation form, can the power be represented as another curve for which the equation can be found? If so, calculate this equation. Hence or otherwise calculate the power which is absorbed at intervals of 5 l/s between zero flow and 140% of BEP flow. Draw the graph of these results and interpret your results.

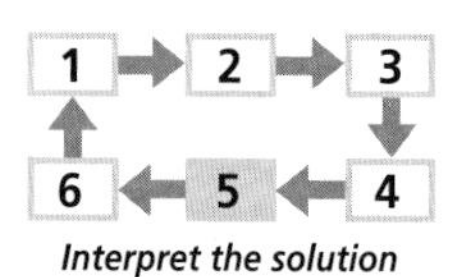

Interpret the solution

Refinement of the model

- Additional efficiency points may be specified through which the efficiency curve must pass (for instance $n = 15$, $q = 15$). Investigate the effect this has on the modelling process.
- The head–flow curve of a pump is often best represented by a straight line at low flow, say linear up to flow 20 l/s and head 24 m, becoming curved at higher flow. How would this affect your modelling process?

CONSOLIDATION EXERCISES FOR CHAPTER 8

1 Lou Zitt has a budget of £2000 to spend on storage units for his office. The storage units must not cover more than 50 m^2 of floor space, and Lou wants to maximise the storage capacity. The three types of storage unit that he can choose from are shown in the following table.

Type	Storage Capacity (m^3)	Floor space covered (m^2)	Cost (£)
Antique Pine Units	3	1	100
Beech Wood Units	8	4	500
Cedar Wood Units	6	3	200

Suppose that Lou buys a antique pine units, b beech wood units and c cedar wood units.

a) Write down two constraints that must be satisfied by a, b and c other than $a \geq 0$, $b \geq 0$, $c \geq 0$.
b) Write down the objective function for this problem.
c) Set up the problem as an LP formulation. (You are not expected to solve the problem.)
d) Identify which aspect of the original problem has been overlooked in the LP formulation.

(OCR Discrete Mathematics 1 Specimen Paper 2000)

2 In order to supplement his daily diet Harry wishes to take some Xtravit and some Yeastalife tablets. Their contents of iron, calcium and vitamins (in milligrams per tablet) are shown in the table.

Tablet	Iron content	Calcium content	Vitamin content
Xtravit	6	3	2
Yeastalife	2	3	4

a) By taking x tablets of Xtravit and y tablets of Yeastalife Harry needs to receive at least 18 milligrams of iron, 21 milligrams of calcium and 16 milligrams of vitamins. Write these conditions down as three inequalities in x and y.

b) In a coordinate plane illustrate the region of those points (x, y) which simultaneously satisfy $x \geq 0$, $y \geq 0$ and the three inequalities in **a)**.

c) If the Xtravit tablets cost 10p each and the Yeastalife tablets cost 5p each, how many tablets of each should Harry take in order to satisfy the above requirements at the least cost?

(AEB Question 9, Discrete Mathematics, Specimen Paper, 1996)

3 The Bonzo Manufacturing Company makes model cars and lorries. Each car sells at a profit of £2.50 and each lorry sells at a profit of £3.00. Three departments: Manufacturing (Dept A); Assembly (Dept B); Finishing (Dept C) are involved in the production of the models. The times, in hours, that the cars and lorries are in each department are shown in the table.

	Car	Lorry
Dept A	1.50	3.00
Dept B	2.00	1.00
Dept C	0.25	0.25

In a given week, 45 hours are available in Department A, 35 hours in Department B and 5 hours in Department C. The manufacturer wishes to maximise his profit £P in this week.

Let x be the number of cars made, and y be the number of lorries made.

You may assume that all models made can be sold.

a) Model this situation as a linear programming problem, giving each inequality in its simplest form with integer coefficients.

b) Display the inequalities on a graph and identify the feasible region.

c) Use the vertex method to obtain the maximum profit and the corresponding values of x and y.

d) State which department has unused time and calculate this time.

(Edexcel Decision Mathematics D1 Specimen Paper 2000)

4 A craftworker makes cushions and rag dolls to sell to a shop. Each cushion uses £15 worth of materials and takes 40 minutes to make. Each rag doll uses £4.50 worth of materials and takes one hour to make. The craftworker has only enough material to make at most four rag dolls.

a) Given that the craftworker has £78 to buy materials and 6 hours in which to make his products, show that this situation can be modelled by the following constraints:

$10x + 3y \leq 52$
$2x + 3y \leq 18$
$y \leq 4, x \geq 0, y \geq 0$
where x is the number of cushions made and y the number of rag dolls made.
The craftworker makes £7 profit for each cushion he sells and £8 profit for each rag doll he sells.

b) Write down an expression for the profit P.

c) Given that the craftworker can only sell completed work, use a graphical method to obtain his most profitable strategy.

(ULEAC Question 5, Decision Mathematics Specimen Paper D1, 1996)

5 A baker has 8.5 kg of flour, 5.5 kg of butter and 5 kg of sugar available at the end of a working day. With these ingredients, biscuits and/or buns can be made. The recipes are as follows.

30 biscuits	200 g flour 120 g butter 100 g sugar	40 buns	200 g flour 200 g butter 200 g sugar

a) Biscuits are sold for 5p each and buns for 7p each. Let x be the number of biscuits baked and y be the number of buns baked. Assuming that any number of biscuits and/or buns can be baked, show that the amount of flour used, in g, is given by $6\frac{2}{3}x + 5y$. Hence write down and simplify an inequality constraining the values of x and y.

b) Produce two further inequalities relating to the availability of butter and of sugar.

c) Assuming that all biscuits and buns can be sold, produce a function which gives the income from the sales.

d) Formulate and solve graphically the linear program: 'Maximise income subject to the availability of flour, butter and sugar.' State the maximum income expected, and the numbers of biscuits and buns baked.

(UODLE Paper 4, 1991)

6 A paper manufacturer has a roll of paper to cut up. It is 40 cm wide and 200 metres long, and is to be cut along its length to produce widths of 11 cm for toilet rolls, and widths of 24 cm for kitchen rolls. The diagram shows two possible cutting plans. Some of the roll may be cut to plan A, some to plan B, and some may be left uncut.

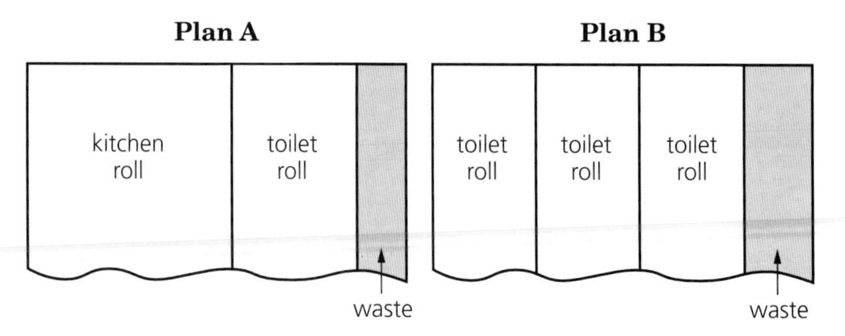

Kitchen roll paper sells for 7p a metre. Toilet roll paper sells for 4p a metre. Any of the roll that remains uncut can be sold for 8p a metre. The manufacturer is keen to ensure that no more than 15 per cent of the roll is wasted.

a) Let a metres be the length that is cut to plan A. Let b metres be the length cut to plan B. Let c metres remain uncut.
Values of a, b and c are to be found so as to produce the maximum income within the given constraints. The following linear programme is formulated to find this.

Maximise $11a + 12b + 8c$
Subject to $a + b + c = 200$
$5a + 7b < 1200$
$a > 0, b > 0, c > 0$

i) Explain how the objective function and each of the first two constraints were produced.
ii) Use the constraint $a + b + c = 200$ to reduce the problem to one in two variables only. (Make c the subject, and then substitute the result for c into the objective function, and into the constraint $c > 0$.)

b) Solve the problem graphically.

(OCR MEI Decision and Discrete Mathematics 1 Specimen Paper 2000)

7 James Bond is very particular about his cocktails. He has them mixed from gin and martini and he insists that they satisfy the following constraints:

Dryness
Gin has a dryness rating of 1. Martini has a dryness rating of 3. Dryness blends linearly, i.e. a mixture of x ml of gin and y ml of martini has a dryness of $\frac{x+3y}{x+y}$.
James insists that the dryness rating of his cocktail is ≤ 2.

Alcohol
Gin is 45 per cent alcohol by volume. Martini is 15 per cent. Alcohol also blends linearly. James insists that his cocktail is between 18 per cent and 36 per cent alcohol.

James has ordered a 200 ml cocktail. Let the amount of gin in it be x ml, and let the amount of martini be y ml, so that $x + y = 200$.

a) Explain why the dryness constraint may be expressed as $x + 3y \leq 400$.
b) In a similar way produce and simplify two constraints relating to alcohol content.
c) Graph all three constraints.
d) Given that gin is more expensive than martini, and remembering $x + y$ must equal 200, give the cheapest and most expensive cocktails that will satisfy James's requirements.
e) An alternative approach to this problem is to let p be the proportion of gin in the cocktail, so the $1 - p$ is the proportion of martini. Using this approach the first constraint becomes: $p + 3(1 - p) \leq 2$; that is, $p \geq \frac{1}{2}$.
Express the other constraints in this way and compare the acceptable values for x with those which you obtained for x in **d)**.

(UODLE Paper 4, 1994)

8 James T Squirt and Co., toy manufacturers, produce two different model spaceships, the Alpha and the Beta. There are 160 units of materials, 120 robot hours and 90 hours for hand-finishing are available. The Alpha uses 15 units of materials, 20 robot hours and takes 12 hours to finish. The Beta uses 20 units of materials, 5 robot hours and takes 9 hours to finish. The profits on the Alpha and Beta are £8.50 and £11.75 respectively. How many of each model should Squirt and Co. produce?

9 A company needs a production plan for one of its products for the next three months. The demand for the product in each month is known and must be satisfied. Production in excess of demand in months 1 and 2 may be stored until needed in subsequent months – at a cost. Demand, production and storage costs are as follows.

	Month 1	**Month 2**	**Month 3**
Demand (units)	5	4	6
Production costs (per unit)	1	5	2

Storage costs = 3 per unit per month

Let p_1 be the number of units produced during the first month and p_2 be the number of units produced during the second month. Then the company's requirements are modelled by the following five inequalities.

i) $p_1 \geq 5$
ii) $p_1 + p_2 \geq 9$
iii) $p_1 + p_2 \leq 15$
iv) $p_1 \geq 0$
v) $p_2 \geq 0$

a) Explain the meaning of each of inequalities (i), (ii) and (iii).
b) Draw the five inequalities on a graph. What production plan is represented by each of the vertices of the feasible region?
c) Specify an objective function for this production problem.
d) Give the optimal production plan and its cost.

(AQA Discrete Mathematics D1 Specimen Paper 2000)

10 A company blends its own brand of fruit drink, "Golden Special", by mixing two types of fruit juice, type X and type Y, which are bought in bulk from their suppliers.

Each week they buy x litres of type X and y litres of type Y which are then mixed in large tanks to produce $(x + y)$ litres of 'Golden Special'.

Type X contains 40% orange juice, no lemon juice and 60% other flavourings and water.

It costs 30p per litre and may be ordered in any quantity.

Type Y contains 15% orange juice, 10% lemon juice and 75% other flavourings and water.

It costs 20p per litre and the minimum acceptable order is 40 000 litres per week.

The 'Golden Special' label states that the product must contain at least 20% orange juice. For a pleasing flavour the proportion of lemon juice must not be below 5%.

The company must produce at least 60 000 litres per week. They wish to plan their weekly purchase of bulk supplies to minimise the total weekly cost £C.

a) Show that the limit on the proportion of orange juice implies the constraint $y \leq 4x$.
b) Formulate the problem of minimising weekly costs as a linear programme with further constraints.
c) Draw a suitable diagram to enable the problem to be solved graphically, indicating the feasible region and the direction of the objective line.
d) Use your diagram to find the minimum total weekly cost £C.

(AQA Mathematics A Discrete 1 Specimen Paper 2000)

Summary

After working through this chapter you should know that:

- in a linear programming problem we find the maximum or minimum of a linear function (called the objective function) which is subject to a set of constraints which are linear inequalities or equalities
- the steps in solving a linear programming type problem are:
  Step 1: Formulate the linear programming problem:
  a) define variables
  b) form objective function
  c) form the constraints.
  Step 2: Write down the equations of the boundary of the feasible region.
  Step 3: Determine the vertices of the feasible region.
  Step 4: Evaluate the objective function at each vertex and determine the optimal value.

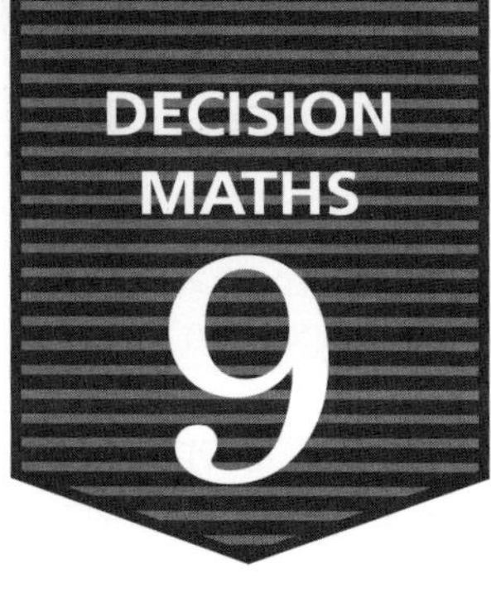

Linear programming 2: the simplex method

In this chapter we continue the investigation of linear programming problems and introduce:

- *the method of Gaussian elimination for solving simultaneous linear equations*
- *the simplex method for solving linear programming problems which uses the elimination ideas.*

The fundamental theorem of linear programming tells us that an important step in the solution process is the need to find the coordinates of the vertices of the feasible region. The optimal solution occurs at one of these vertices. The graphical approach adopted in Chapter 8, *Linear programming 1: Modelling and graphical solution*, works well for problems with two variables but cannot easily be extended to problems with three or more variables. In these cases we need an algebraic method of solution that can be used in a computer algorithm.

THE GAUSSIAN ELIMINATION METHOD

An algebraic solution

Example 9.1

Find the solution of this pair of linear equations.

$x + 2y = 7$

$2x + 7y = 23$

Solution

Suppose we label the two equations like this.

$x + 2y = 7 \qquad \mathbf{E_1}$

$2x + 7y = 23 \qquad \mathbf{E_2}$

Now subtract $2\mathrm{E}_1$ *from* E_2.

$$\begin{array}{rl} 2x + 7y = 23 & \mathrm{E}_2 \\ -\quad 2x + 4y = 14 & 2\mathrm{E}_1 \\ \hline 3y = 9 & \end{array}$$

Solving for $y \Rightarrow y = 3$

Substitute for y *into* E_1 *then solve for* x *to find* $x = 1$.

The solution is $x = 1$ *and* $y = 3$.

The number 2 that multiplies E_1 in $\mathrm{E}_2 - 2\mathrm{E}_1$ is called the **multiplier**.

Exploration 9.1

The method of elimination

Consider this pair of linear equations.

$3x + 4y = 5$ **E_1**

$2x - 3y = 9$ **E_2**

> ***Note:*** *For two linear simultaneous equations:*
> $ax + by = c$
> $dx + ey = f$
> *then* $r = -\frac{d}{a}$

- Find the multiplier r so that $E_2 - rE_1$ is an equation without x.
- Solve for y.
- Solve for x.

Elimination

Choosing $r = \frac{2}{3}$ then $E_2 - \frac{2}{3}E_1$

$\Rightarrow\ -3y - \frac{2}{3} \times 4y = 9 - \frac{2}{3} \times 5$ **E_{2a}**

$\Rightarrow\ -\frac{17}{3}y = \frac{17}{3}$ **E_{2a}**

Solving for y:

$y = -1$

Substituting for y into E_1 and solving for x:

$x = 3$

This method of solving simultaneous linear equations is called the **Gaussian elimination method**, after the mathematician Carl Friedrich Gauss.

The method is a two-stage process.

- The first stage is the process of subtracting a multiple of E_1 from E_2 to obtain a new equation E_{2a} that does not contain x. This is the **elimination stage**.
- The second stage of solving for y and substituting back into E_1 to find x is the **back substitution stage**.

For a pair of linear equations written in more general form:

$a_1x + b_1y = c_1$ E_1

$a_2x + b_2y = c_2$ E_2

in which x and y are the variables and a_1 is not zero, the multiplier is $\frac{a_2}{a_1}$ so that for the elimination stage we calculate:

$$E_2 - \frac{a_2}{a_1}E_1$$

to give an equation without a term in x.

Carl Friedrich Gauss (1777–1855)

The Gaussian elimination method can be applied to sets of equations containing more than two variables.

This forms an important part of the method of solving linear programming problems introduced in the next two sections.

Example 9.2

Solve these equations by the Gaussian elimination method.

$3x + y - z = 1$ **E_1**

$5x + y + 2z = 6$ **E_2**

$4x - 2y - 3z = 3$ **E_3**

Solution

The elimination stage:

$E_2 - \frac{5}{3}E_1 \Rightarrow 0x - \frac{2}{3}y + \frac{11}{3}z = \frac{13}{3}$ $\mathbf{E_{2a}}$

$E_3 - \frac{4}{3}E_2 \Rightarrow 0x - \frac{10}{3}y - \frac{5}{3}z = \frac{5}{3}$ $\mathbf{E_{3a}}$

We now have three equations but two do not contain x.

$3x + y - z = 1$ $\mathbf{E_1}$

$-\frac{2}{3}y + \frac{11}{3}z = \frac{13}{3}$ $\mathbf{E_{2a}}$

$-\frac{10}{3}y - \frac{5}{3}z = \frac{5}{3}$ $\mathbf{E_{3a}}$

Now we eliminate y using equations E_{3a} and E_{2a}.

$$E_{3a} - 5E_{2a} \Rightarrow 0y - \frac{60}{3}z = -\frac{60}{3}$$

$$\Rightarrow 20z = 20$$

Now we apply the back substitution stage.

Solve equation E_{3b} for z:

$z = 1$

Substitute into E_{2a} for z:

$-\frac{2}{3}y + \frac{11}{3} = \frac{13}{3} \Rightarrow -\frac{2}{3}y = \frac{2}{3}$

Solve for y:

$y = -1$

Substitute into E_1 for z and y:

$3x - 1 - 1 = 1 \Rightarrow 3x = 3$

Solve for x:

$x = 1$

The solution for the equations is $x = 1$, $y = -1$, $z = 1$.

In this example note the systematic method of approach. Equation E_1 remains unchanged; equation E_{2a} does not contain x and equation E_{3b} does not contain x or y. We then work backwards through the equations to find the variables: E_{3b} then E_{2a} then E_1.

EXERCISES

9.1 CLASSWORK

1 Use the method of Gaussian elimination to solve the following equations. In each case write down the multiplier.

a) $x + 3y = 11$
$3x + y = 9$

b) $x - y = 8$
$4x + y = 23$

c) $2x - y = 15$
$x + 5y = 19$

d) $3x - 5y = 8$
$4x + 7y = 11$

e) $2x + y = 3$
$3x + 2y = 5$

f) $0.5x - 0.3y = -0.5$
$0.1x + 0.4y = 2.2$

2 Use Gaussian elimination to solve the following equations. In each case write down the multiplier for each of the elimination processes.

a) $3x - 6y + 9z = 0$
$4x - 6y + 8z = -4$
$-2x - y + z = 7$

b) $3x - y + z = 6$
$6x - z = 3$
$y + z = 0$

c) $x - 3y + 4z = 1$
$4x - 10y + 10z = 4$
$-3x + 9y - 5z = -6$

d) $x + 2y + 2z = 11$
$2x + 5y + 9z = 39$
$x - y - z = -4$

3 Solve the following using Gaussian elimination.

a) $x_1 - 2x_2 + 5x_3 = 6$
$x_1 + 3x_2 - 4x_3 = 7$
$2x_1 + 7x_2 - 12x_3 = 12$

b) $2x_1 - 3x_2 + x_3 + x_4 = 10$
$x_1 + x_2 + 4x_3 - x_4 = 5$
$3x_1 - x_2 - x_3 + 2x_4 = 8$
$4x_1 + 2x_2 - 3x_3 + 5x_4 = 11$

4 Solve the pair of simultaneous linear equations:
$x + 4y = 9$
$-2x - \mu y = 7$
for the two cases: **a)** $\mu = 1$ **b)** $\mu = 4$.
Interpret your results in terms of the graphs of appropriate linear functions.

5 A stone is thrown vertically upwards and its height h metres above the ground is given by the equation:
$h = a + bt - ct^2$
where a, b and c are constants and t is the time, in seconds, the stone has been in the air.
Given that h = 27, 42, 47 after 1, 2, 3, seconds respectively, determine the values of a, b and c. From what height was the stone launched?

EXERCISES

9.1 HOMEWORK

1 Use the method of Gaussian elimination to solve the following equations. Sketch the graphs of each pair of equations and confirm geometrically the validity of your solution.

a) $x + y = 3$
$2x + y = 4$

b) $3x + 4y = 6$
$x - y = 2$

c) $3x + y = x + 2y - 1 = 6$

d) $-4x + y = -9$
$x - 2y = 4$

e) $-2x - 3y = 1$
$5x - y = 6$

f) $4x - y = 7a + 4b$
$5x - 4y = 6a + 5b$
where a and b are constants

2 Sketch the graphs of each pair of equations.

a) $4x - 6y = 5$
$6x - 9y = 7.5$

b) $4x - 6y = 5$
$6x - 9y = 10$

How many solutions do **a)** and **b)** have?
Use the method of Gaussian elimination in the algebraic solution of these equations, explicitly stating equation E_{2a}. Use these results to confirm your conclusions from the sketches.

3 Solve the pair of simultaneous linear equations:
$2x - 3y = 4$
$-4x + \lambda y = -7$
for the two cases: **a)** $\lambda = 1$ **b)** $\lambda = 6$.
Interpret your results in terms of the graphs of appropriate linear functions.

4 Solve the following using Gaussian elimination.

a) $x + y - z = 3$
$2x - 2y - 3z = 3$
$4x - 3y - 2z = 3$

a) $-2x_1 + x_2 - 3x_3 = 7$
$3x_1 + 2x_2 - x_3 = -5$
$2x_1 + x_2 = -4$

5 It is required to find the real coefficients a, b, c and d of the cubic equation:

$$y = ax^3 + bx^2 + cx + d$$

If it is known that the graph of y against x cuts the y-axis at 8, evaluate one of the coefficients immediately.
If we further know that the graph passes through the points (–1, 0), (1, 6) and (2, 6) obtain and solve three linear simultaneous equations in the other three coefficients.

ORGANISING THE WORK IN A TABLE

The important part of the Gaussian elimination method is the elimination stage of the process. We can lay out the information and equation manipulations in the form of a tabular block called a **matrix**. The coefficients of the linear equations are the elements of the matrix.

Consider the pair of equations in Example 9.1.

$$x + 2y = 7$$
$$2x + 7y = 23$$

We repeat the solution of these equations showing on the left-hand side the manipulation of the equations and on the right-hand the equivalent operation on the rows of the matrix.

Equation manipulations		**Row operations**
$x + 2y = 7$	$\mathbf{E}_1$	$\left[\begin{array}{cc|c} 1 & 2 & 7 \\ 2 & 7 & 23 \end{array}\right] \begin{array}{l} R_1 \\ R_2 \end{array}$
$2x + 7y = 23$	$\mathbf{E}_2$	

Coefficients of x and y — Right-hand side of linear equations

$E_1 \Rightarrow x + 2y = 7$	$\mathbf{E}_1$	$R_1 \Rightarrow \left[\begin{array}{cc|c} 1 & 2 & 7 \\ 0 & 3 & 9 \end{array}\right] R_1$
$E_2 - 2E_1 \Rightarrow 3y = 9$	$\mathbf{E}_{2a}$	$R_2 - 2R \Rightarrow R_{2a}$

The back substitution is carried out in the same way as before. Operating on the rows of the matrix turns out to be less tedious for the linear programming problems than writing out the full equations.

Exploration 9.2

Row operations

Consider the equations in Example 9.2.

$$3x + y - z = 1$$
$$5x + y + 2z = 6$$
$$4x - 2y - 3z = 3$$

- Write the equations in tabular form.
- Write the equation manipulations as row operations on the table.

Using the matrix approach

Equation manipulations | **Row operations**

$3x + y - z = 1$ $\quad \mathbf{E}_1$

$5x + y + 2z = 6$ $\quad \mathbf{E}_2$

$4x - 2y - 3z = 3$ $\quad \mathbf{E}_3$

$$\left[\begin{array}{ccc|c} 3 & 1 & -1 & 1 \\ 5 & 1 & 2 & 6 \\ 4 & -2 & -3 & 3 \end{array}\right] \begin{array}{l} R_1 \\ R_2 \\ R_3 \end{array}$$

$\mathbf{E}_2 - \frac{5}{3}\mathbf{E}_1 - \frac{2}{3}y + \frac{11}{3}z = \frac{13}{3}$ $\quad \mathbf{E}_{2a}$

$\mathbf{E}_3 - \frac{4}{3}\mathbf{E}_1 - \frac{10}{3}y + \frac{5}{3}z = \frac{5}{3}$ $\quad \mathbf{E}_{3a}$

$$\begin{array}{r} \\ R_2 - \frac{5}{3}R_1 \\ R_3 - \frac{4}{3}R_1 \end{array} \left[\begin{array}{ccc|c} 3 & 1 & -1 & 1 \\ 0 & -\frac{2}{3} & \frac{11}{3} & \frac{13}{3} \\ 0 & -\frac{10}{3} & -\frac{5}{3} & \frac{5}{3} \end{array}\right] \begin{array}{l} R_1 \\ R_{2a} \\ R_{3a} \end{array}$$

$\mathbf{E}_{3a} - 5\,\mathbf{E}_{2a} - \frac{60}{3}z = -\frac{60}{3}$ $\quad \mathbf{E}_{3b}$

$$\begin{array}{r} \\ \\ R_{3a} - 5R_{2a} \end{array} \left[\begin{array}{ccc|c} 3 & 1 & -1 & 1 \\ 0 & -\frac{2}{3} & \frac{11}{3} & \frac{13}{3} \\ 0 & 0 & -\frac{60}{3} & -\frac{60}{3} \end{array}\right] \begin{array}{l} R_1 \\ R_{2a} \\ R_{3b} \end{array}$$

Note that in the example and in Exploration 9.2 the final table has 0 in each position below the main diagonal.

EXERCISES

9.2 CLASSWORK

For each question in Exercises 9.1 CLASSWORK:

a) write the equations in tabular form

b) carry out the row operations on the table to give zeros in all the positions below the diagonal.

EXERCISES

9.2 HOMEWORK

1 Write question 2 of Exercises 9.1 HOMEWORK in tabular form and carry out the row operations on the table to obtain zeros in all positions below the diagonal.
Note the form of the bottom line in the final table. In which way will this line indicate whether there is a unique solution, an infinity of solutions or no solutions?

2 For question 1 of Exercises 9.1 HOMEWORK:

a) write the equations in tabular form

b) carry out the row operations on the table to give zeros in all the positions below the diagonal.

3 In each of three sets of simultaneous equations in x, y and z, Gaussian elimination has been performed in tabular form. After reduction to diagonal form we have these arrays.

a) $\left[\begin{array}{ccc|c} 4 & -1 & -2 & 0 \\ 0 & -\frac{3}{4} & \frac{1}{2} & -2 \\ 0 & 0 & \frac{1}{3} & \frac{2}{3} \end{array}\right]$ **b)** $\left[\begin{array}{ccc|c} 2 & 1 & 4 & 2 \\ 0 & 3 & 2 & 5 \\ 0 & 0 & 0 & 1 \end{array}\right]$ **c)** $\left[\begin{array}{ccc|c} 1 & 2 & 4 & 5 \\ 0 & 1 & 3 & 3 \\ 0 & 0 & 0 & 1 \end{array}\right]$

For cases **a)** and **b)**, state the number of solutions and determine the solution where it exists.
For case **c)** state why there are infinitely many solutions. To obtain a solution in general form, let $z = t$ say, where t can take any real value. Determine y and x in terms of t.

4 Solve the pairs of simultaneous equations using Gaussian elimination in tabular form. In each case, determine the value of which will give:

i) a unique solution

ii) an infinity of solutions

iii) no feasible solutions.

a) $x + y = 1$

$x + \lambda^2 y = \lambda$

b) $x + y + z = 3$

$4x + 5y + 2z = 11$

$2x + y + (\lambda^2 + 6)z = \lambda + 12$

For those cases where solutions exist determine the solutions.

LINEAR PROGRAMMING AND THE SIMPLEX TABLEAU

Exploration 9.3

Maximising profit

A manufacturer makes three products A, B and C. Each product requires time in the cutting, assembly and finishing departments. The times in each department, total time available and profit-margin per item are shown in the table.

		Product			
		A	B	C	Time available (hours per week)
Time (hours)	Cutting	2	3	5	204
	Assembling	4	3	6	260
	Finishing	3	1	10	304
	Profit (£)	10	12	16	

How many of each product should be made each week in order to maximise profits?

Slack variables

If we set up the exploration as a linear programming problem it becomes:

Maximise $P = 10A + 12B + 16C$

subject to: $2A + 3B + 5C \leq 204$ (cutting)

$4A + 3B + 6C \leq 260$ (assembling)

$3A + B + 10C \leq 304$ (finishing)

It would not be possible to solve this problem graphically, as it contains three variables (A, B and C). So we need to consider an alternative approach.

We can use the row operation process to solve linear programming problems. The first step is to convert the inequalities in the constraints into a system of linear equations. We do this by introducing **slack variables**.

Example 9.3

Consider this problem.

Maximise	$f = 80x + 70y$	**1**
where	$2x + y \leq 32$	**2**
	$x + y \leq 18$	**3**
	$x \geq 0, y \geq 0$	**4**

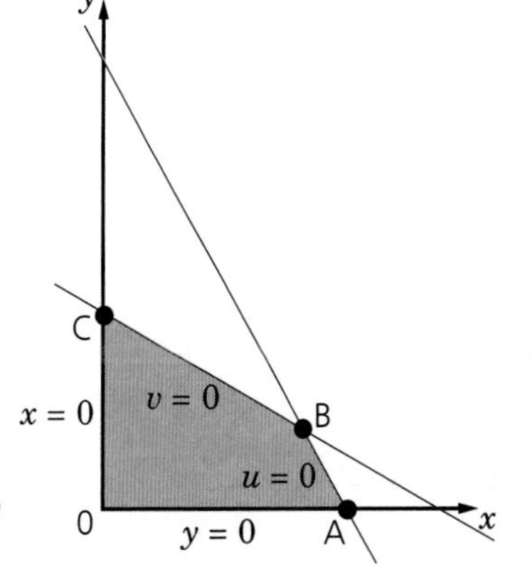

If we were to graph this problem we would obtain a feasible region bounded by four lines. We know that the optimal solution lies at a vertex, so we need only consider the value of f at the four vertices O, A, B and C.

Consider the solutions at vertices A, B, and C. What is the optimal solution to this problem?

Each vertex is at the intersection of two lines, for example O is where $x = y = 0$.

We introduce the slack variables, u and v, so that we can define AB as $u = 0$ and BC as $v = 0$. In this way we say that each vertex corresponds to
two of the variables being zero:

at O $x = y = 0$
at A $y = u = 0$
at B $u = v = 0$
at C $v = x = 0$

In effect, u and v represent the slack between the total available and what is actually being used, and they enable us to replace the inequalities with linear equations.

$2x + y + u = 32$ **2A**
$x + y + v = 18$ **3A**

Because of the way they are defined $u \geq 0$ and $v \geq 0$ at the optimal solution. The linear programming problem can be redefined as follows.

> Among all the solutions of the system of the linear equations:
> $f = 80x + 70y$
> $2x + y + u = 32$
> $x + y + v = 18$
>
> find the one for which $u \geq 0, v \geq 0, x \geq 0, y \geq 0$ and f is as large as possible.

This is an example of **the standard form of linear programming** problem in which:

1 the objective function is to be maximised
2 each variable (actual and slack) ≥ 0.

The simplex method

We are now ready to introduce the **simplex method** of solving linear programming problems, which uses the elimination steps of the Gaussian elimination method. In demonstrating the method the graphs show a comparison with the graphical method of Chapter 8, *Linear programming 1: Modelling and graphical solution.*

Consider the problem in standard form:

Maximise $f = 80x + 70y$
where $2x + y + u = 32$
$x + y + v = 18$
$x \geq 0, y \geq 0, u \geq 0, v \geq 0$

Note: The objective equation is found by rearranging the objective function to give $f = 80x - 70y = 0$.

We write the equations in a tabular form called the **simplex tableau**.

f	x	y	u	v		
1	−80	−70	0	0	0	Objective equation
0	2	1	1	0	32	Constraint equations
0	1	1	0	1	18	

The idea behind the simplex method is that we move round the feasible region, travelling along the edges and visiting the vertices in turn until a maximum value of f is reached. A row operation on the simplex tableau is equivalent to jumping from one vertex to another. The edges of the feasible region are defined by $x = 0$, $y = 0$, $u = 0$ and $v = 0$. A vertex is a solution of two of these equations.

Consider the top row of the tableau:
$f - 80x - 70y = 0$

If we start at $x = 0$, $y = 0$ then $f = 0$.

We can increase f by increasing x or y (or both but we choose one or the other).

Since increasing x leads to a larger change in f than increasing y, we choose to increase x and keep $y = 0$.

The effect is to move along the x-axis. From the graph we see that the vertex (16, 0) is visited first. This is the intersection of $y = 0$ and $u = 0$.

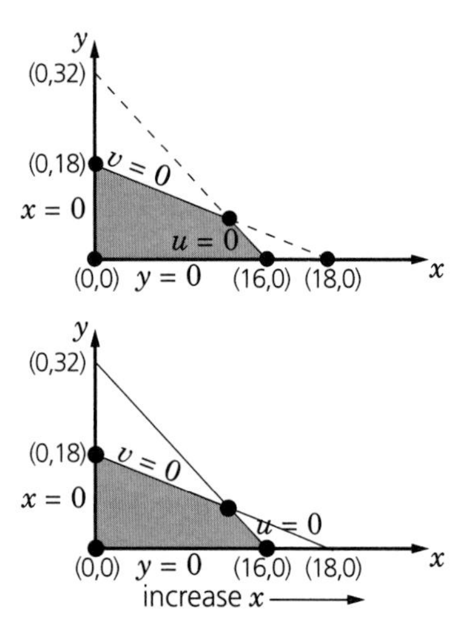

We can find this information from the tableau by dividing the right-hand column by the x-column entries.

$$\begin{array}{ccccc|c} f & x & y & u & v & \\ 1 & -80 & -70 & 0 & 0 & 0 \\ 0 & 2 & 1 & 1 & 0 & 32 \\ 0 & 1 & 1 & 0 & 1 & 18 \end{array} \quad \begin{array}{l} \\ \\ 32 \div 2 = 16 \\ 18 \div 1 = 18 \end{array}$$

$\uparrow$

This is called the **pivot column**.

In this case row 1 is the pivot row. The element in the intersection of the pivot row and pivot column is the **pivot element**. We make this element 1 and then carry out row operations to make each other element in the pivot column equal to zero.

What happens to the equations

$$\begin{array}{c} \\ \\ \\ R_2 \div 2 \end{array} \begin{array}{ccccc|c} f & x & y & u & v & \\ 1 & -80 & -70 & 0 & 0 & 0 \\ 0 & 1 & \frac{1}{2} & \frac{1}{2} & 0 & 16 \\ 0 & 1 & 1 & 0 & 1 & 18 \end{array}$$

$$f - 80x - 70y = 0 \quad \mathbf{E_1}$$
$$x + \tfrac{1}{2}y + \tfrac{1}{2}u = 16 \quad \mathbf{E_2}$$
$$x + y + v = 18 \quad \mathbf{E_3}$$

$$\begin{array}{c} \\ R_1 + 80R_2 \\ \\ R_3 - R_2 \end{array} \begin{array}{ccccc|c} f & x & y & u & v & \\ 1 & 0 & -30 & 40 & 0 & 1280 \\ 0 & 1 & \frac{1}{2} & \frac{1}{2} & 0 & 16 \\ 0 & 0 & \frac{1}{2} & -\frac{1}{2} & 1 & 2 \end{array} \leftarrow \text{new value of } f$$

$$p - 30y + 40u = 1280 \quad (\mathbf{E_1} + 80\mathbf{E_2})$$
$$x + \tfrac{1}{2}y + \tfrac{1}{2}u = 16$$
$$\tfrac{1}{2}y + \tfrac{1}{2}u + v = 2 \quad (\mathbf{E_3} - \mathrm{E_2})$$

The new value of f is 1280 and occurs at the vertex where the lines $y = 0$, $u = 0$ meet. In the new simplex tableau the value of f has increased.

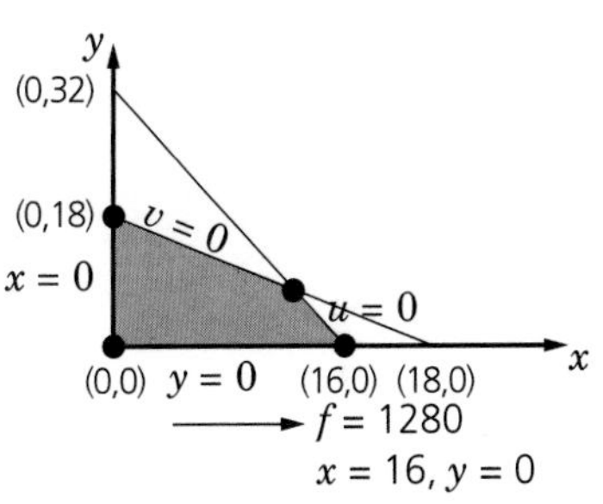

Increasing *y*

From the top row:

$f - 30y + 40u = 1280$

To increase f further we hold $u = 0$ and increase y. Graphically this means that we are moving along the edge $u = 0$.

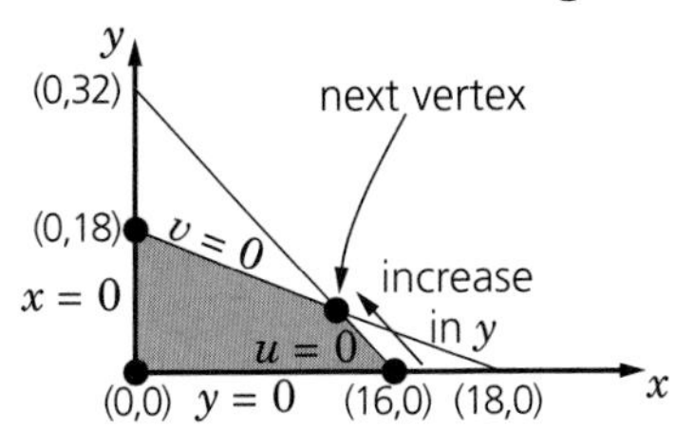

The pivot column is the y-column.

$$\begin{array}{ccccc|c} f & x & y & u & v & \\ \hline 1 & 0 & -30 & 40 & 0 & 1280 \\ 0 & 1 & \frac{1}{2} & 0 & 0 & 16 \\ 0 & 0 & \textcircled{\frac{1}{2}} & -\frac{1}{2} & 1 & 2 \end{array} \quad \begin{array}{l} 16 \div \frac{1}{2} = 32 \\ \\ 2 \div \frac{1}{2} = 4 \leftarrow \text{pivot} \end{array}$$

↑ pivot column

Arrive at $y = 4$ before $y = 32$.
$\frac{1}{2}$ is called the pivot element

$$R_3 \times 2 \quad \begin{array}{ccccc|c} f & x & y & u & v & \\ \hline 1 & 0 & -30 & 40 & 0 & 1280 \\ 0 & 1 & \frac{1}{2} & 0 & 0 & 16 \\ 0 & 0 & 1 & -1 & 2 & 4 \end{array}$$

Now pivot about the pivot element using row operations.

$$\begin{array}{l} R_1 + 30R_3 \\ \\ R_1 - \frac{1}{2}R_3 \end{array} \begin{array}{ccccc|c} f & x & y & u & v & \\ \hline 1 & 0 & 0 & 10 & 60 & 1400 \\ 0 & 1 & 0 & \frac{1}{2} & -1 & 14 \\ 0 & 0 & 1 & -1 & 2 & 4 \end{array} \quad \begin{array}{rl} f + 10u + 60v = 1400 & (\mathbf{E_1} + 30\mathbf{E_3}) \\ x + u - v = 14 & (\mathbf{E_2} - \frac{1}{2}\mathbf{E_3}) \\ y - u + 2v = 4 & \mathbf{E_3} \end{array}$$

The equation for f is now:

$f + 10u + 60v = 1400$

Increasing u and v will decrease f, so we have found the maximum value for f.

It is $f = 1400$ and occurs when $u = 0$ and $v = 0$.

From the table we can read off the values of x and y for the maximum value of f.

It is where $x = 14$ and $y = 4$.

We will now look at a two variable, two constant problem which sets out both the graphical and simplex methods side by side so that you can compare how they obtain the solution.

Example 9.3

This example shows the graphical and simplex methods side-by-side.

A small factory produces two types of toys: trucks and bicycles. In the manufacturing process two machines are used: the lathe and the assembler. The table shows the length of time needed for each toy:

	Lathe	Assembler
Bicycle	2 hours	1 hour
Truck	1 hour	1 hour

The lathe can be operated for 16 hours a day and the assembler for 9 hours a day. Each bicycle gives a profit of £16 and each truck gives a profit of £14. Formulate and solve a linear programming problem so that the factory maximises its profit.

1 Formulate the problem

Let x be number of bicycles made.
Let y be number of trucks made.

Objective function

Maximise $P = 16x + 14y$

$P - 16x - 14y = 0$

Subject to constraints:

Introduce slack variables

Lathe: $2x + y \leq 16$ — $2x + y + s1 = 16$

Assembler: $x + y \leq 9$ — $x + y + s2 = 9$

2 Solve the problem

Graphical solution

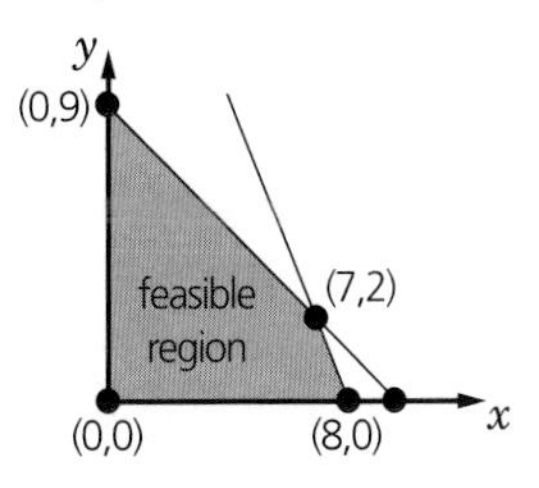

Simplex tableau

p	x	y	s1	s2		
1	−16	−14	0	0	0	
0	2	1	1	0	16	1
0	1	1	0	1	9	
1	0	−6	8	0	128	
0	0	0.5	0.5	0	8	2
0	0	0.5	−0.5	1	1	
1	0	0	2	12	140	
0	1	0	1	−1	7	3
0	0	1	−1	2	2	

1 Choose to increase x (largest profit) in row 2 because $\frac{16}{2} < \frac{9}{1}$

2 Choose to increase y in row 3 because $\frac{1}{0.5} < \frac{1}{8.5}$

3 top row all positives so optimal solution

3 Interpret solution

By considering vertices

(8, 0) $P = 16 \times 8 = £128$

(7, 2) $P = (16 \times 7) + (14 \times 2) = £140$

(0, 9) $P = 14 \times 9 = £126$

Factory should make 7 bicycles and 2 trucks each day.
Profit £140

Reading from the tableau

$P = £140$, $x = 7$, $y = 2$
Providing $s1 = 0$ and $s2 = 0$

Factory should make 7 bicycles and 2 trucks each day.
Profit £140

Now let us return to Exploration 9.3. If we set this out in standard form it becomes

$P = 10A + 12B + 16C$
$2A + 3B + 5C + s = 204$
$4A + 3B + 6C + t = 260$
$3A + B + 10C + u = 304$

which we can put into the simplex tableau to solve. (You will notice that we have put each single iteration on one matrix – this takes less time and space, but you may prefer to use more working until you are confident.)

	P	*A*	*B*	*C*	*s*	*t*	*u*		
R_1	1	–10	–12	–16	0	0	0	0	choose to increase x
R_2	0	2	3	5	1	0	0	204	(204 ÷ 2 = 102)
R_3	0	4	3	6	0	1	0	260	(260 ÷ 4 = 65)
R_4	0	3	1	10	0	0	1	304	(304 ÷ 3 = $103\frac{1}{3}$)
$R_1 + 10R_3$	1	0	–4.5	–1	0	2.5	0	650	choose to increase y
$R_2 - 2R_3$	0	0	1.5	2	1	–0.5	0	74	
$R_3 \div 4$	0	1	0.75	1.5	0	0.25	0	65	
$R_4 - 3R_3$	0	0	–1.25	5.5	0	–0.75	1	109	
$R_1 + \frac{9}{2}R_2$	1	0	0	5	3	1	0	872	
$R_2 = \frac{3}{2}$	0	0	1	$1.\dot{3}$	$0.\dot{6}$	$-0.\dot{3}$	0	49.333	
$R_3 - \frac{3}{4}R_2$	0	1	0	0.5	–0.5	0.5	0	28	
$R_4 + \frac{5}{4}R_2$	0	0	0	$7.1\dot{6}$	$0.8\dot{3}$	$-1.1\dot{6}$	1	170.666	

At this stage we stop because all the entries in the top row are positive. If we increase any of the other values, we will decrease P, so we know that this is the optimal solution.

Reading from the final tableau this gives

$$P + 5C + 3s + t = 872$$
$$B + \tfrac{4}{3}C + \tfrac{2}{3}s - \tfrac{1}{3}t = 49\tfrac{1}{3}$$
$$A + \tfrac{1}{2}C - \tfrac{1}{2}s + \tfrac{1}{2}t = 28$$
$$7\tfrac{1}{6}C + \tfrac{5}{6}s - 1\tfrac{1}{6}t + u = 170\tfrac{2}{3}$$

This can only give a valid solution if C = 0 s = 0 and t = 0 giving

$P = 872$, $B = 49\frac{1}{3}$, $A = 28$

However, since we are dealing with products, the answers must be integer values – we would not make $\frac{1}{3}$ of an item – so we must interpret our solution correctly. Thus we would make 28 of product A, 49 of product B and none of product C. This would give a profit of

$$P = 10(28) + 12(49) + 16(0)$$
$$P = £868.$$

Example 9.4

Use the simplex tableau to solve this linear programming problem.

Maximise $f = 9x + 4y$
where $3x + 4y \leq 48$
$2x + y \leq 17$
$3x + y \leq 24$
$x \geq 0, y \geq 0$

Solution

Introducing slack variables u, v and w the system of linear equations is:

$3x + 4y + u = 48$
$2x + y + v = 17$
$3x + y + w = 24$
$-9x - 4y + f = 0$

The simplex tableau is:

Pivot about one of the x-values in order to increase f.

f	x	y	u	v	w			
1	−9	−4	0	0	0	0		
0	3	4	1	0	0	48	$48 \div 3 = 16$	
0	2	1	0	1	0	17	$17 \div 2 = 8.5$	
0	3	1	0	0	1	24	$24 \div 3 = 8$	Pivot about row 3

	f	x	y	u	v	w	
	1	−9	−4	0	0	0	0
	0	3	4	1	0	0	48
	0	2	1	0	1	0	17
$R_1 \div 3$	0	1	$\frac{1}{3}$	0	0	$\frac{1}{3}$	8

Pivot about y.

	f	x	y	u	v	w			
$R_4 - 9R_3$	1	0	−1	0	0	3	72		
$R_1 - 3R_3$	0	0	3	1	0	−1	24	$24 \div 3 = 3$	
$R_2 - 2R_3$	0	0	$\frac{1}{3}$	0	1	$-\frac{2}{3}$	1	$1 \div \frac{1}{3} = 3$	Pivot about row 2
	0	1	$\frac{1}{3}$	0	0	$\frac{1}{3}$	8	$8 \div \frac{1}{3} = 24$	

	f	x	y	u	v	w	
	1	0	−1	0	0	3	72
	0	0	3	1	0	−1	24
$R_2 \div \frac{1}{3}$	0	0	1	0	3	−2	3
	0	1	$\frac{1}{3}$	0	0	$\frac{1}{3}$	8

	f	x	y	u	v	w	
$R_4 + R_2$	1	0	0	0	3	1	75
$R_1 - 3R_2$	0	0	0	1	−9	5	15
	0	0	1	0	3	−2	3
$R_3 \div \frac{1}{3}R_2$	0	1	0	0	−1	1	7

Since each entry in the last row is positive we have achieved the maximum value for f. If we increase v and w then we decrease f, since the bottom row gives:

$f = 75 - 3v - w$

The maximum value of f is 75 and occurs when $x = 7$ and $y = 3$.

Example 9.5

This is a three variable, two constraint problem.

Maximise the objective function $f = -x + 8y + z$

where $x + 2y + 9z \leq 10$

$y + 4z \leq 12$

$x \geq 0, y \geq 0, z \geq 0$

Solution

Introduce two slack variables u *and* v *since we have two constraints.*

$x + 2y + 9z + u = 10$

$y + 4z + v = 12$

The simplex tableau is:

Pivot about y.

f	x	y	z	u	v			
1	1	−8	−1	0	0	0		
0	1	2	9	1	0	10	$10 \div 2 = 5$	Pivot about row 1.
0	0	1	4	0	1	12	$12 \div 1 = 12$	

First divide the pivot row by the pivotal element.

	f	x	y	z	u	v	
	1	1	−8	−1	0	0	0
$R_1 \div 2$	0	$\frac{1}{2}$	1	$\frac{9}{2}$	$\frac{1}{2}$	0	5
	0	0	1	4	0	1	12

	f	x	y	z	u	v	
$R_3 + 8R_1$	1	5	0	35	4	0	40
	0	$\frac{1}{2}$	1	$\frac{9}{2}$	$\frac{1}{2}$	0	5
$R_2 - R_1$	0	$-\frac{1}{2}$	0	$-\frac{1}{2}$	$-\frac{1}{2}$	1	7

The bottom row contains positive elements and so the maximum value for f has been reached. The bottom row gives:

$f = 40 - 5x - 35z - 4u$

Hence the maximum value is $f = 40$ *and occurs when* $x = 0$, $z = 0$ *and* $u = 0$ *and from the tableau* $y = 5$.

You can now use the simplex tableau to solve the problem which was posed at the beginning of this section on page 179.

The answer is Item A – 14
Item B – 12 P = £724
Item C – 25

EXERCISES

9.3 CLASSWORK

1 Use the simplex method to solve the following linear programming problems.

a) Maximise $f = 2x + 3y$
subject to $x + 3y \le 120$
$7x + 2y \le 154$
$x \ge 0, y \ge 0$

b) Maximise $c = x + \frac{1}{2}y$
subject to $2x + y \le 36$
$x + y \le 15$
$x \ge 0, y \ge 0$

c) Maximise $f = 8x + 13y$
subject to $2x + 3y \le 350$
$5x + y \le 500$
$x \ge 0, y \ge 0$

d) Maximise $g = 3x + 4y$
subject to $x + y \le 20$
$x + 2y \le 25$

e) Maximise $M = x + 3y$
subject to $-x + y \le 160$
$3x - y \le 3$
$x \ge 0, y \ge 0$

2 Use the simplex method to solve the following linear programming problems.

a) Maximise $f = x + 3y$
subject to $5x + y \le 30$
$3x + 2y \le 60$
$x + y \le 50$
$x \ge 0, y \ge 0$

b) Maximise $f = x + y$
subject to $2x + y \le 6$
$2x + 3y \le 12$
$4x + y \le 10$
$x \ge 0, y \ge 0$

c) Maximise $c = x + \frac{1}{2}y$
subject to $2x + y \le 24$
$x + y \le 15$
$y \le 2x$
$x \ge 3, y \ge 3$

d) Maximise $c = 10x + 7y - 1500$
subject to $3x + 2y \le 3600$
$x + 4y \le 1800$
$y \le 4x$
$x \ge 0, y \ge 0$

You may have seen this question in Chapter 8. You should now use the simplex method to tackle it.

3 A factory produces two types of toys: trucks and bicycles. In the manufacturing process of these toys three machines are used. These are a moulder, a lathe and an assembler. The table shows the length of time needed for each toy.

	Moulder	Lathe	Assembler
bicycle	1 hour	3 hours	1 hour
truck	0 hours	1 hour	1 hour

The moulder can be operated for 3 hours per day, the lathe for 12 hours per day and the assembler for 7 hours per day. Each bicycle made gives a profit of £15 and each truck made gives a profit of £11. Formulate a linear programming problem so that the factory maximises its profit.

You may have seen this question in Chapter 8. You should now use the simplex method to tackle it.

4 Chris is making coffee cakes and chocolate cakes for a cake stall at the village fete. Each chocolate cake needs 2 eggs and 250 g of margarine. Each coffee cake needs 3 eggs and 150 g of margarine. Chris buys three dozen eggs and a 3 kg tub of margarine. Formulate and solve a linear programming problem so that Chris can make as many cakes as possible.

You may have seen this question in Chapter 8. You should now use the simplex method to tackle it.

5 A food manufacturer makes Alma margarine from vegetable and non-vegetable oils. The raw oils are refined by the manufacturer and blended to form the margarine. One objective of the manufacturer is to make as large a profit as possible; however, there are various constraints. It is important that the raw oils are refined separately to avoid contamination and the final blend of oils must be soft enough to spread, but not too runny. To achieve this quality each ingredient and the final product has a hardness factor. The following table gives the cost and hardness of the raw materials used to make Alma margarine.

	Oil	Cost £ per kg	Hardness factor
Vegetable	Groundnut	0.7	1.2
	Soya bean	1.05	3.4
	Palm	1.05	8.0
Non-vegetable	Lard	1.30	10.8
	Fish	1.10	8.3

Alma margarine sells for £1.55 per kg and to ensure good spreading, its hardness must be between 5.6 and 7.4. The manufacturers can refine up to 40 000 kg of vegetable oil and up to 32 000 kg of non-vegetable oil per day. Formulate the problem of maximising the profit as a linear programming problem.

EXERCISES

9.3 HOMEWORK

1 Use the simplex method to solve the following linear programming problems.

a) Maximise $f = 20x + 30y$
subject to $7x + 10y \leq 80$
$x + 2y \leq 12$
$x \geq 0, y \geq 0$

b) Maximise $g = x + \frac{1}{10}y$
subject to $2x + 3y \leq 6$
$5x + 2y \leq 10$
$x \geq 0, y \geq 0$

c) Maximise $h = 2x + 7y$
subject to $x + 2y \leq 30$
$2x + 3y \leq 35$
$x \geq 0, y \geq 0$

d) Maximise $M = 13x + 25y$
subject to $-8x + 5y \leq 35$
$3x \leq y$
$x \geq 0, y \geq 0$

e) Maximise $N = 11x + 9y$
subject to $5x - 4y \leq 10$
$-x + 2y \leq 10$
$x \geq 0, y \geq 0$

2 Use the simplex method to solve the following linear programming problems.

a) Maximise $f = 17x + 3y$
subject to $5y + 8x \leq 100$
$3y + 5x \leq 61$
$2y + 7x \leq 70$
$x \geq 0, y \geq 0$

b) Maximise $g = 8x + 5y$
subject to $x + y \leq 100$
$5x + 4y \leq 420$
$3x + 2y \leq 240$
$x \geq 0, y \geq 0$

c) Maximise $h = 9x + 2y + 100$
subject to $5x \leq 5y$
$6x + y \leq 44$
$3x + y \leq 38$
$x \geq 0, y \geq 0$

d) Maximise $M = 4x + y$
subject to $2x + 7y \leq 114$
$5x + y \leq 54$
$2x \leq 5y$
$x \geq 1, y \geq 2$

3 A shop sells a mixture of two types of coffee beans designated A and B. Bean A costs £1.50 per kilogram, Bean B costs £2 per kg and £1000 has been set aside for purchasing the beans. For a palatable mixture, the weight of bean B plus twice the weight of bean A must not exceed 550 kg. If the shop sells the mixture at £6 per kg, how many kilograms of each type of coffee bean should be purchased to maximise the income based on the assumption that the entire mixture can be sold?

4 A haulage firm is negotiating a contract with a company which uses two sizes of containers: large 0.4 m^3 containers weighing 15 kg and small 0.2 m^3 containers weighing 4 kg. The haulage firm will use a vehicle that can handle a maximum load of 1640 kg and a cargo size of up to 100 m^3. For stability of load the vehicle should carry no more than 300 small containers. if the companies have agreed on a transport rate of £1 for each large container and 75p for each small container, how many containers of each type should the haulage company place on a vehicle to maximise the income?

5 A company manufactures two types of PCs: a 16-bit and 32-bit machine. In the manufacturing cycle the company can produce no more than 200 16-bit machines and 300 32-bit machines. The company production cycle means the total number cannot exceed 400. Chip requirements entail that twice the number of 16-bit plus three times the number of 32-bit machines made has a maximum of 1050 machines. The profit is £300 on each 16-bit machine and £400 on each 32-bit machine. Determine the number of each machine that should be manufactured to maximise the profit.

CONSOLIDATION EXERCISES FOR CHAPTER 9

1 A linear programming problem gives the following LP information.

Maximise $P = 3x + 4y + 5z$

subject to $2x + 8y + 5z \leq 3$

$9x + 3y + 6z \leq 2$

and $x \geq 0, y \geq 0, z \geq 0$

a) Set up an initial simplex tableau for this problem. Perform two iterations, choosing first to pivot on an element chosen from the z column.

b) State the values of x, y, z and P that the result from each of the two iterations carried out in part **(i)**.

c) Explain how you know whether or not the optimal solution has been achieved.

(OCR Discrete Mathematical Specimen paper)

2 A coach company has 20 coaches. At the end of a given week, 8 coaches are at depot A, 5 coaches are at depot B and 7 coaches are at depot C. At the beginning of the next week, 4 of these coaches are required at depot D, 10 of them at depot E and 6 of them at depot F. The table below shows the distances, in miles, between the relevant depots.

	D	E	F
A	40	70	25
B	20	40	10
C	35	85	15

The company needs to move the coaches between depots at the weekend. The total mileage covered is to be a minimum. Formulate this information as a Linear Programming Problem.

a) State clearly your decision variables.
b) Write down the objective function in terms of your decision variables.
c) Write down the constraints, explaining what each constraint represents.

(Edexcel Decision Mathematics D2 Specimen paper)

3 Lose Weight Foods Ltd produce a slimming drink consisting of vitamin-enriched milk. To satisfy dietary requirements the drink should contain 300 units of vitamin A, 150 units of vitamin B and 200 units of vitamin C. Three additives are available to provide the vitamins.

Each millilitre of Additive I contains 1 unit of vitamin A, 2 units of vitamin B, 2 units of vitamin C and costs 1p.
Each millilitre of Additive II contains 3 units of vitamin A, 1 unit of vitamin B, 2 units of vitamin C and costs 2p.
Each millilitre of Additive III contains 2 units of vitamin A, 3 units of vitamin B, 5 units of vitamin C and costs 3p.
Formulate and solve by the simplex method the problem as a linear programming problem in which the manufacturer wishes to minimise the cost of the additives.

You met this family in Exploration 8.1. Now you are required to solve the problem using the simplex method.

4 The parents of a two-year-old child want to ensure that they include at least the minimum satisfactory amount of protein, energy and vitamin C. The daily amounts recommended by good health guides are 32 g of protein, 5500 kilojoules (kJ) of energy and 30 mg of vitamin C. Suppose that the parents decide to achieve the daily amounts through milk and orange juice. The following table shows the amount of protein, energy and vitamin C available in these drinks.

Item	Measure of drink	Protein (g)	Energy (kJ)	Vitamin C (mg)
milk	100 ml	3.0	300	1
juice	100 ml	0.2	40	10

Milk costs 38p per litre and orange juice costs 48p per litre. The parents want to find the minimum costs to achieve the recommended daily intake.

If x and y are the number of measures of milk and juice, write down:

a) the total costs of the drinks
b) three inequalities for the amount of protein, energy and vitamin C so that the minimum satisfactory amounts are achieved.

Solve the linear programming problem using the simplex method.

You met Ben in Example 8.1. Now you are required to solve the problem using the simplex method.

5 Ben has two evening and weekend jobs, mowing lawns and cleaning cars. He can spend a maximum of 15 hours per week on these jobs. He has several regular clients for whom he spends six hours per week mowing lawns and three hours per week cleaning cars. Ben is paid £2.50 per hour for mowing lawns and £3.50 per hour for cleaning cars. Formulate and solve a linear programming problem for Ben to maximise his earnings.

You may have seen this question in Chapter 8. You should now use the simplex method to tackle it.

6 A house builder builds two type of home.

The first type requires one plot of land, £40 000 capital, 150 worker days of labour and is sold for £8000 profit.
The second type requires two plots of land, £105 000 capital, 200 worker days of labour to build and is sold for £11 500 profit.
The house builder owns 150 plots of land, has available £9 600 000 and 24 000 worker days of labour.
Formulate and solve a linear programming problem to advise the builder on a strategy to maximise the profit.

7 In the course of an investigation of the following problem arose.

Maximise $f = 3x + 2y + 7z$
subject to $x + y + 2z \leq 50$
$x + 4z \leq 20$
$x \geq 0, y \geq 0, z \geq 0$

Use the simples algorithm to determine the maximum value of P and the values of x, y and z for which this occurs.

8 A production unit makes two types of product, X and Y. Production levels are measured in tonnes and are constrained by availability of finance, staff and storage space. Requirements for, and daily availability of, each of these resources are summarised in the table.

	Finance required (£/tonne)	**Staff time required (hours/tonne)**	**Storage space requirements (m^3/tonne)**
Product X	200	8	1
Product Y	100	8	3
Resource availability each day	£1000	48 hours	15 m^3

Profits on these products are £160 per tonne for X and £120 per tonne for Y.

a) Express the three resource constraints as inequalities, and write down two further inequalities indicating that production levels are non-negative.
b) Given that the objective is to maximise profit, write down the objective function.
c) Illustrate your inequalities graphically, and use your graph to find the best daily production plan.
d) Set up an initial tableau for the problem and use the simplex algorithm to solve the problem. Relate each stage of the tableau to its corresponding point on your graph.

(UODLE Question 22, Paper 4, 1993)

9 In an executive initiative course, participants are asked to travel as far as possible in three hours using a combination of moped, car and lorry. The moped can be carried in the car and the car can be carried on the lorry.

The moped travels at 20 miles per hour (mph) with a petrol consumption of 60 miles per gallon (mpg). The car travels at 40 mph with a petrol consumption of 40 mpg. The lorry travels at 30 mph with a petrol consumption of 20 mpg. $2\frac{1}{2}$ gallons of petrol are available.

The moped must not be used for more than 55 miles, and a total of no more than 55 miles must be covered using the car and/or lorry.

The problem is formulated as the linear programming problem

Maximise $D = m + c + l$
subject to $6m + 3c + 4l \leq 360$
$2m + 3c + 6l \leq 300$
$m \leq 55$
$c + l \leq 55$

where m is the number of miles covered on the moped, c is the number of miles covered in the car and l is the number of miles covered in the lorry.

a) Explain how the constraints were produced.
b) Set up a simplex tableau to solve the problem.
c) Perform two iterations of the simplex algorithm.
d) State whether or not the tableau resulting from part **c)** represents an optimal solution.
Describe the strategy that is indicated by the tableau.
Write down how much time is used and how much petrol is left?

(OCR MEI Decision and Discrete Mathematics D2 Specimen paper)

10 A diet planning problem involves mixing two foodstuffs, X and Y, to satisfy three dietary constraints at minimum cost, C. The amounts of X and Y used are x and y kg respectively. The costs are £1 per kg for each. The problem is formulated as a linear programme.

Minimise $C = x + y$
subject to the constraints $2x + y \geq 8$
$3x + 2y \geq 14$
$x + 2y \geq 6$
$x \geq 0$
$y \geq 0$

a) Draw a graph to illustrate the inequalities and the feasible region.
b) The first three inequalities are converted into equalities by introducing slack variables, s_1, s_2 and s_3, and are laid out in a tableau.

C	x	y	s_1	s_2	s_3	RHS
1	−1	−1	0	0	0	0
0	2	1	−1	0	0	8
0	3	2	0	−1	0	14
0	1	2	0	0	−1	6

Explain why it is not possible to apply the simplex algorithm to this tableau.

c) A modification to the simplex algorithm is applied to produce a feasible solution to the above problem. Expressed in tableau form the solution is:

C	x	y	s_1	s_2	s_3	RHS
1	1	0	–1	0	0	8
0	2	1	–1	0	0	8
0	1	0	–2	1	0	2
0	3	0	–2	0	1	10

i) Interpret this tableau by stating the values given for the variables x and y and the cost C, and show that the constraints are satisfied.

ii) What are the values taken by s_1, s_2 and s_3 in this solution? How do these values relate to the constraints?

d) Starting from the tableau in **c)** two iterations of the simplex algorithm are needed to achieve the optimal solution. With this formulation you will choose to pivot on a column (not the C-column) that has a positive value in the first row of the table. Thus the first iteration will involve increasing the value of x. Perform the two iterations, interpret the result, and mark the point to which it relates on your graph.

(UODLE Question 20, Paper 4, 1995)

Summary

After working through this chapter you should be able to:

- formulate a linear programming problem in standard form
- introduce slack variables so that constraints become linear equations
- set up and use the tabular form of the simplex algorithm for problems with two or three variables
- identify initial, intermediate and final tableaux and know when the solution is optimal
- interpret the solution produced by the Simplex tableaux.

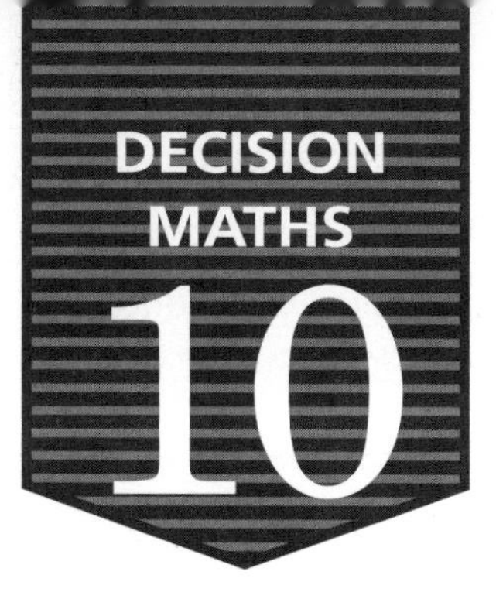

Simulation

By the end of this chapter you should:

- *know that a simulation is a mathematical model that can be used to test a real-life situation*
- *be able to carry out simulations using random numbers.*

Exploration 10.1

Queuing at the post office

A small post office has only one serving counter. Sometimes long queues build up during busy periods, so Mr Simms, the post master, decides to investigate to see whether it is worth installing a second serving counter. How might he go about this?

SIMULATIONS

One way that the post master could investigate the problem is to set up a **simulation**. A simulation is a mathematical model that can be used to test what might happen in situations where an experiment with real subjects may be too long or too dangerous to be practical, for example:

- What system will minimise the waiting time for customers queuing in a bank or post office?
- How does a disease spread through a population (human or animal)?

It may be theoretically possible to investigate these situations by observation and experimentation, but simulation gives us a feasible alternative approach. A good simulation gives the opportunity to ask 'What happens if …?' questions and consider the various outcomes of making changes. Therefore it is a very important tool in business and social situations. Setting up a simulation follows a similar procedure to that involved in setting up any mathematical model.

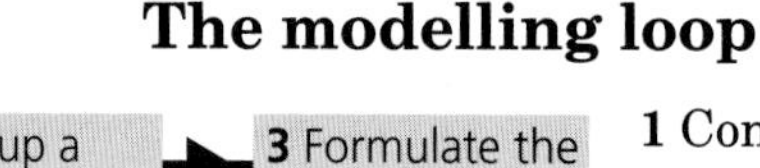

The modelling loop

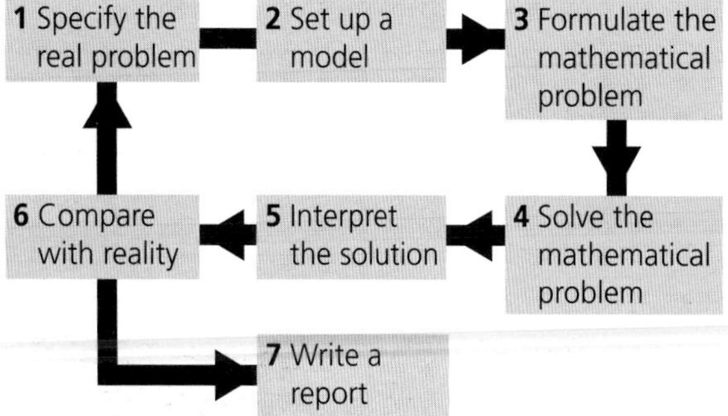

1 Consider the real situation and decide on the important variables and processes (box 1, specify the real problem).
2 Identify the simplifying assumptions that must be made (box 2).
3 Decide on the parameters you need to know. Take a sample of observations in order to estimate the values of these parameters (box 3).
4 Set up the simulation, stating the rules of the simulation clearly, and run it several times (box 4).
5 Compare the results with reality (boxes 5 and 6).

6 If your results are not good enough, amend your simulation to improve results (run the loop again).
7 Interpret the results with reference to the real situation. Use them to predict the likely outcomes if changes are made (box 7).

Notice that point 4 suggests running the simulation several times. This is important in order to generate enough information to give reliable results. However, running a simulation can in itself be a lengthy and repetitive process so computers are normally used. There are two types of simulation model.

The deterministic model

This model assumes that chance events do not occur. It can only generate one answer to a given situation.

Exploration 10.2

Forest fires

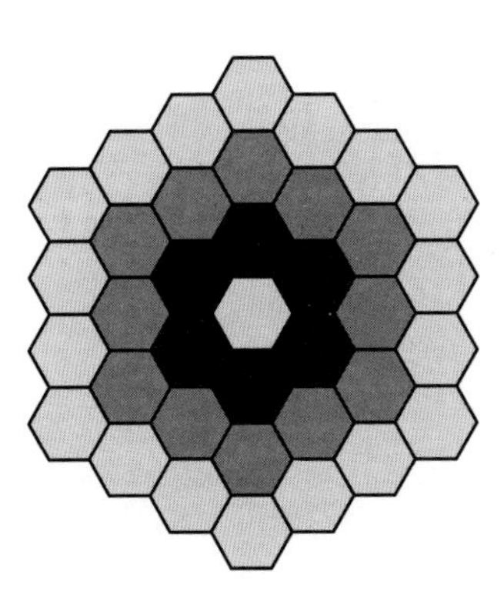

The spread of a forest fire can be modelled as follows. Each tree is represented by a hexagon. The fire is assumed to start at one tree and spreads to all adjacent trees within one minute.

Time	Trees alight	Total number burning
0	1	1
1	6	7
2	12	19
3	18	37

- Can you find an equation that will enable you to solve this model to find the number of trees burning after one hour?
- Is the model realistic?

A realistic model?

This type of model is not always very realistic. For example, an airline may find that a deterministic model suggests that five per cent of passengers do not turn up to claim their seats on flights. On the basis of this information, they may decide to book 105 passengers for every 100 seats available on their flights but, of course, on many flights this assumption will prove to be incorrect.

The stochastic model

This model allows for an element of chance in the outcome, and will therefore generate a different answer each time the simulation is run. These models use some kind of device for introducing a random element such as a dice or random number generator.

The collector's problem

Little plastic toys are being given away with breakfast cereal. If there are six different toys, how many boxes of cereal will a collector need to buy on average in order to gain a full set of toys?

This type of problem lends itself to using a die since each toy can be represented by a number on the die, assuming that there is an equally likely chance of getting each of the six toys. One run through would look like this:

2, 1, 4, 4, 6, 3, 4, 1, 2, 3, 5

The collector needs eleven boxes of cereal before she has all six toys.

*Stochastic methods are often referred to as **Monte Carlo methods**, after the town famous for its casinos.*

We should need to run the simulation several times, each time recording how many throws it takes to collect a set of all six numbers. The average of these will give a guide as to how many boxes of cereal the collector would need to buy.

EXERCISES

10.1 CLASSWORK

1. Use the information from the simulation on forest fires to find predictions for the number of trees that will be burning after:

   a) 10 minutes **b)** 30 minutes **c)** t minutes.

2. Suggest a way in which an element of chance could be introduced into the forest fire simulation.

3. Use a die to perform the collector's simulation ten times. Suggest how many boxes of cereal should be bought to be sure of collecting all six toys.

4. **a)** Suggest how the rules of the collector's simulation could be changed if there were twelve plastic toys instead of six.
   b) Run your new simulation ten times and suggest the number of boxes of cereal that would be needed to collect all twelve toys.

5. Calum wants to find the probability that two people in a class of 30 have the same birthday. He decides to use two dice to simulate the day, and sets up a table like this.

First die (columns), Second die (rows)

	1	2	3	4	5	6
1	1	2	3	4	5	6
2	7	8	9	10	11	12
3	13	14	15	16	17	18
4	19	20	21	22	23	24
5	25	26	27	28	29	30
6	31					

   a) What is his rule for simulation of the day?
   b) How might he simulate the month?
   c) How do you suggest he deals with dates that do not exist, such as 30 February?
   d) Run the simulation for 30 people at least three times. How would you use the results to calculate an estimate for the probability that two people in a class of 30 have the same birthday?

EXERCISES

10.1 HOMEWORK

1 Simulate a gambling game in which you start with £10 and bet £1 at a time on a game. The rules are:

- you receive £2 back (£1 stake plus £1 winnings) with probability $\frac{1}{3}$
- you lose your stake of £1 with probability $\frac{2}{3}$.

Run this game ten times, using a six-sided die with each run lasting for 20 throws. (If you run out of money the run finishes.)

2 A mouse in a maze will take a left turn at a T-junction after having taken a left turn at the previous T-junction with probability $\frac{7}{12}$ and a left turn at a T-junction after having taken a right turn at the previous T-junction with probability $\frac{1}{3}$. Fill in the probabilities in each arc in the diagram below.

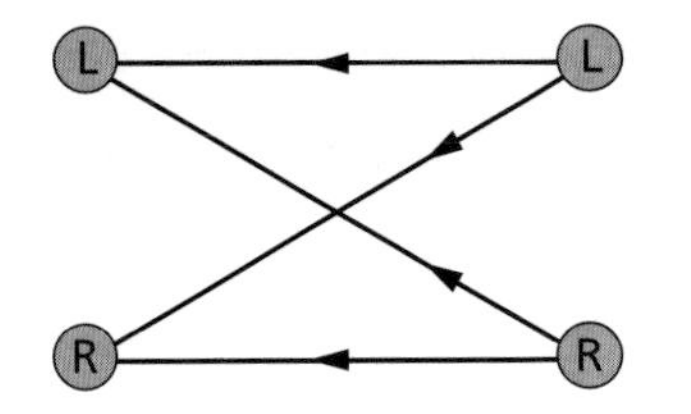

Further moves are to be simulated using a coin and a six-sided die.

a) Give a simulation rule that will generate a right and left turn succeeding:
i) a left turn **ii)** a right turn.

b) If the mouse turns right at a junction, use the rules with the coin and die to simulate the next 20 turns of the mouse.

3 In a bagatelle game, a ball is dropped down a chute so that it hits a pin and may be deflected to the left or right with an equal probability of $\frac{1}{2}$. This occurs until the ball enters a slot, as shown in the diagram below. Determine the probability of a ball finishing in each of the five slots.

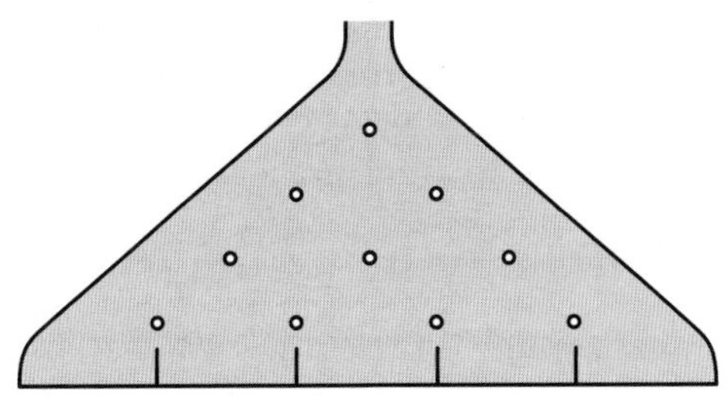

Suggest how this game may be used to simulate the determination of the number of defective items in random sample of four, drawn from a large population, if it is suspected that 50 per cent of this population are defective. Suggest a reason why you may wish to simulate the number of defectives in random samples of four rather than test the whole population.

4 Elderly people are admitted to a hospital and progress through the diagnostic stages of assessment, rehabilitation and long-stay care, as in the diagram. Most are rehabilitated and discharged but a small proportion become long-stay patients and remain in hospital. The situation can be represented diagrammatically like this.

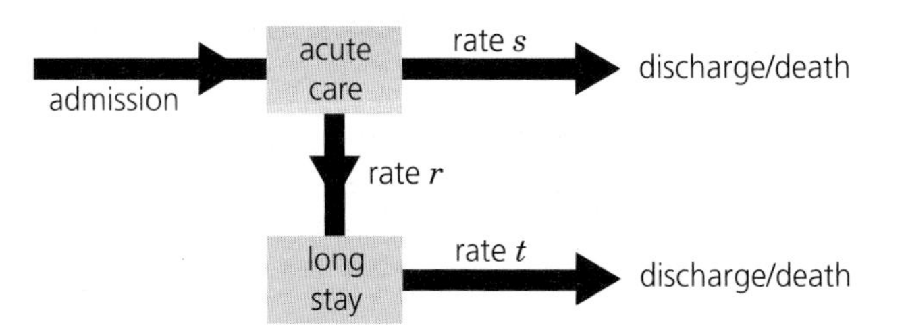

Describe how a simulation could be used by the hospital administration to plan the number of beds needed in this part of the hospital.

5 Find a situation that can be modelled by a simulation. Devise a method of simulating the situation you are suggesting.

USING RANDOM NUMBERS

When there are many possible outcomes, or if the outcomes are not all equally likely, then a die, coin or similar method will not be suitable for generating the simulation. When this is the case we often use random numbers.

Most calculators will generate random numbers. A scientific calculator usually has the symbol RAN# or RAND#. Pressing this key will produce a series of three-digit numbers between 0 and 1. Graphic calculators usually have the random number function on the probability menu and most will also have a facility for setting the number of digits required.

To demonstrate a simulation that uses random numbers, let us rejoin the queue in the post office in Exploration 10.1, to see whether it is worth installing a second serving counter. How might Mr Simms proceed?

1 Important variables and processes

Since Mr Simms is concerned with the length of queues, he needs to find out how frequently customers arrive at the post office and how long it takes to serve each one. This will enable him to calculate the average waiting time and length of queues.

2–3 Simplifying assumptions

First, it is important to establish the parameters that you need to know.

Mr Simms needs to find out how often a new customer comes into the post office. This is called the **inter-arrival time** and is an important feature of simulations of this type. Mr Simms performs a small survey and finds that the inter-arrival times at peak periods are usually between 30 seconds and 3 minutes, with roughly the distribution shown in this table.

Inter-arrival time (minutes)	0.5	1	1.5	2	2.5	3
Probability (%)	5	10	30	35	15	5

His counter assistant says that serving times usually vary between one and four minutes, with this pattern.

Serving time	1	1.5	2	2.5	3	3.5	4
Probability (%)	5	10	25	25	20	10	5

This provides sufficient information for Mr Simms to set up his simulation.

To avoid the simulation becoming too cumbersome, these simplifying assumptions will be made.

- All times are taken to the nearest 30 seconds.
- Any inter-arrival and serving times which fall outside the normal range will be ignored.
- The simulation will be performed for 25 customers, starting at a time 0.

4 Rules of the simulation

The simulation is to be generated using random numbers. There are two variables in this simulation, and so Mr Simms will need to generate two sets of random numbers, one for the inter-arrival times and one for the service times. Since the probabilities are given in percentages, Mr Simms can use random numbers from 0 to 99 (or 0.00 to 0.99 if using a calculator) to represent the probabilities. The numbers will be allocated like this.

For inter-arrival times

Inter-arrival time (minutes)	0.5	1	1.5	2	2.5	3
Probability (%)	5	10	30	35	15	5
Random numbers	0–4	5–14	15–44	45–79	80–94	95–99

For serving times

Serving time	1	1.5	2	2.5	3	3.5	4
Probability	5	10	25	25	20	10	4
Random numbers	0–4	5–14	15–39	40–64	65–84	85–94	95–99

Note: it is very important to ensure that you allocate the random numbers correctly – it is easy to make a mistake at this stage of a simulation.

Solution

Having done this, Mr Simms can now run the simulation. The easiest way to record the results is in a table like this.

Customer	RND	Inter-arrival time	Arrives (after start time)	Service start (after start time)	RND	Service time (minutes)	Service end (after start time)	Wait time (minutes)	Queue length (maximum)
1		0	0	0	72	3	3	0	0
2	93	2.5	2.5	3	61	2.5	5.5	0.5	1
3	14	1	3.5	5.5	46	2.5	8	2	1
4	66	2	5.5	8	77	3	11	2.5	1
5	74	2	7.5	11	41	2.5	13.5	3.5	2
6	82	2.5	10	13.5	98	4	17.5	3.5	2
7	8	1	11	17.5	29	2	19.5	6.5	2
8	90	2.5	13.5	19.5	97	4	23.5	6	2
9	35	1.5	15	23.5	0	1	24.5	8.5	3
10	92	2.5	17.5	24.5	51	2.5	27	7	3
11	81	2.5	20	27	9	1.5	28.5	7	3
12	89	2.5	22.5	28.5	57	2.5	31	6	4
13	89	2.5	25	31	96	4	35	6	3
14	17	1.5	26.5	35	2	1	36	8.5	4
15	63	2	28.5	36	43	2.5	38.5	7.5	3
16	45	2	30.5	38.5	4	1	39.5	8	4
17	68	2	32.5	39.5	23	2	41.5	7	4
18	43	1.5	34	41.5	19	2	43.5	7.5	5
19	22	1.5	35.5	43.5	17	2	45.5	8	5
20	72	2	37.5	45.5	69	3	48.5	8	5
21	5	1	38.5	48.5	36	2	50.5	10	5
22	41	1.5	40	50.5	87	3.5	53.5	10.5	5
23	7	1	41	53.5	28	2	55.5	12.5	6
24	21	1.5	42.5	55.5	78	3	58.5	13	6
25	57	2	44.5	58.5	3	1	59.5	14	6

Average wait time	6.94 minutes ≈ 7 minutes
Maximum wait time	14 minutes
Average queue length	3.4
Maximum queue length	6

For the purpose of the example, we have only performed the simulation once. In reality, simulations need to be performed several times so that patterns can be seen and useful conclusions drawn.

This gives the average waiting time to be around 7 minutes, though it rises to a maximum of 14 minutes and was still rising after 25 customers. The queue length reached a maximum of 6 at the end, with an average of 3.4, showing that an extra service point would definitely be used during this time.

5 Compare results with reality

To verify his simulation Mr Simms would then compare his results with the situation he observed in the shop each day.

6 Amend your model

In order to improve the model Mr Simms could do a more accurate survey of the inter-arrival times and service times and he may also wish to take smaller time intervals to improve accuracy. Another possibility would be to consider how these factors change once the busiest period is over, to see how quickly the queues decline.

7 Interpret the results

On the basis of this simulation, it would appear that queues will inevitably build up during peak periods. If this continues for more than an hour, Mr Simms may well need to install another service point during this time. If, however, the busy period is fairly short, he may feel that it is not necessary as the queues will decline fairly quickly once things slow down again.

Exploration 10.3

Queuing

Mr Simms decides to open his new service point. Run another simulation to see which is the most effective method for queuing.

One method is the type usually encountered in a supermarket, where each service point has a separate queue and people join the queue of their choice.

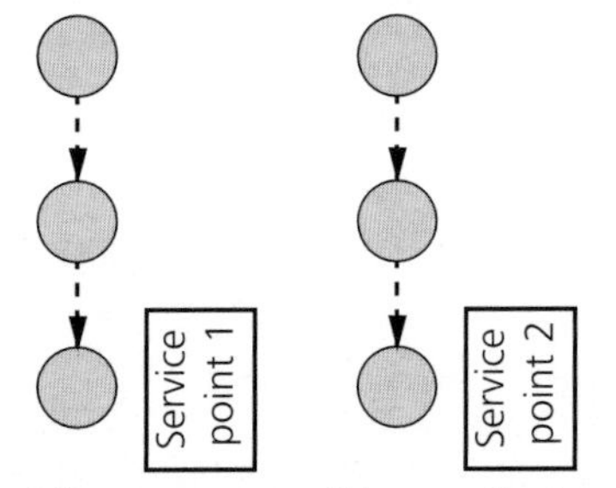

Of course, we all know that, under this system, everyone *feels* that their queue is moving more slowly than others! The alternative is to use the system operated by many post offices and banks, which is to have a single queue feeding into several service points.

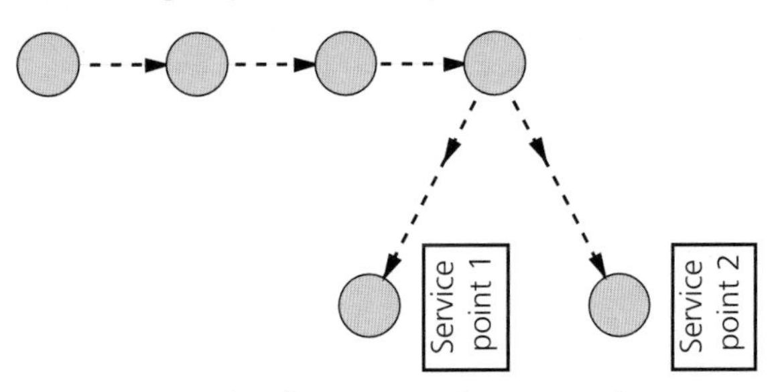

As you try to simulate a queuing procedure, you may well find it helpful to draw the service points on a piece of paper and use counters to represent the people in the queue.

Devise a method for simulating the two queuing procedures given above and state which you think would be the more efficient.

EXERCISES

10.2 CLASSWORK

1 Run the simulation for Mr Simms' queuing problem again, this time for 50 customers. On the basis of this extra information, would you advise him to install a second service point?

2 Quicksnack take-away claims to provide fast food for local workers during lunch time. The inter arrival times of its customers on a typical weekday are summarised here.

Inter-arrival time (minutes)	1	2	3	4	5
Frequency (%)	55	30	10	4	1

The average service times are given in this table.

Service time (minutes)	1	1.5	2	2.5	3
Frequency (%)	20	24	28	18	10

If the take-away opens at midday, simulate the waiting times for the first 20 customers. Find estimates for the average waiting time and maximum queue length. Does Quicksnack really provide fast food?

3 The time taken to serve a customer at a payment till in a department store varies as shown in the following table.

Time taken to serve (minutes)	1	2	5
Probability	$\frac{1}{4}$	$\frac{1}{2}$	$\frac{1}{4}$

Customers arrive at the till according to the following table.

Time between customers (minutes)	1	2	3	4
Probability	$\frac{1}{6}$	$\frac{1}{3}$	$\frac{1}{3}$	$\frac{1}{6}$

a) Give a rule for generating customer service times from two-digit random numbers.

b) Use your rule and the following random number stream to simulate the service times for the next twelve customers.

87 64 67 35 15 52 04 57
91 83 05 23 76 22 41

c) Give a rule for generating times between customers from two-digit random numbers.

d) Use your rule and the following random number stream to simulate the time intervals between the arrivals of the next twelve customers. (The first interval should represent the time interval before the first customer arrives.)

37 15 09 35 13 97 65 69
86 90 85 07 04 55 41

e) Using your service times from **b)** and your inter-arrival times from **d),** simulate the arrival and serving of twelve customers at the till. Record the events of your simulation in the following way (the numerical entries are for illustration).

Services times			5	2	2	2		...
Inter-arrival times	2	1	1	2	1	3	3	...

Time (minutes)	Customer arriving	Time of next arrival	Customer starting service	Time of end of service	Length of queue
0		2			0
1					0
2	first	3	first	7	0
3	second	4			1
4	third	6			2
5					2
6	fourth	7			3
7	fifth	10	second	9	3
8					3
9			third	11	2

f) Showing your working, find the mean queue length between the time of the arrival of the fourth customer and the time of the arrival of the ninth customer.
Why is this more representative of the performance of the system than the mean queue length throughout the simulation?

g) Give any times during which the server is not occupied between the time of the arrival of the fourth customer and the time of the arrival of the ninth customer.
Find the percentage of the time between the time of the arrival of the fourth customer and the time of the arrival of the ninth customer for which the server is occupied.

h) What reliance would you place on your simulation for estimating the mean queue length and the server utilisation? How could you improve its reliability?

(UODLE Question 12, Decision Mathematics AS, 1994)

4 Twenty companies each have a bay of five reserved spaces in a multi-storey car park (100 spaces in all). Records indicate that at midday, on average, fifteen per cent of spaces are empty.

a) Use the random digits below to simulate the presence (P) or absence (A) of a car at midday in each of the 100 spaces. State your simulation rules clearly.

30 31 02 83 22   29 82 73 04 79   47 62 79 84 38   97 08 59 27 07
02 28 56 83 24   78 07 99 41 32   01 87 20 92 75   88 00 06 08 68
84 35 93 65 79   13 97 92 79 99   01 90 23 41 76   77 26 98 00 56
68 39 19 39 25   29 05 01 09 91   33 11 71 21 94   82 72 42 82 48
76 33 84 92 71   50 10 91 58 66   37 76 29 02 51   20 01 58 77 04

b) Group your results in sets of five, and thus count how many cars are present for each company. Record this number for each company.

c) Using your results from **b)**, estimate:
  i) the probability that a randomly selected company has all of its spaces occupied
  ii) the mean number of spaces occupied per company.

d) Calculate the theoretical probability that a randomly selected company has at midday:
  i) all of its spaces occupied
  ii) four of its spaces occupied.

(UODLE Question 2, Decision Mathematics AS, 1991)

5 Simulation is often used in population modelling. A pair of birds will produce from four to ten eggs in a clutch each spring, with this distribution.

Eggs per clutch	4	5	6	7	8	9	10
Frequency (%)	8	10	16	25	30	7	4

Of these, 0.4 will either not hatch or die in the first three months. Through the winter 0.5 of the adults and 0.7 of the surviving chicks will also die due to cold and food shortage.

a) Starting with a population of 20 pairs, set up a simulation to estimate the number of birds that will be alive at the beginning of the next spring.

b) Assuming that the young can reproduce after one year, explain how you could continue the simulation. What simplifying assumptions would you need to make?

EXERCISES

10.2 HOMEWORK

1 Allocate random numbers m in the range 00–99 for the following situations.

a) X is distributed as illustrated by the histogram.

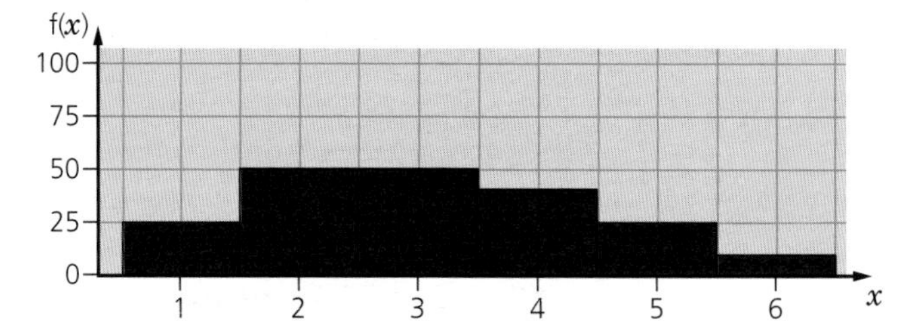

Parts (b), (c) and (d) should only be attempted by students who have or who are studying A-level Statistics.

b) The random variable X has a Poisson distribution of mean 2 i.e:

$$P(X = j) = \frac{e^{-2}.2^j}{j!}, j = 0, 1, 2, 3, \ldots$$

c) The continuous random variable X has a probability density function which is triangular, as illustrated below.

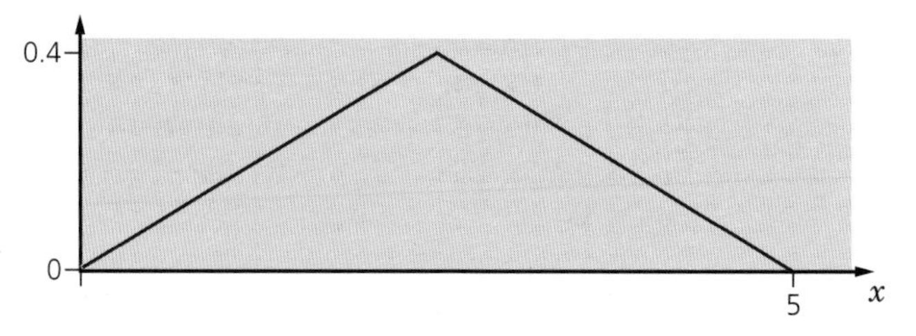

(Divide the range $0 \le x \le 5$ into five classes of equal width.)

d) The continuous random variable X is distributed as a negative exponential distribution of mean 0.5 so its probability density function is $f(x) = 2e^{-2x}, x \ge 0$.
(Divide the range $x \ge 0$ into half-unit intervals for five intervals and then consider $x \ge 2.5$. **Note:** The area under the curve $y = 2e^{-2x}$ from a to b, where $b > a \ge 0$, is given by $e^{-2a} - e^{-2b}$.)

> *The numbers generated are called **pseudo-random numbers** and when such generating methods are devised for computers a and m are large, for example one recommended choice is a = 16 807, b = 0, m = $2^{31} - 1$.*

2 Simulations are usually run on computers and to use random numbers there must be some deterministic way of generating them (i.e. by a fixed rule). Numbers may be generated by the rule:

$$x_{n+1} = ax_n + b \pmod{m}$$

i.e. x_{n+1} is the remainder after $ax_n + b$ has been divided by m.

To keep the numbers in the range 00–99, we take $m = 100$. Generate ten numbers by taking:

a) $a = 23$, $b = 3$, $x_0 = 7$ **b)** $a = 30$, $b = 10$, $x_0 = 7$.

Comment on your results.

3 Consider the simulation of a game of tennis between two weak players X and Y. Player X has a 50:50 chance of getting the service in play and only a chance of 0.01 of winning from a valid serve without a rally taking place. Player Y has a 60:40 chance of getting the service in play and a probability of 0.1 of winning directly from a valid serve. If the game goes to a rally, X has a probability 0.6 of winning and hence Y has a probability 0.4 of winning a rally, independent of who is serving. Draw a flow diagram for the model of this game of tennis.

Simulate a game of tennis in which

a) X serves **b)** Y serves.

4 The local health trust has a short-stay ward for minor illnesses in one of their hospitals. There are five beds in the ward, but they wish to remove one of the beds to cut costs. From previous records, the distribution of the number of patients who became ill each day is summarised in this table.

Number	0	1	2	Total
Frequency	35	50	25	100

The length of stay of patients has the distribution shown in this table.

Length of stay (days)	1	2	3	4	5	6	7	Total
Frequency	5	10	20	40	16	5	4	100

As the local hospital manager, you have to determine the demand on the five beds. Assume that on day 0 two beds are occupied and count day 0 as the first day of these patients' stay; they are in bed for three and four days respectively. Simulate the situation for 28 days and repeat the simulation with four beds. For each simulation run, find the percentage of patients who are accepted and the mean number of empty beds per day. Find the distribution of these quantities over as many runs of the simulation as time allows you.

5 A computer company receives calls for technical support which are dealt with on one direct telephone line. If the line is engaged a customer will try at a later stage. The calls arrive with the following patterns.

Time between calls (minutes)	0–4	5–9	10–14	15–19	20–39
Percentage of calls	15	20	25	20	20

The calls are answered as politely as possible and take the following times to be dealt with.

Time to satisfy a customer (minutes)	0–4	5–9	10–19	20–29	30–89
Percentage of calls	50	10	15	20	5

A customer who finds the phone engaged will call back. The effect on the frequency of calls has been measured and may be simulated by a ten per cent reduction in the spread of inter-call times for each customer unable to get through.

Simulate the above situation and produce a short report for the management on the suitability of the current position, indicating in particular the number of customers unable to get through when they ring, and the time that the support system is idle.

MATHEMATICAL MODELLING ACTIVITY

Lights or filters

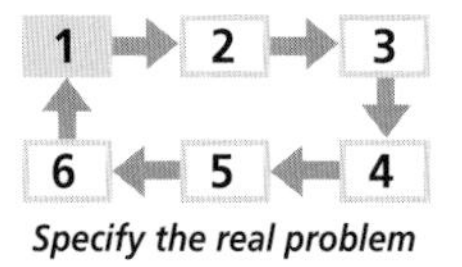

Specify the real problem

Specify the real problem

On the island of Guernsey you would encounter a familiar looking road sign bearing the unfamiliar instruction, '*Filter in turn*'.

Consider the junction below where cars are approaching along roads A and B. They then take it in turns to enter the box and cross the junction, one from A followed by one from B and then one from A again, filtering in turn. Investigate how quickly cars can move through this type of junction.

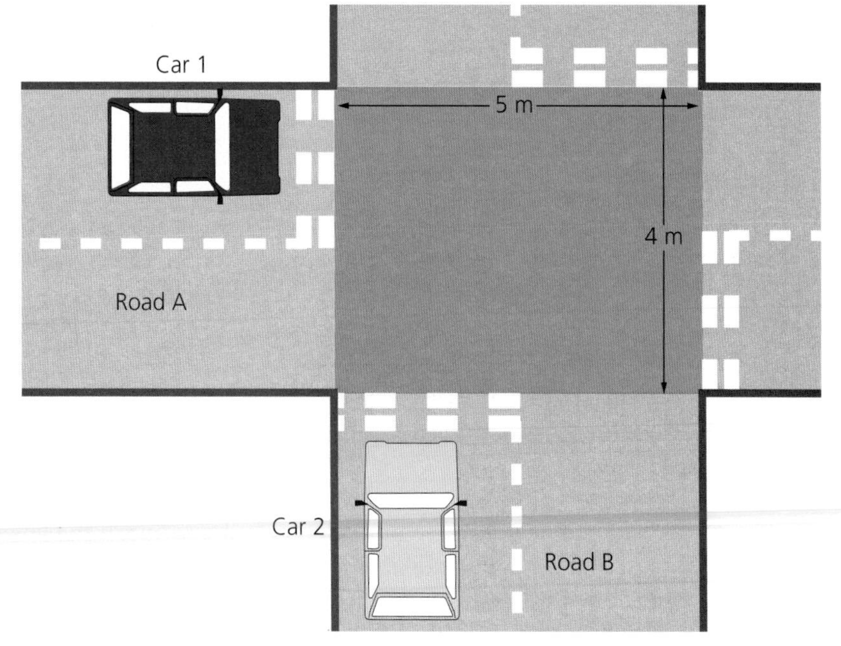

Imagine you are doing a study into whether traffic lights or a 'Filter in turn' would be a more effective way of relieving conjestion at a busy road junction during peak traffic flow times.

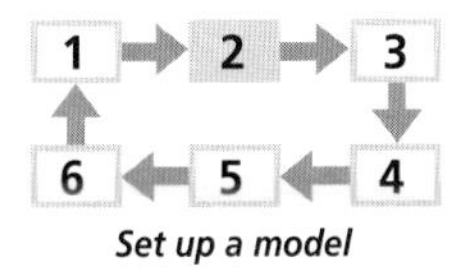

Set up a model

Set up a model

Decide on the simplifying assumptions which must be made. You may wish to consider these ideas.

- All lengths will be measured to the nearest 25 cm.
- All times will be to the nearest second.
- All cars move off from being stationary with the same constant acceleration (e.g. $2\,\text{m s}^{-2}$).

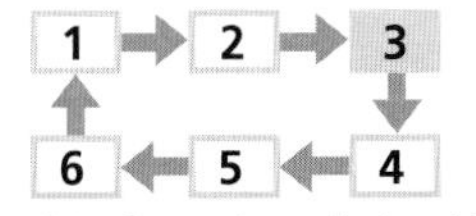

Formulate the mathematical problem

Formulate the mathematical problem

Identify the important variables. These might include:

- the inter-arrival times of vehicles at the junction during a busy time
- the distribution of lengths of vehicles which are likely to use the roads (Do you include buses and lorries? If not you may need to amend your simplifying assumptions.)
- How long the traffic lights stay green.

You may wish to conduct surveys to obtain the information. You will probably end up with tabulated data like this.

Inter-arrival time (seconds)	1	2	3	4	...
Percentage of vehicles (%)	60	18	12	...	...

Length of vehicle (nearest 25 cm)	2.5	2.75	3	...
Percentage of vehicles (%)	10	15	20	...

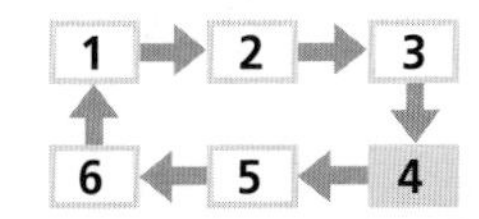

Solve the mathematical problem

Solve the mathematical problem

Run the simulation. State your rules clearly. Run the simulation for both traffic lights and a filter in turn as many times as you think feasible.

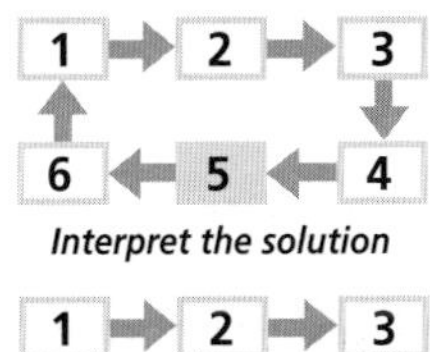

Interpret the solution

Interpret the solution

Use your simulation to calculate queue length and average waiting times over a given period.

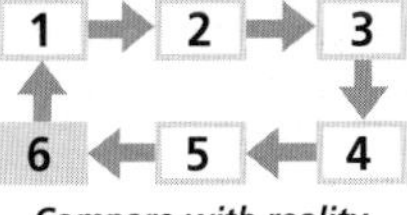

Compare with reality

Compare with reality

See whether your traffic light simulation agrees with observations made at the lights you originally used. You cannot directly check a filter, but if the lights simulation is realistic there is a good chance the filter simulation will also be realistic.

If the simulation does not compare well with reality, you will need to amend your model. Then follow through the procedure again, and interpret your results.

Write a report

- Which system seems to be more efficient?
- Are your distributions for car lengths and inter-arrival times adequate?
- Did you omit large vehicles from the original model?
- What effect does changing the delay on traffic lights have?

CONSOLIDATION EXERCISES FOR CHAPTER 10

1 A variety of sweet pea produces flowers which are either pink, blue or white in the proportion $\frac{1}{5}$, $\frac{3}{10}$ and $\frac{1}{2}$ respectively. It is not possible to distinguish flower colour by examining sweet pea seeds.

a) A seed is chosen at random and sown. The colour of the resulting flower is recorded. Give a rule for using two-digit random numbers to simulate this experiment.

A similar variety of sweet pea has pink, blue and white flowers in the proportions $\frac{1}{7}$, $\frac{3}{14}$ and $\frac{9}{14}$ respectively.

b) A seed is chosen at random and sown. The colour of the resulting flower is recorded. Give a rule for using two-digit random numbers to simulate this experiment.

2 The diagram shows part of a pinball machine which has tracks along which balls roll. The tracks divide at points at which separators (labelled 1–6) deflect the balls one way or the other with equal probabilities.

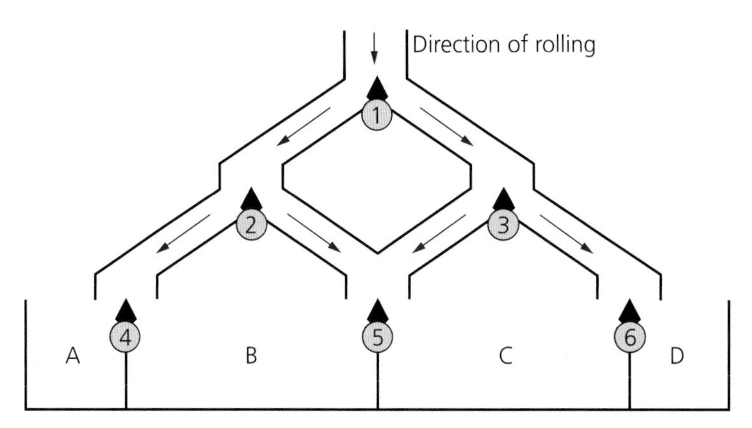

The balls which end up in A or D score five points each. Those which end up in B or C score ten points each.

a) You are now to simulate the progress of 24 balls, either by using the random numbers listed beneath this question, or by using the RAN# (or equivalent) key on your calculator.
You must indicate clearly:

- how many random numbers you use to simulate the progress of each ball
- the rules which you use to simulate the progress of each ball
- for each of the first five balls, a list of the random numbers used and a clear description of the path taken.

b) **i)** What was the mean score per ball from your simulation?
ii) What would you have expected the mean score per ball to be?

Random numbers:

72824	54409	53048	30383
04212	87273	25027	43581
87018	05989	36415	96027
31624	65384	40563	17496

(UODLE Question 6, Decision Mathematics AS, 1995)

3 When two friends arrive at an airport there are two check-in desks operating for their flight. There is a separate queue for each desk. There are three groups of people in the queue on the left and two groups in the queue on the right, and in each case the first group in the queue has just begun the check-in procedure.

The distribution of check-in times for groups at a check-in desk is shown in the table.

Time taken to check-in (mins)	1	2	3	4	5
Probability	0.1	0.4	0.2	0.2	0.1

a) Complete the table to produce a rule for simulating group check-in times.

Random numbers	Time to check in (mins)
00–09	1
	2
	3
	4
	5

b) Complete the following table by simulating six times what might happen if the two friends were to join the queue on the right, i.e. the shorter queue.
The friends' waiting time is the total time taken for the two groups in front to check-in.

Simulation number	Random number	Time for first group to check in	Random number	Time for second group to check in	Total waiting time
1	73		91		
2	38		07		
3	03		64		
4	24		10		
5	53		70		
6	38		70		

Give the mean waiting time from your six simulations.

c) Instead of both joining the shorter queue the two friends decide to use the following strategy. They will join one queue each. The first to be served will then check **both** in.

i) Complete the table below by simulating six times the progress of the longer queue.

Simulation number	Random number	Time for first group to check in	Random number	Time for second group to check in	Random number	Time for third group to check in	Total waiting time
1	21		26		67		
2	32		73		85		
3	86		28		74		
4	60		77		95		
5	35		41		04		
6	32		93		45		

ii) Combine the results with those from the simulation in part **b)** to obtain six values for the friends' waiting time if they use this strategy.

Simulation number	1	2	3	4	5	6
Waiting time (minutes)						

Give the mean of the six waiting times.

d) Give three ways in which this simulation experiment could be improved or made more realistic.

(OCR MEI Decision and Discrete June 1998)

4 Random number lists for this question are printed at the end of the question.
The random numbers are to be used as two-digit random numbers, and are to be read in order from the top left across the page, a row at a time, as required. there are separate lists for part **a)** and for part **b)** of the question.

a) A small post office has one server. Customer inter-arrival times follow the following distribution.

Inter-arrival time (minutes)	2	3	4
Probability	$\frac{1}{4}$	$\frac{1}{2}$	$\frac{1}{4}$

Using two-digit random numbers from the list marked A, simulate five inter-arrival times. Describe the rule that you used to generate your inter-arrival times.

b) Service times follow the following distribution.

Length of service (minutes)	1	2	5
Probability	$\frac{1}{3}$	$\frac{1}{2}$	$\frac{1}{6}$

Using two-digit random numbers from the list marked B, simulate service times for six customers. Describe the rule that you used to generate your service times.

c) Assuming that the service of the first customer has just begun, simulate the service of six customers. Number the customers 1 to 6, and record your results in the table.

Customer number	Arrival time	Inter-arrival time	Start of service	Service time	End of service
1	0		0		
2					
3					
4					
5					
6					

d) Compute the mean queue time (i.e. the mean of the times for which each customer queues), the mean length of queue and the server utilisation (i.e. the percentage of time for which the server is busy).

e) Say how the reliability of the results from your simulation could be improved.

Random number list A
42 22 03 80 19 37 62 93 55 12
12 07 46 91 63 28 20 37 92 83
45 37 64 58 55 07 14 12 75 43
45 94 74 21 56 64 39 54 64 19
23 36 78 53 46 85 10 19 09 05

Random number list B
35 92 78 98 16 06 31 34 82 17
64 54 38 03 17 19 20 91 76 73
36 49 54 84 87 69 46 21 09 70
75 24 15 13 96 50 02 04 56 87
56 62 39 53 81 04 24 25 71 61

(O & C MEI Question 4, Decision & Discrete Mathematics Paper 19, June 1995)

5 Two tills are in operation at a supermarket.

a) The time intervals between customers arriving to pay have the following probability distribution.

Time interval (seconds)	10	30	50	70	90
Probability	0.3	0.2	0.1	0.1	0.3

i) Complete the table to give a rule for using two-digit random numbers to simulate time intervals between customer arrivals.

Random numbers	00-29				
Time interval (seconds)	10	30	50	70	90

ii) Use the following five two-digit random numbers to simulate the inter-arrival interval times for five customers. (The first random number is to simulate the time from the start of the experiment to the arrival of the first customer.). Show the actual times, after the start of the experiment, at which customers arrive to pay.
94 06 14 39 46

Customer number	1	2	3	4	5
Time interval (seconds)					
Arrival time (seconds)					

b) The time taken to serve a customer at a till is given by the following probability distribution.

Time (seconds)	15	45	80
Probability	$\frac{1}{2}$	$\frac{1}{8}$	$\frac{3}{8}$

i) Complete the table to give a rule for using two-digit random numbers to simulate times for serving a customer.

Random numbers	00–47		
Time (seconds)	15	45	80

ii) Use the following two-digit random numbers to simulate the service times for five customers.
07 75 98 73 77 59

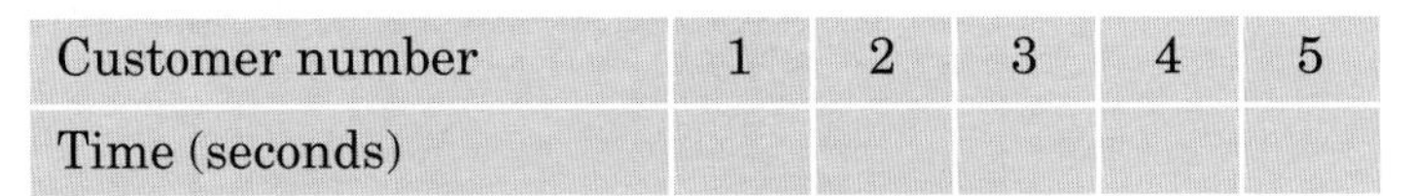

Customer number	1	2	3	4	5
Time (seconds)					

c) Use your times from parts (a) and (b) to simulate the processing of five customers through the tills. There is a separate queue for each till, and customers join the shorter queue (or that for till 1 if both queues are the same length). Start with both tills free.

Customer number	Arrival time (seconds)	Till number	Time of start of service (seconds)	Time of end of service (seconds)	Total time queuing and paying (seconds)
1					
2					
3					
4					
5					

i) Give the mean total time for queuing and paying.
Server utilisation is defined as

$$\frac{(\text{time server 1 busy}) + (\text{time server 2 busy})}{2 \times \text{duration of experiment}} \times 100\%$$

ii) Give the server utilisation.

d) It is suggested that a single queue be set up to serve both tills. Using the same times, simulate the processing of five customers through the tills using a single queue system. Again, start with both tills free.

Customer number	Arrival time (seconds)	Till number	Time of start of service (seconds)	Time of end of service (seconds)	Total time queuing and paying (seconds)
1					
2					
3					
4					
5					

i) Give the mean total time for queuing and paying.
ii) Give the server utilisation.

e) i) Would you expect a single queue system to be more or less efficient than having separate queues? Why?
ii) Give an advantage of a single queue system which is not related to the efficiency of the system.

(OCR MEI Decision and Discrete June 2000)

6 A chocolate factory produces several types of chocolate. The chocolates are mixed together and randomly packed into cartons of 20. Fifteen per cent of the chocolates have nut centres.

a) You are to simulate the process of packing chocolates into cartons of 20, distinguishing between nut centres and other chocolates. You are to use a table of two-digit random numbers from 00 to 99 to generate ten cartons of chocolates. For each carton you are to count the number of nut centres in the carton.

i) Give your simulation rule.
ii) Now apply your rule using the table of two-digit random numbers below. Indicate those numbers which represent nut centres. Count the number of nut centres in each simulated carton.
iii) Use your *results* to estimate the probabilities of a carton containing:
- exactly three nut centres
- fewer than three nut centres.

iv) Suppose that the fifteen per cent figure quoted is only an approximation, and in fact one seventh of the chocolates have nut centres. How would you change your simulation rule to simulate the process using two digit random numbers?

b) A quality control manual states that if p% of items from a large batch are faulty, then the probability of obtaining fewer than three faulty items in a random sample of size 20 is given by:

$$f(p) = \left(1 - \frac{p}{100}\right)^{20} + 20\frac{p}{100}\left(1 - \frac{p}{100}\right)^{19} + 190\left(\frac{p}{100}\right)^{2}\left(1 - \frac{p}{100}\right)^{18}$$

i) Say how this result can be used to compute the probability of obtaining fewer than three nut centres in a carton of 20 chocolates.
ii) Calculate f(15), giving your answer correct to three decimal places. Compare it with the results of your simulation.
Say why the results might differ, and suggest how to improve the simulation.

	Carton number									
	1	**2**	**3**	**4**	**5**	**6**	**7**	**8**	**9**	**10**
Number of nut centres	65	06	55	72	04	87	31	29	39	56
	29	93	95	65	90	95	99	87	46	66
	36	07	93	49	20	02	59	48	54	35
	73	34	68	72	44	28	87	44	81	09
	77	19	52	52	52	65	29	15	82	81
	23	56	99	82	21	01	62	81	98	14
	56	32	69	71	27	29	74	87	24	79
	42	66	10	50	75	40	87	08	26	35
	84	64	56	47	44	11	22	93	84	75
	65	06	91	47	67	25	97	25	08	35
	68	76	98	45	28	80	46	57	74	80
	62	57	51	32	33	42	06	56	17	81
	94	25	05	63	58	62	21	99	86	58
	90	78	87	05	96	57	38	14	37	35
	05	51	87	25	87	71	56	03	65	03
	00	51	60	44	72	59	53	94	22	10
	74	38	54	43	43	45	29	91	74	43
	58	08	72	99	89	09	38	66	75	45
	49	00	47	42	75	47	88	59	25	21
	04	61	07	14	40	73	42	68	67	25

(O & C MEI Question 2, Decision & Discrete Mathematics Paper 19, January 1996)

7 The weather bureau in a particular country defines each day to be either wet or dry. Records show that if the weather today is dry then the probability that it will be dry tomorrow is $\frac{4}{5}$, and the probability that it will be wet tomorrow is $\frac{1}{5}$. If the weather today is wet then the probability that it will be wet tomorrow is $\frac{2}{7}$, and the probability that it will be dry tomorrow is $\frac{5}{7}$.

Future weather is to be simulated. Each day a two-digit random number is to be used, together with a simulation rule based on that day's weather, to generate the weather for the next day.

When the weather is dry the rule to be used is as follows.

00 – 79 $\Rightarrow$ weather tomorrow is dry

80 – 99 $\Rightarrow$ weather tomorrow is wet.

a) Give a simulation rule to generate the weather for a day following a wet day.

b) The weather today is dry. Use the rules, together with the following random numbers to simulate the weather for the next 14 days. Read the two-digit numbers from left to right.

Random numbers:
39 16 44 89 01 56 90 99 11 37 47 84 29 52 21

06 39 43 06 42 82 52 16 39 89 58 61 74 93 82

c) Use the fourteen results from your simulation to calculate an estimate of the overall proportion of wet days.
d) Use the fourteen results from your simulation to calculate an estimate of the probability of the weather tomorrow being the same as the weather today.
e) Give two ways in which the simulation of the weather could be improved.

(O & C MEI Question 1, Decision & Discrete Mathematics Paper 19, June 1996)

8 A circular coin of diameter 3 cm is used in a fairground game. Competitors flick the coin along a board marked across with ten parallel lines at 5 cm intervals, and with end sections of width 4 cm. To be valid the coin must end entirely within the scoring region (shown unshaded in the diagram below), and it will then count as a win if it does not cross a line. Otherwise it will count as a loss. If it does not land entirely within the scoring area then another attempt is made until it does. (The numbers in the diagram indicate the distances in cm between the lines.)

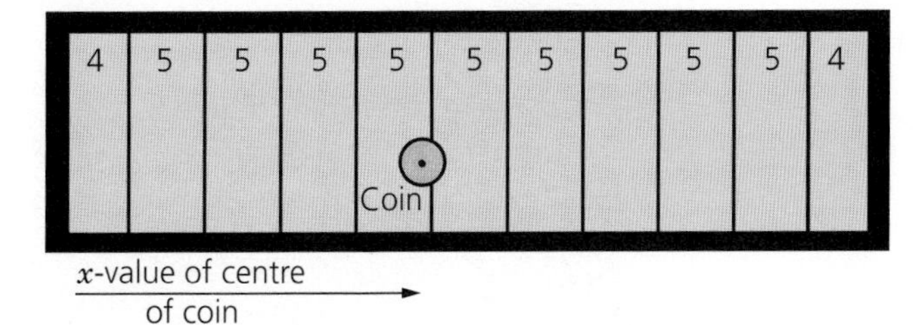

Assume that a valid coin is equally likely to land in any position within the scoring region. Thus the x-value of the centre of a valid coin will be between 1.5 and 51.5.

a) i) Give the ranges of x-values which correspond to the centres of winning coins.
ii) Use the following list of five-digit random numbers to simulate the x-values of the centres of ten randomly-flicked and valid coins. Explain your method for performing the simulation and give the x-value of the centre of each of your coins.
iii) State whether or not each of your simulated coins is in a winning position.

Random numbers:
65236 80077 82581 41085 87273
06567 65104 73943 01674 36415

b) By considering the regions of the board within which the centre of a winning coin must lie, calculate the probability of winning.
c) A new board is to be constructed with ten equally-spaced lines 6 cm apart, but with the width of the end sections each being e cm. Find the value of e for the board to be fair (i.e. one in which the probability of winning is 0.5).

(UODLE Question 11, Decision Mathematics AS, 1995)

9 A 'drive-through' fast food restaurant has a single lane for cars, alongside which are three windows. Two are for taking orders and one is for serving food. Once cars have entered the system no overtaking is possible.

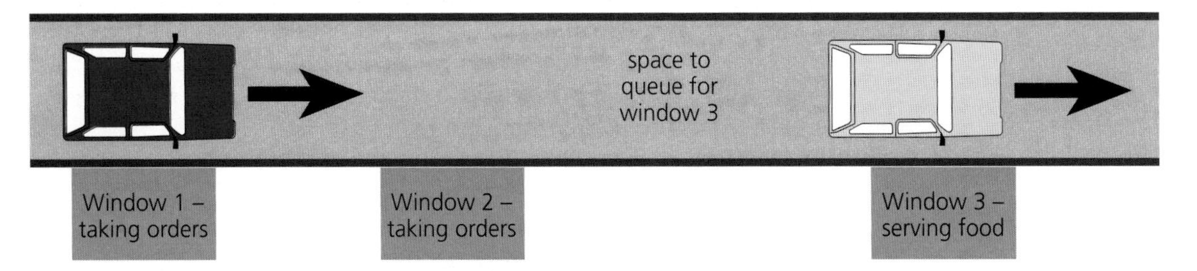

A car arrives, queues if necessary, and is then sent either to window 1 or window 2, where the driver places an order and pays. Having paid, the driver then moves forward to window 3 as soon as possible. At window 3 the food has to be collected, which may involve a wait.

a) The time taken to order and pay (either at window 1 or window 2) has the following probability distribution.

Time (minutes)	$\frac{1}{2}$	1	$1\frac{1}{2}$	2	$2\frac{1}{2}$
Probability	$\frac{1}{10}$	$\frac{4}{10}$	$\frac{2}{10}$	$\frac{2}{10}$	$\frac{1}{10}$

i) Compete the table to give a rule for using two-digit random numbers to simulate times for ordering and paying.

Random numbers	00-				
Time (minutes)	$\frac{1}{2}$	1	$1\frac{1}{2}$	2	$2\frac{1}{2}$

ii) Use the following two-digit random numbers to simulate order and paying times for three cars.
73 98 14

Car number	1	2	3
Time			

b) The time taken from paying until the food is ready for collection at window 3 has the following probability distribution.

Time (minutes)	$\frac{1}{2}$	1	$1\frac{1}{2}$
Probability	$\frac{1}{3}$	$\frac{1}{2}$	$\frac{1}{6}$

i) Complete the table to give a rule for using two-digit random numbers to simulate these times. (Your rule should use as many two-digit random numbers as possible.)

Random numbers	00-		
Time (minutes)	$\frac{1}{2}$	1	$1\frac{1}{2}$

ii) Use the following two-digit random numbers to simulate three of these times.
30 24 98 49

Order number	1	2	3
Time			

c) Use your times from parts **a)** and **b)** to simulate the time it takes to process three cars through the system using two different rules. The arrival times of the cars have been simulated for you. Start with all windows free. (The time taken for a car to move between windows should be ignored.)

Rule 1
If both window 1 and window 2 are free the next car is sent to window 2.
If window 1 is free but window 2 is occupied the next car is sent to window 1.

Car	Arrival time (hrs:mins)	Window 1 or 2?	Time of arriving at window 1 or 2	Time of completing payment	Time of leaving window 1 or 2	Time of leaving system
1	12:00					
2	12:01$\frac{1}{2}$					
3	12:02					

Rule 2
If both window 1 and window 2 are free the next car is sent to window 2.
If window 1 is free and window 2 is occupied, it is sent to window 1 if the car at window 2 has been there for 1 minute or less; otherwise the next car is made to wait until window 2 is free.

Car	Arrival time (hrs:mins)	Window 1 or 2?	Time of arriving at window 1 or 2	Time of completing payment	Time of leaving window 1 or 2	Time of leaving system
1	12:00					
2	12:01$\frac{1}{2}$					
3	12:02					

d) Comment on your results from part **c)**.

(OCR MEI Decision and Discrete June 1999).

10 A dentist wishes to analyse the time that her patients spend waiting for their appointments. She begins surgery at 8.30 a.m. and appointments are made every 15 minutes, the last appointment being at 12.15 p.m. She estimates that roughly 33 per cent of her appointments are check-ups which take an average of ten minutes, twelve per cent are extractions which take around fifteen to twenty minutes, and the rest are usually fillings which can take from ten to twenty minutes. Devise a simulation for the dentist's morning surgery, stating your assumptions clearly. Do you think the dentist is correct in making appointments every fifteen minutes?

Summary

After working through this chapter you should:

- *understand that a simulation is a model set up to represent a real situation*
- *be able to set up and perform simple simulations, stating the rules you use and the simplifying assumptions, which are the assumptions you make to enable you to carry out the simulation*
- *be able to interpret the results of your simulation in the context of the original problem.*

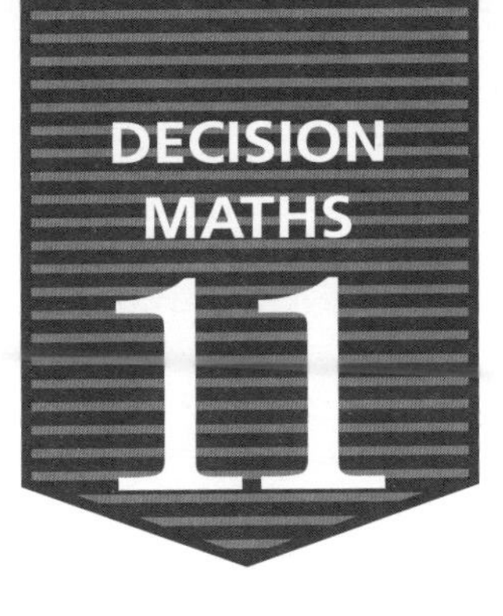

Boolean algebra

By the end of this chapter you should:

- *understand the basic principles of Boolean algebra*
- *be able to construct truth tables and apply these to the output for some types of logic gates in electronics.*

Exploration 11.1

A long corridor in a school is lit by strip lighting which needs to have switches at each end and one in the middle where the side corridor goes off. How would you design a circuit so that the light is independently controlled by any one of the three light switches?

switch

switch

switch

BOOLEAN ALGEBRA

When John switches on his graphic calculator, the screen shows the same information that was on it when he switched it off. Electronic circuits inside the machine enable it to retain information (provided the batteries don't go flat).

Throughout this book we have frequently talked of the importance of computer technology in the application and development of algorithms. The branch of Mathematics we now call *Decision mathematics* has developed rapidly since the advent of computers and we could argue that without this development, many of the algorithms that are now widely used in business would have remained little more than recreational maths puzzles, simply because of the time it takes to apply such algorithms by hand. If a technique is to be viable in business, its cost-effectiveness is all important and a method that takes a long time to apply would not be viable.

'There is no branch of mathematics, however abstract, which may not some day be applied to the phenomena of the real world.' (Nikolai Lobachevsky 1792 – 1856)

Computers are one of the many electronic devices that we now take for granted, yet their development would not have been possible without one of the most abstract branches of mathematics: logic. A lesson to be learned: never assume a branch of mathematics is useless!

'Pure mathematics was discovered by Boole in a work which he called The Laws of Thought. *His work was concerned with formal logic, and this is the same thing as mathematics.' (Bertrand Russell,* International Monthly, *1901)*

Machines such as computers handle discrete data and almost always operate in binary (base 2). The circuits and networks which describe their operation are called **switching circuits**. These were developed using the work of the British logician George Boole. He is considered to have been the founder of symbolic logic, in which mathematical notation is applied to questions of logic, and Boolean algebra is named after him.

George Boole (1815–64)

Boolean variables can only take two values, often given as true or false, 0 and 1 or off and on.

Logical propositions

A proposition is a statement which is either true or false, but cannot be both. For example:

$(x + 4)(x - 4) = x^2 - 16$ **true**
$2 + 2 + 2 = 0$ **false**

This means that questions such as 'How are you?' are not acceptable since no truth value can be assigned to them.

In language we do not construct every sentence as a list of statements, but link things together using connectives or conjunctions:

- Jill lives in Birmingham *and* Jenny lives in Leeds.

This contains two propositions and, for the statement to be true, both of the constituent statements must be true. If we consider the ways of linking statements that we use in everyday language, we find we use the same concepts in logic, and we have symbols to represent them.

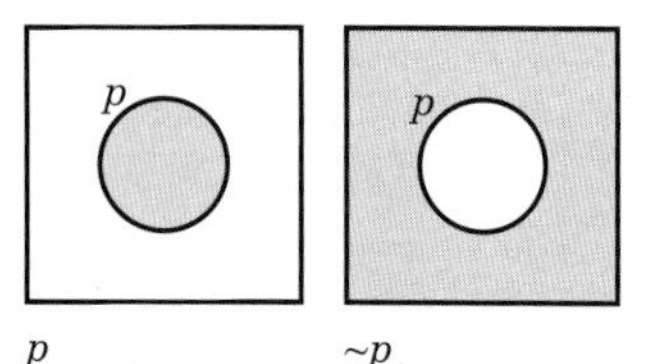

p $\sim p$

Note: This is often called complement

Negation

Let p denote a simple proposition, then the negation of p is 'not p' and is written $\sim p$. If p is true, then $\sim p$ will be false, and if p is false, then $\sim p$ will be true. For example if p denotes the statement 'crocuses flower in spring' (true) then $\sim p$ denotes 'crocuses do not flower in spring' (false).

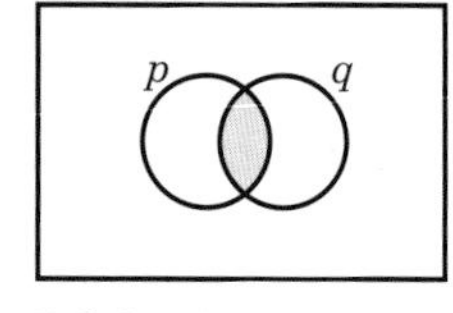

$p \cap q = p \wedge q$

Conjunction

The most common conjunction used in speech is ***and***. This is represented in symbolic logic by $\wedge$. If p and q denote two simple statements then $p \wedge q$ is true if and only if both p and q are true.

This can be represented in a **truth table**, like this.

p	q	$p \wedge q$
T	T	T
T	F	F
F	T	F
F	F	F

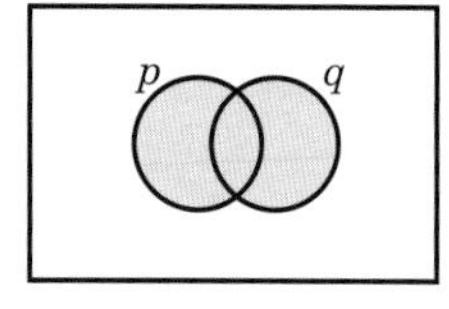

$p \cup q = p \vee q$

Disjunction

A disjunction is represented in speech by ***or*** and in symbolic logic by $\vee$. If p and q denote two simple statements then $p \vee q$ is true if p or q is true. In this case the truth table looks like this. Note the situation where both are true.

p	q	$p \vee q$
T	T	T
T	F	T
F	T	T
F	F	F

Condition

A condition is a proposition of the form 'p implies q' and is represented by the symbol $\Rightarrow$. The truth table for this is like this.

p	q	$p \Rightarrow q$
T	T	T
T	F	F
F	T	T
F	F	T

Note: $p \Rightarrow q$ is only false if a false conclusion q is drawn from a true statement p.

Double condition

The double condition $\Leftrightarrow$ is a way of saying 'p implies q which implies p'. This is rather cumbersome so we usually say 'if and only if' which is abbreviated to ***iff***.

Consider these statements.

1 School sports day will be cancelled if it rains.
2 School sports day will be cancelled only if it rains.
3 School sports day will be cancelled if, and only if, it rains.

Let p be the statement 'School sports day will be cancelled.'

Let q be the statement 'It rains.'

Statement **1** says rain will lead to the cancellation of sports day, so we can say $q \Rightarrow p$ (or $p \Leftarrow q$, 'p is implied by q').

We say rain is a **sufficient condition** for cancellation of sports day.

Statement **2** says if sports day is cancelled, then it must be raining since no other reason for cancellation is given, so $p \Rightarrow q$.

We say rain is a necessary condition for cancellation.

Statement **3** says both of these, so $p \Leftrightarrow q$.

We say rain is a **necessary and sufficient condition** for the cancellation of sports day.

The truth table for $\Leftrightarrow$ looks like this.

p	q	$p \Leftrightarrow q$
T	T	T
T	F	F
F	T	F
F	F	T

Compound propositions

When more than one connective is needed to express the proposition symbolically we can draw combined truth tables.

Example 11.1

Draw the truth table for $(p \wedge \sim q) \Rightarrow q$.

Solution

p	q	$\sim q$	$p \wedge \sim q$	$(p \wedge \sim q) \Rightarrow q$
T	T	F	F	T
T	F	T	T	F
F	T	F	F	T
F	F	T	F	T

Laws of Boolean algebra

The most obvious way to simplify Boolean expressions is to manipulate them in the same way as you would manipulate algebraic expressions. A set of rules for is needed for this and these are shown in the table.

In many cases, the operations used in Boolean algebra are defined as:
p + q = p ∨ q
(this is often written pq*)*
p . q = p ∧ q.

$p \vee \sim p = 1$ $p \wedge \sim p = 0$	Complement laws
$\sim(\sim p) = p$	Double complement
$p \vee p = p$ $p \wedge p = p$	Indempotent laws
$p \vee 0 = p$ $p \wedge 1 = p$	Identity laws
$p \vee 1 = 1$ $p \wedge 0 = 0$	Dominance laws
$p \vee q = q \vee p$ $p \wedge q = q \wedge p$	Commutative laws
$p \vee (q \wedge r) = (p \vee q) \wedge (p \vee r)$ $p \wedge (q \vee r) = (p \wedge q) \vee (q \wedge r)$	Distributive laws
$p \wedge (p \vee q) = p$	Absorption law
$\sim(p \vee q) = \sim p \wedge \sim q$ $\sim(p \wedge q) = \sim p \vee \sim q$	DeMorgan's laws

The distributive law is one with which you will probably be familiar from number work e.g: $7(5 + 3) = 7 \times 8 = (7 \times 5) + (7 \times 3)$.

Example 11.2

Simplify $p \vee (\sim p \wedge q)$

$$p \vee (\sim p \wedge q) = (p \ \sim q) \wedge (p \ q)$$
$$= 1 \wedge (p \ q)$$
$$= p \vee q$$

Example 11.3

Prove that $x \wedge (x \vee y) = x$ *(the Absorption law)*

$$
\begin{aligned}
LHS &= (x \vee 0) \wedge (x \wedge y) && \textit{(since } x = x \vee 0\textit{)}\\
&= (x \wedge x) \vee (x \wedge y) \vee (0 \wedge x)\ (0 \wedge y) &&\\
&= (x \wedge x) \vee (x \wedge y) && \textit{(distributive law)}\\
&= x \wedge (1 \vee y) && \textit{(distributive law)}\\
&= x \wedge 1 &&\\
&= x &&
\end{aligned}
$$

Electronics

Logic gates are widely used in computers and other electronic devises. In this case, the variables p and q represent inputs which can be either ON (1) or OFF (0). A simple logic gate is shown in the diagram.

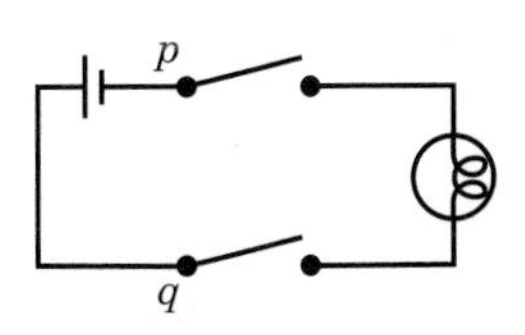

For the bulb to light, both switches must be on, so this is an AND gate which means that p and q must both be on to get an output. We can draw a truth table which describes the behaviour of this system.

p	q	$p \wedge q$
0	0	0
0	1	0
1	0	0
1	1	1

The circuit can be represented on a simplified diagram like this.

This is called an AND gate and is expressed symbolically as $p \wedge q$.

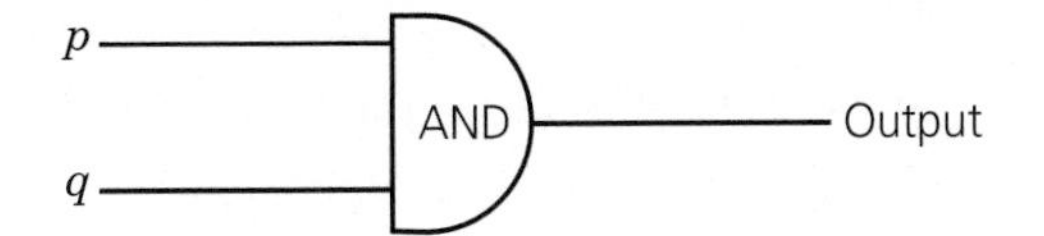

Other types of gates, corresponding to the connectives used in symbolic logic also have circuit symbols.

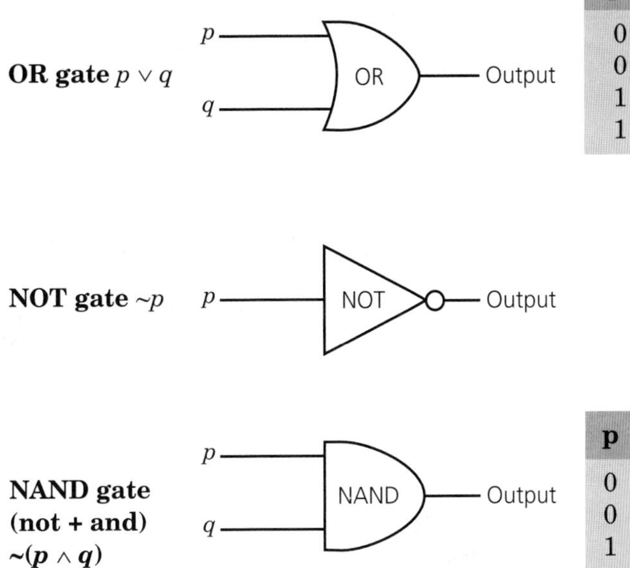

p	q	$p \vee q$
0	0	0
0	1	1
1	0	1
1	1	1

p	q	$\sim(p \wedge q)$
0	0	1
0	1	1
1	0	1
1	1	0

A personal computer contains about half a million NOR gates.

NOR gate (not + or) $\sim(p \vee q)$

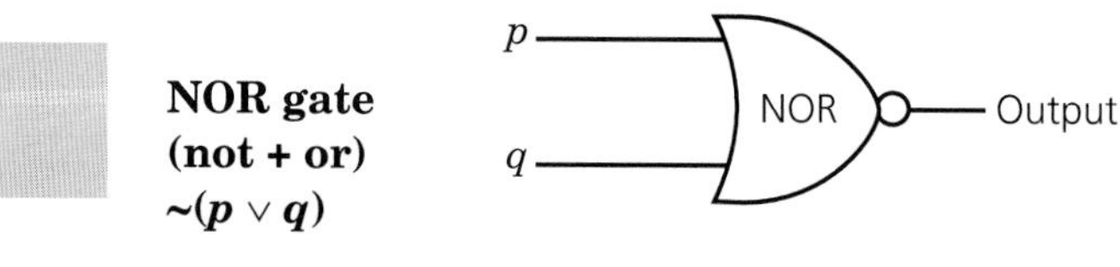

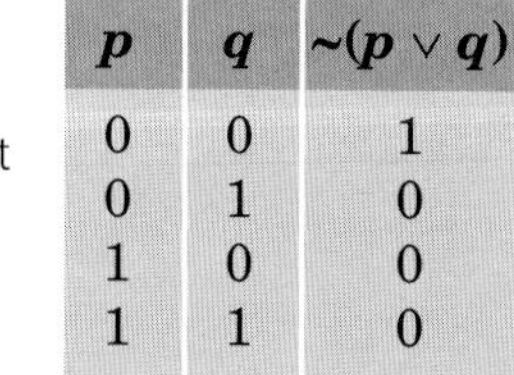

p	q	$\sim(p \vee q)$
0	0	1
0	1	0
1	0	0
1	1	0

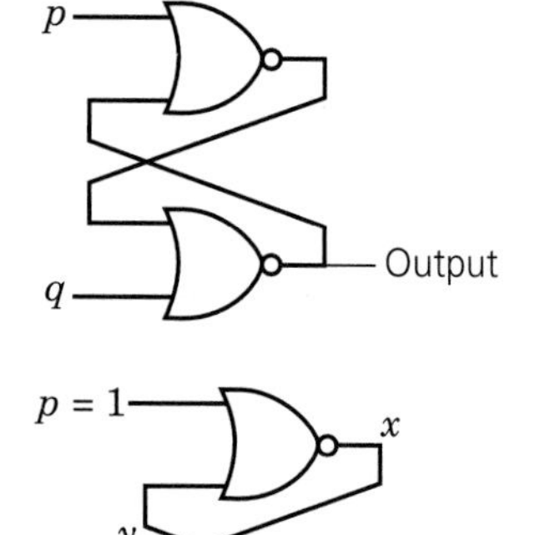

These gates can be connected together in many ways. One of these is called a **bistable**, which can be made of two NOR gates as shown in the diagram. (Remember, NOR is an abbreviation of 'not or', which would be stated symbolically as $\sim(p \vee q)$. Its output will only be 1 (T) if both inputs are 0).

The bistable forms the basis for computer memories, the sequence of diagrams shows how a bistable based on two NOR gates works.

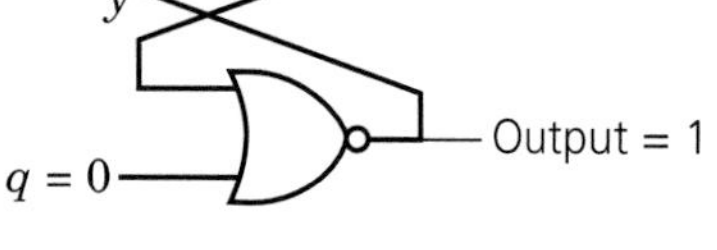

If the input at p is 1 then x must be 0 regardless of the value of y, so if q is 0, the output will be 1 (and hence y will also be 1).

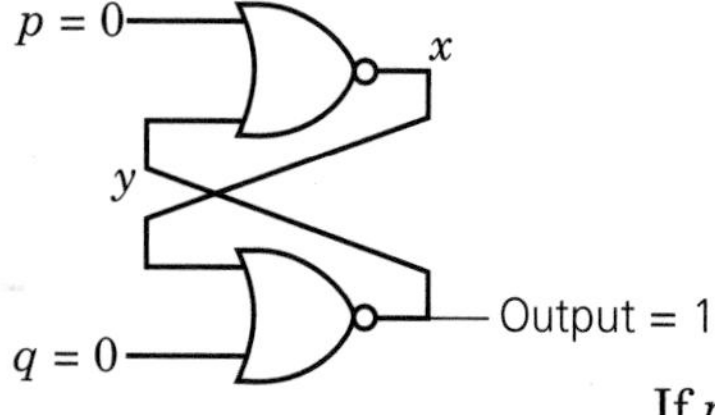

If the input at p is 0 and the output at x is 0, then y must be 1 (if y were 0, then x would also be 0), so q must be 0 and the output is also 1.

In other words, changing the state of p alone does not alter the output.

If p is left at 0 and q is now 1, x becomes 1 and y is 0, so the output is 0. The bistable has been cleared.

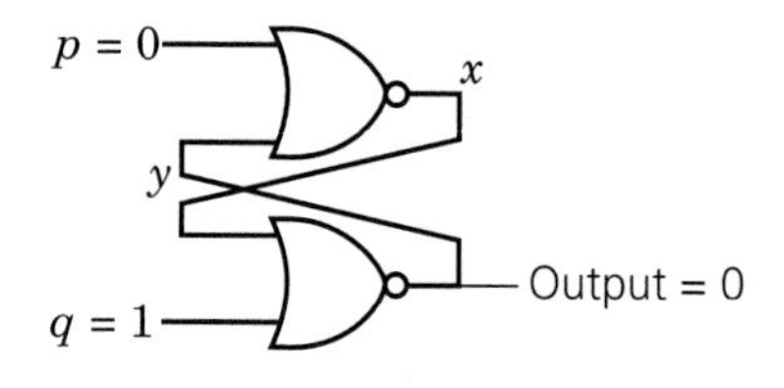

If we switch q back to 0, it will not alter the output which is still 0.

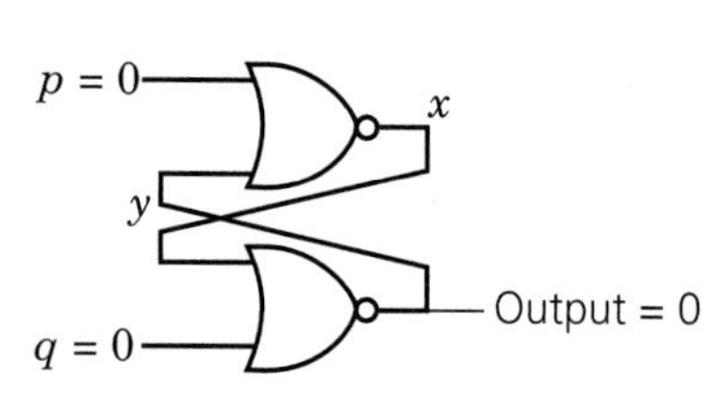

The important point about the bistable is that it 'remembers' whether p or q was last in the logic state 1 and the outcome depends on this. The truth table for a bistable is like this.

p	q	output
1	0	1
0	0	1
0	1	0
0	0	0

The output is stable in two states (hence the name bistable) and we say the bistable remembers one **binary digit**, or **BIT** of information.

John's graphic calculator has several thousand bytes of memory (a byte is 1024, or 2^{10} bits), so as long as there is a source of power, it will retain the information he wants it to. A computer has several million bytes of memory which it uses while the machine is on, though it will lose any information not stored when the power is switched off.

Switching circuits

Exploration 11.1 is about switches in a lighting circuit. A switch is either on or off, giving an analogy with the boolean logic we have been using thus if a switch is x when it is on, $\sim x$ describes the switch when it is off.

Two switches in series give $x \wedge y$.

Two switches in parallel give $x \vee y$.

Thus this circuit represents $(x \vee y) \wedge [(\sim x \wedge \sim y) \vee z]$.

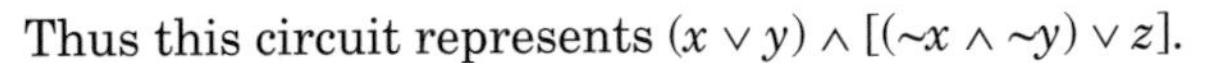

This diagram shows an equivalent circuit.

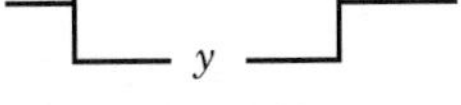

We can show analytically that the two circuits are equivalent like this.

$(x \vee y) \wedge [z \vee (\sim x \wedge \sim y)]$
$= [(x \vee y) \wedge z] \vee [(x \vee y) \wedge (\sim x \wedge \sim y)]$
$= [(x \vee y) \wedge z] \vee [(x \wedge \sim x) \wedge (x \wedge \sim y) \vee (y \wedge \sim x) \wedge (y \wedge \sim y)]$
$= [(x \vee y) \wedge z] [\, 0 \wedge (x \wedge \sim y) \vee (y \wedge \sim x) \wedge 0]$
$= (x \vee y) \wedge z$
$= z \wedge (x \vee y)$

EXERCISES

11.1 CLASSWORK

1 Let p be the proposition 'Ali studies Mathematics'.
Let q be the proposition 'Ali studies German'.
Let r be the proposition 'Ali studies Art'.
Write the following propositions in words.

a) $p \wedge q$ **b)** $p \wedge \sim r$
c) $p \wedge q \wedge \sim r$ **d)** $p \wedge \sim(q \vee r)$

2 Construct a truth table to show $p \Rightarrow \sim q$.

3 Construct a truth table to show $(p \vee q) \Leftrightarrow (\sim p \wedge q)$.

4 The diagram below represents the logic function $\mathrm{F}(p, q) = (\sim p \vee q)$.

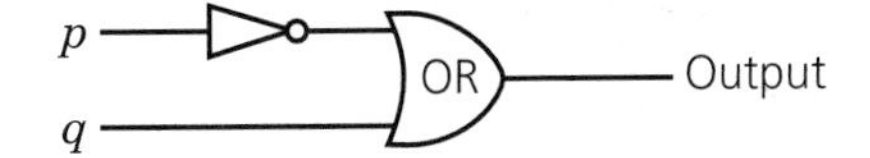

Draw a table to show the possible outputs for the circuit.

5 A bistable can also be constructed from NAND gates. Draw a bistable with two NAND gates and construct the truth table for it.

EXERCISES

11.1 HOMEWORK

1 Given that x is a real number, consider the statements
$r : x = 2 \qquad s : x^2 = 4.$

a) Discuss whether the following are true or not.
r is a necessary condition for s.
r is a sufficient condition for s.

b) Repeat the analysis for new statements
$r : x < 1 \quad s : x^2 < 1$

$\sim q \Rightarrow \sim p$ is called the ***contra-positive.***

2 By constructing truth table show that $p \Rightarrow q$ is not equivalent to its converse, $q \Rightarrow p$, but is equivalent to $\sim q \Rightarrow \sim p$.

3 In a switching circuit, p' represents 'not p' so when p is open p' is closed and vice versa.

a) Draw a circuit represented by $(a \vee b \vee c') \wedge (a' \vee b \vee c) \wedge (b' \wedge a) \vee c)$.

b) Derive an algebraic expression for the switching circuit below. Discuss why it is intuitively acceptable that it could be simplified to $b \wedge c$.

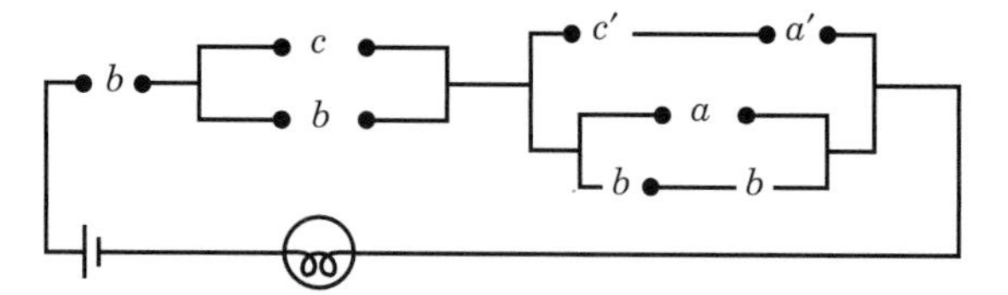

Construct a truth table for the above and show it is logically equivalent to $b \wedge c$.

c) Using AND, OR and NOT gates, find an equivalent diagram to the switching circuit above.

4 Consider the network below with inputs x_1 and x_2 and outputs s and c.

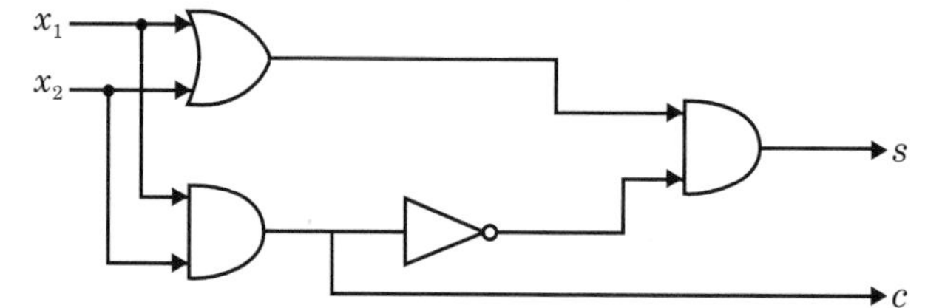

a) Write down Boolean expressions c and s and use truth tables to show s is equivalent to $x_1{}' x_2 \wedge x_1 x_2{}'$.

The device is called a half-adder.

b) If x_1 and x_2 are binary digits, by considering the truth table for s and c, what do you think s and c represent?

5 **a)** A corridor is illuminated by a strip lighting element. Design a circuit so that the light is independently controlled by any one of three light switches placed respectively at the two ends and in the middle of the corridor. Write down the equivalent Boolean expression and the corresponding truth table.

b) Repeat the above stages for a circuit in which a light is lit when a majority occurs when a panel of three people vote by pressing a switch.

MATHEMATICAL MODELLING ACTIVITY

Logic and proof

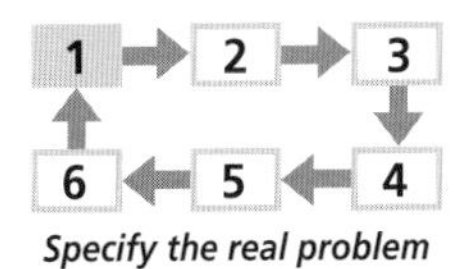

Specify the real problem

Problem statement

We have the following statements:

If Jones met Smith last night then Smith was the murderer. The murder took place before midnight if Jones is telling the truth. If Jones did not meet Smith last night the inescapable conclusion is that Jones is telling the truth. The murder did take place at or after midnight hence Smith was the murderer.

The problem is to determine whether the above constitutes a valid argument.

Set up a model

Set up the model

The first step is to identify the simple statements such as:
  p : Jones met Smith last night.

Define each simple statement as above and express the sentences in the introduction in terms of your defined symbol:
  $\Rightarrow$ ~ (not)

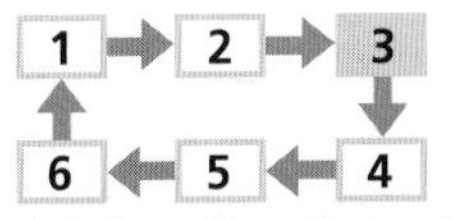

Formulate the mathematical problem

Mathematical solution

A mathematical argument is the process of starting with statements called premises P and deducing another called the conclusion C. For an argument to be valid the implication $P \Rightarrow C$ must be true for all possible outcomes. A valid argument is a proof.

A **tautology** is a statement that, if the truth values are always, true shows that:

$p \wedge (p \Rightarrow q) \Rightarrow q$,

For example, $(p \Rightarrow q) \wedge (q \Rightarrow r) \Rightarrow (p \Rightarrow r)$ are tautologies.

Two of the more common standard methods of proof are therefore **detachment** where p and $p \Rightarrow q$ are premises and the valid conclusion is q, and **syllogism** where $p \Rightarrow q$ and $q \Rightarrow r$ are premises and the valid conclusion is $p \Rightarrow r$.

Using the methods of proof above analyse the statements in the introduction (you may find the result of Question 2 of Exercises 13.3B useful).

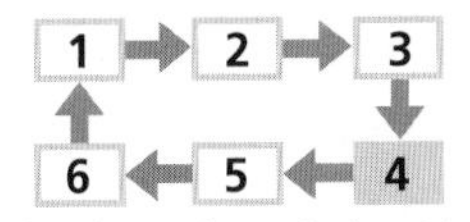

Solve the mathematical problem

Continuation

Consider the following argument by Lord Bridgewater who lived in the 18th century and was a big supporter of canal transport.

The development of steam locomotion will surely result in a vast rail system. But for the continued prosperity of the canals we must do without such a rail system. Surely the love of Lord Bridgewater does guarantee the ever-continuing prosperity of the canals. Thus, we may conclude that steam locomotion will not develop further.

Show that there is nothing invalid in the above argument. Discuss why the above was not patently true.

CONSOLIDATION EXERCISES FOR CHAPTER 11

1 Use a truth table to prove that the following two statements are equivalent

- 'If the north wind blows then we shall have snow.'
- 'If we have snow then the north wind does not blow.'

(AQA B Discrete Mathematics D1 Specimen Paper 2000)

2 Consider the following switching circuit.

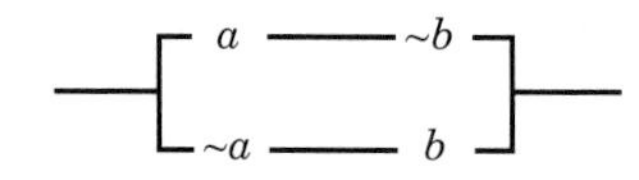

a) Write down a Boolean expression for the circuit.
b) Use the distributive rules to prove that the Boolean expression for the circuit is equivalent to $(a \vee b) \wedge (\sim a \vee \sim b)$.
c) Use part **b)** to draw an alternative, equivalent switching circuit to that given in part **a)**.

(AQA B D1 Specimen 2000)

3 **a)** The following is a switching circuit for a half adder.

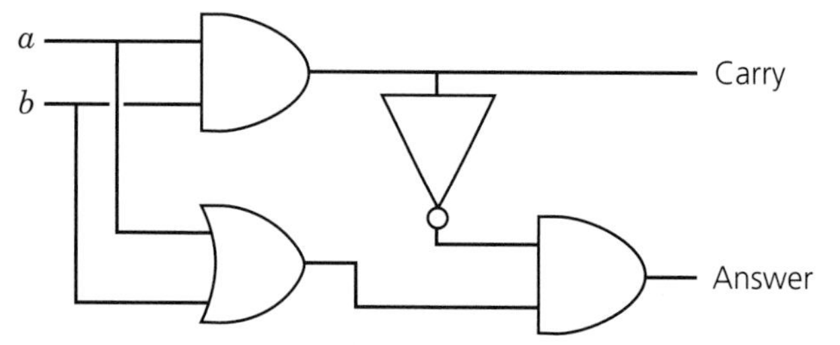

Draw a table of inputs and associated outputs.
b) The following circuit uses two half adders:

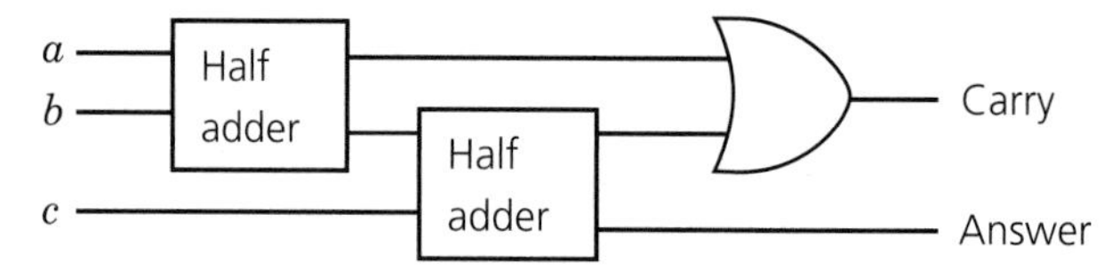

Draw a table of inputs and associated outputs, and explain briefly the purpose of the circuit.

(AQA B D1 Specimen 2000)

4 The figure represents a two-way switch. The arm of the switch is shown between the 'up' and 'down' positions (marked with broken lines), but it is in fact always in one or other of those two positions. When a switch is operated the arm is either moved from the down position to the up position, or vice versa.

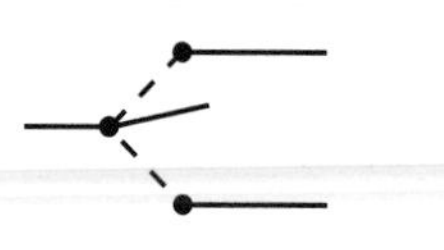

i) Draw a circuit diagram connecting two such switches so that when a switch is operated current starts to flow if it was not flowing, and stops if it was flowing.

ii) Draw a truth table for the expression $(a \vee \sim b) \wedge (b \vee \sim a)$. Explain how this expression models the circuit in part **i)**. Produce an alternative Boolean expression which also models the circuit.

iii) In the circuit diagram in the figure below the two switches labelled B are 'ganged'. That is they are constructed to operate together, so that they are either both up or both down.

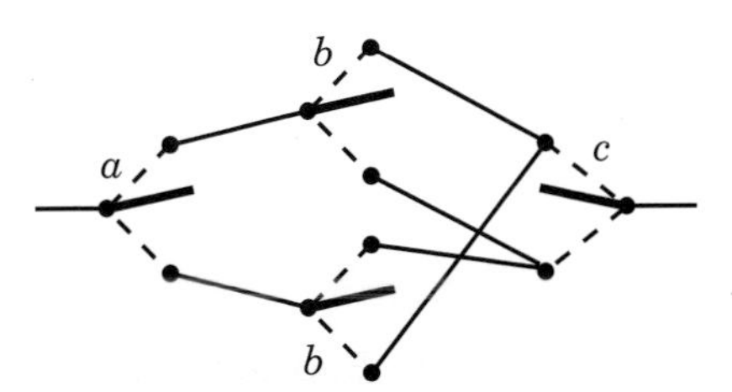

Draw up a table to show when current flows and when it does not, and describe what happens when the state of a switch is changed.

(OCR MEI Decision and Discrete D2 Specimen Paper)

5 For the following circuit find a Boolean expression and construct the truth table.

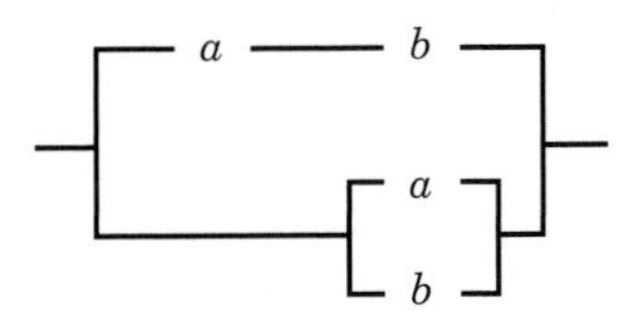

6 Construct circuits for the following Boolean expressions.

a) $(a \wedge \sim b) \vee (\sim a \wedge b)$
b) $(a \wedge a) \vee (a \wedge b) \vee (a \wedge c) \vee (b \wedge c)$

7 **a)** Draw the circuit represented by the Boolean expression:

$[(a \vee b) \wedge (c \vee \sim a)] \vee [b \wedge (\sim c \vee b)]$

b) Show that the circuit can be simplified to

$b \vee (a \wedge c)$

8 Let p represent 'I work hard' and q 'I pass my exams'.

a) Express in terms of p and q the proposition 'I work hard so I will pass my exams'.
b) Interpret the proposition $\sim p \wedge q$.
c) If r represents 'I take regular exercise', interpret the proposition: $(p \wedge \sim r) \Rightarrow \sim q$.

9 Construct truth tables for the following propositions.

a) $\sim(p \Rightarrow (q \Rightarrow p))$ **b)** $p \Rightarrow (p \wedge \sim(q \wedge r))$

10 **a)** Show, by constructing truth tables or otherwise, that the following statements are equivalent.
$p \Rightarrow q$ and $\sim(\sim(p \wedge q) \wedge p)$
b) With the aid of **a)**, or otherwise, construct combinatorial circuits consisting only of NAND gates to represent the following functions.
$f(x, y) = x \Rightarrow y$ and $g(x, y) = \sim(x \Leftrightarrow y)$

(AEB Question 6, Specimen Paper, 1996)

Summary

After working through this chapter you should

- *understand the ideas of Boolean logic, know the notation and be able to construct simple truth tables*
- *know how the ideas of Boolean algebra can be applied to electronic circuits*
- *know the symbols for AND, OR, NOT, NOR and NAND gates and be able to construct tables of outputs for two inputs.*

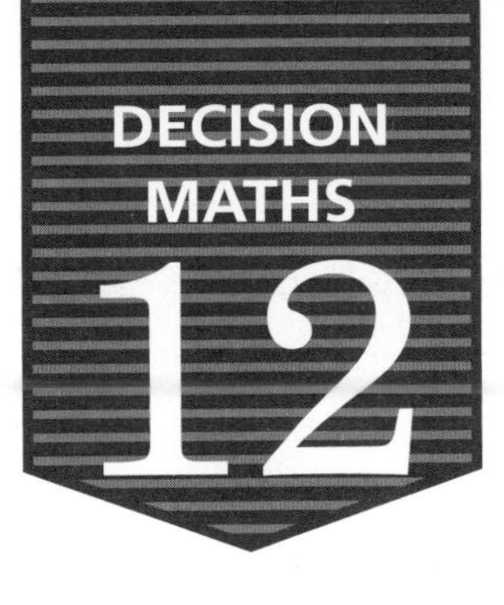

Network flows

By the end of this chapter you should be able to:

- *find out how to model flows with networks and use the associated vocabulary*
- *find the maximum flow using the labelling procedure*
- *use the maximum flow – minimum cut theorem.*

Exploration 12.1

Grid lock!

The centre of a small town suffers from traffic congestion at rush hour. The local council decides to introduce a one-way system to try to solve the problem, so the planning officer develops some plans. Her plan for part of the town centre is shown below; the arrows indicate the direction of traffic flow and the figures give the maximum number of vehicles per hour that can pass along each road.

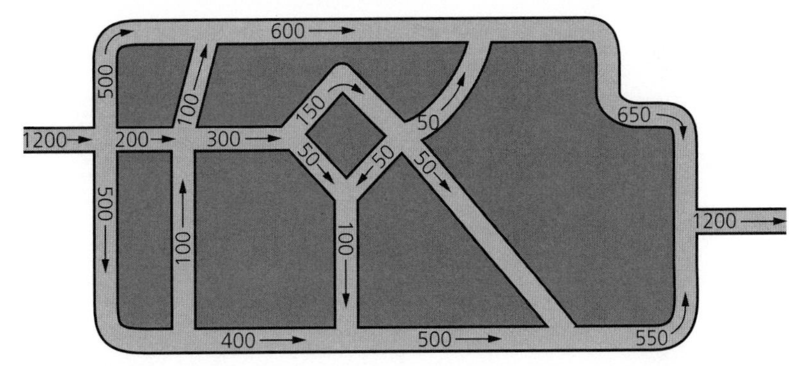

A survey has shown that at peak times this section of the town centre has a flow of 1200 vehicles per hour. Will the planning officer's design cope with this amount of traffic?

In this chapter, we shall consider this problem.

NETWORK FLOWS

Networks can be used to model situations involving the flow of a substance or single items, such as water or oil in a pipe, traffic in a road system, electricity through a circuit. In these situations the edges of the network represent the routes along which the items can flow.

In most of the networks we have met so far, we have been able to travel in either direction along an edge. When modelling real situations this is not always the case. Sometimes the movement can only be in one direction. This is usually the case when we are dealing with flows such as water in a pipe or traffic in a one-way system. Then we can use a **directed network** to model the situation.

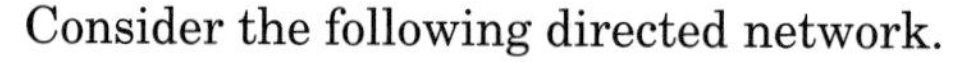

Consider the following directed network.

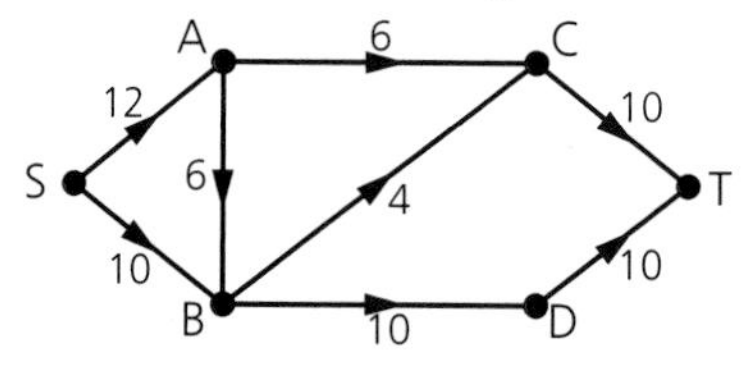

An edge which is directed is called an **arc**; arrows on the arcs indicate the direction of flow. Thus we would refer to the arc AC as being the route along which the substance flows with direction from A to C.

A flow must go from one point to another so a network modelling a flow must have:

- a starting point – called the **source**, denoted by S
- a finishing point – called the **sink**, denoted by T

and since there is usually a limit on the magnitude of the flow in any arc, the weights on the arcs show the **maximum capacity** of that arc.

Thus in this network arc SA has a maximum capacity of 12, arc AB has a maximum capacity of 6 and so on.

The flow along an arc does not have to be maximum, but when it is we say the arc is **saturated**.

A maximum flow is the largest possible flow through a network from S to T. An example of a possible flow through the network above is given by this diagram.

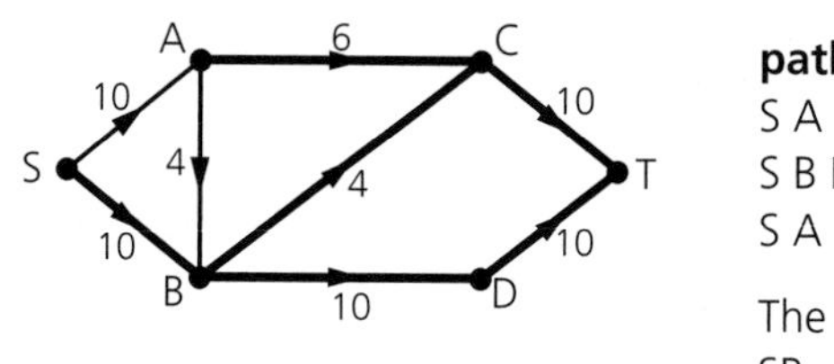

path	flow
S A C T	6
S B D T	10
S A B C T	4

The saturated arcs are SB, AC, BC, BD, CT, DT.

EXERCISES

12.1 CLASSWORK

1 Suggest what these directed networks could represent.

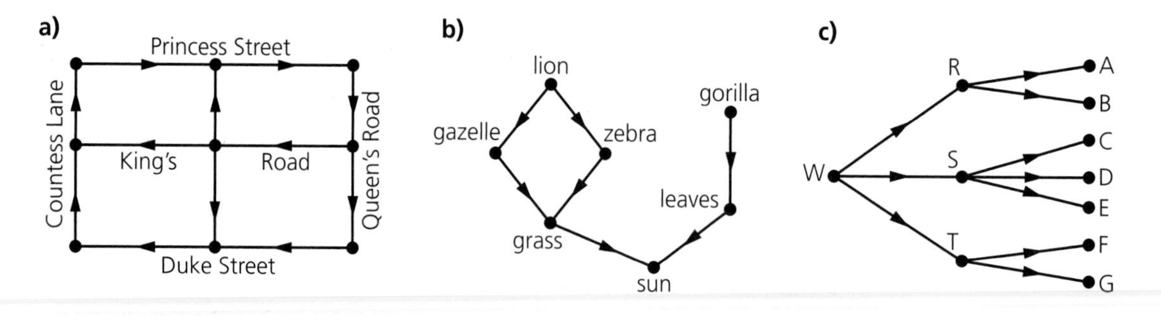

2 State, with reasons, which two of the following directed graphs are isomorphic.

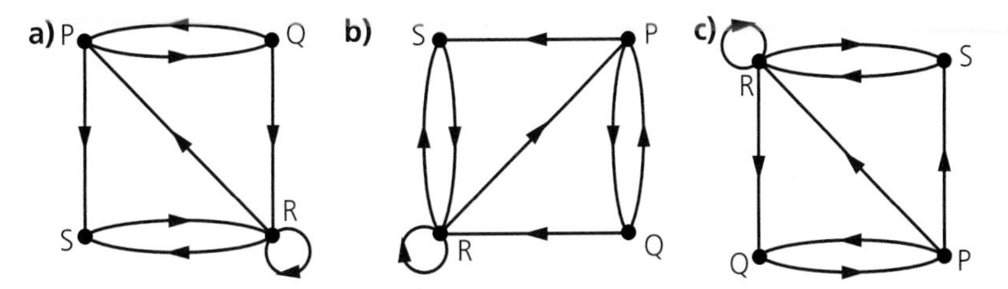

3 Draw the directed network that is represented by this matrix.

From

To		S	A	B	C	T
	S	—	—	—	—	—
	A	4	—	—	2	—
	B	3	—	—	—	—
	C	5	—	2	—	—
	T	—	6	2	8	—

4 Find a possible flow from S to T through this network. Is it a maximum flow? Justify your answer by considering the arcs which are saturated.

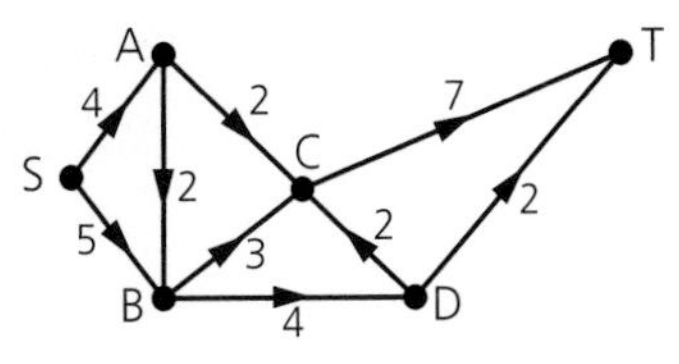

5 Find, by inspection, the maximum flow from S to T in this network.

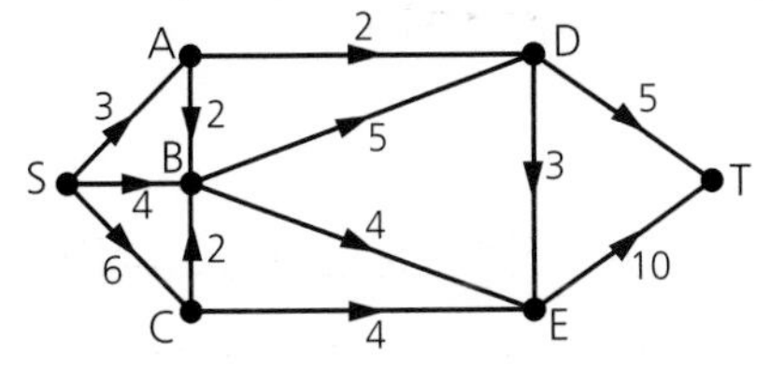

State the arcs in your solution which are saturated.

EXERCISES

12.1 HOMEWORK

1 Draw a directed network for the following situations.

a) In a squash ladder, A had beaten B and D, B had beaten C, C had beaten A and D, D had beaten B.

b) A talks to B, C and F. C talks to A, E and D. E talks to F, F talks to E, and A and D talk to C.

c) Communication links between five centres are: a can pass information to d, b to c and d, c to a, d to e and e to a and b.

Express each of the above in a matrix representing an arc between two vertices by 1, and by 0 if no directed edge exists.

2 Use a tree diagram to find all the paths from S to T in the directed graph below.

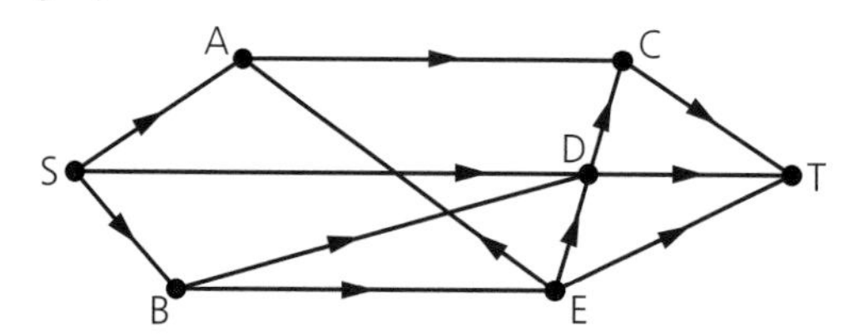

3 Draw a directed network represented by this matrix.

From

To		S	A	B	C	D	E	J
	S	–	–	–	–	–	–	–
	A	5	–	–	–	–	4	–
To	**B**	6	–	–	–	–	–	–
	C	–	10	–	–	5	–	–
	D	4	5	4	–	–	5	–
	E	–	–	8	–	–	–	–
	J	–	–	–	7	3	4	–

4 Which of the following directed graphs are isomorphic? Justify your conclusions.

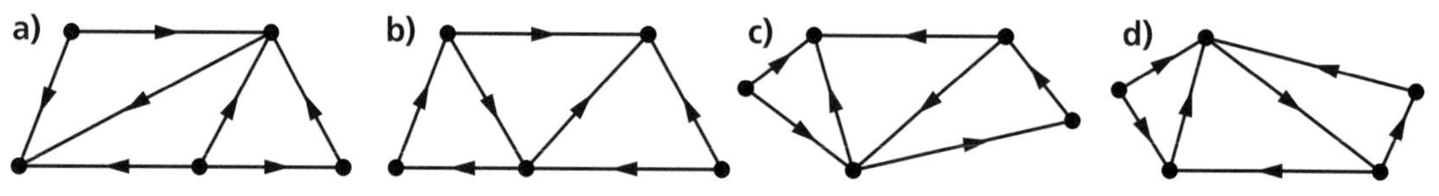

5 Consider the network below in which each arc is labelled with its capacity. Find a flow of value 7 from S to T. Draw a diagram with the flows in each arc.

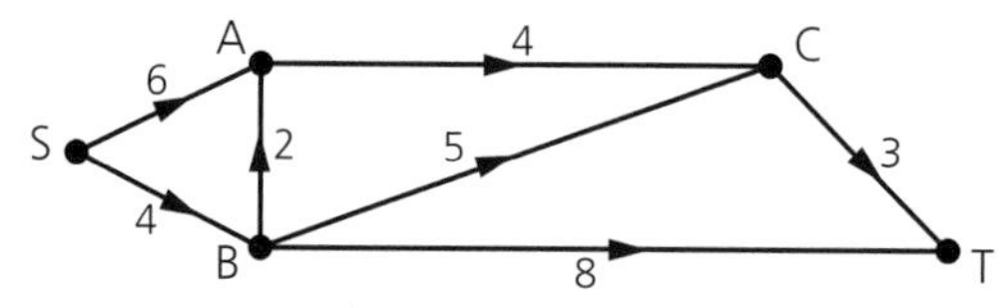

LABELLING PROCEDURE

To solve a problem such as the one in Exploration 12.1, we need a systematic way of ensuring that the flow we have found is the best one possible. The basic idea when looking for a maximum flow is to find a flow by inspection, then to increase its value step by step until we cannot increase it any further. We shall develop such a method using this directed network.

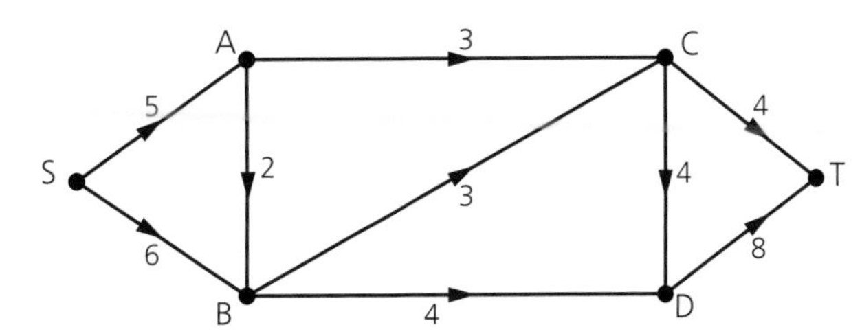

The first step is to look for a **feasible flow** through the network. We do this by finding routes through the network from S to T, for example SACT, and considering the capacities of the arcs along this route.

- SA has capacity 5.
- AC has capacity 3.
- CT has capacity 4.

So the maximum flow along SACT is 3, that is the capacity of the arc with the minimum capacity.

Arrows in the direction S to T indicate flow is still available in that arc.

Arc AC is now saturated, but arc SA and CT still have some capacity available, which we call the **excess capacity**. We can show this on the network, by means of arrows alongside the arcs showing their excess capacity.

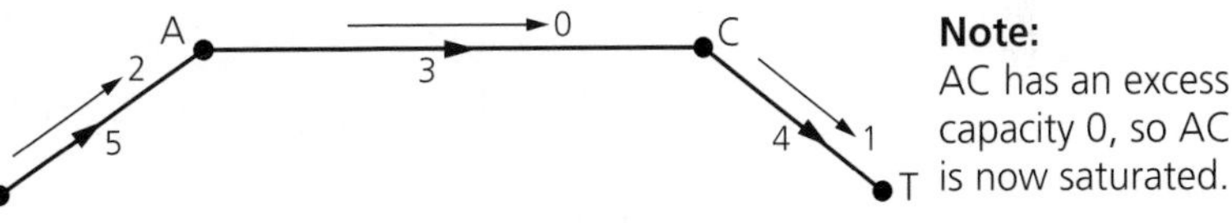

It is also useful to record the flow that we are sending down each arc by means of arrows directed in the opposite direction.

Arrows in the direction T to S indicate flow so far along that arc.

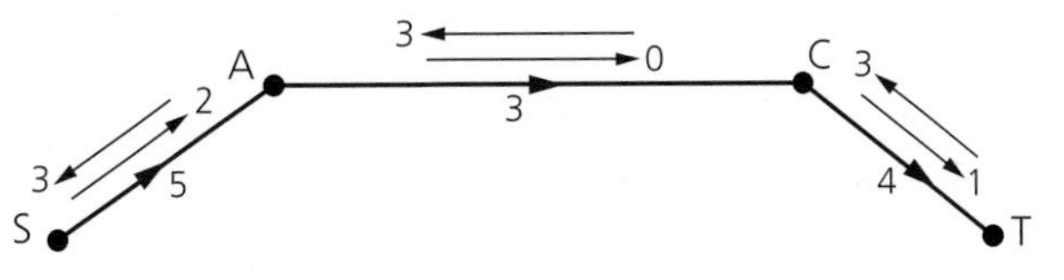

These values are called **artificial backward capacities**. As well as recording the flow in that arc, they also offer a means of altering the flow in that arc if, at some stage, we see a better solution and wish to reduce the flow along that arc and direct it along a different arc.

It is unlikely that our initial feasible flow would consist of just one path; we should normally start with a flow that takes in two or three of the most direct routes from S to T, for example an initial flow in this network might be like this.

Step 1:

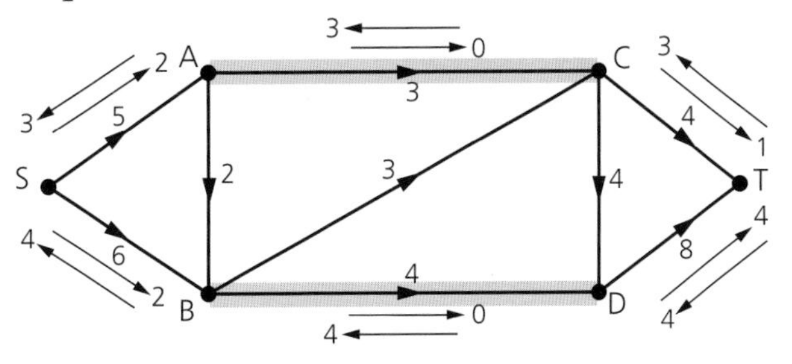

We write this as:

Path	Flow	Saturated arcs	
SACT	3	AC	
SBDT	4	BD	Total flow 7

giving an initial flow of 7.

Notice that the saturated arcs have been recorded both on the network, by listing them, and on the diagram, by marking them in a coloured highlighter. It is useful to do this as you perform each step, because it helps to decide when you have an optimal solution.

Now we need to try to improve on this flow if possible. We do this by looking for other paths from S to T which consist entirely of unsaturated arcs; these are called **flow augmenting paths**. One such is SBCT which has a capacity of 1.

Step 2:

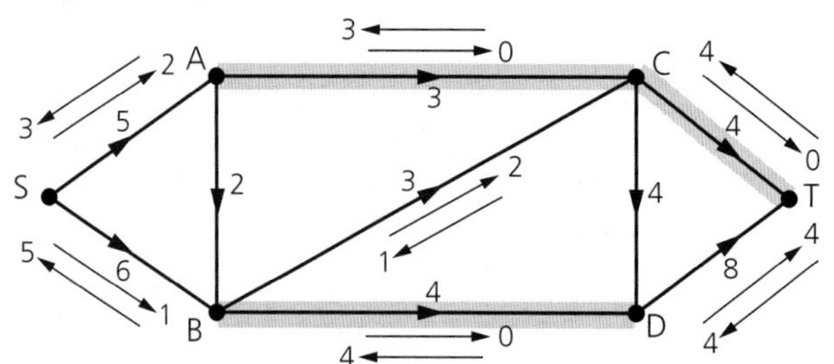

We add this to our network, amending the figures for the artificial backward capacities and excess flows. Arc CT is now saturated.

Path	Flow	Saturated arcs	
SACT	3	AC	
SBDT	4	BD	
SBCT	1	CT	Total flow 8

We repeat this process until there are no possible improvements. The next path could be SBCDT which has a maximum flow of 1.

Step 3:

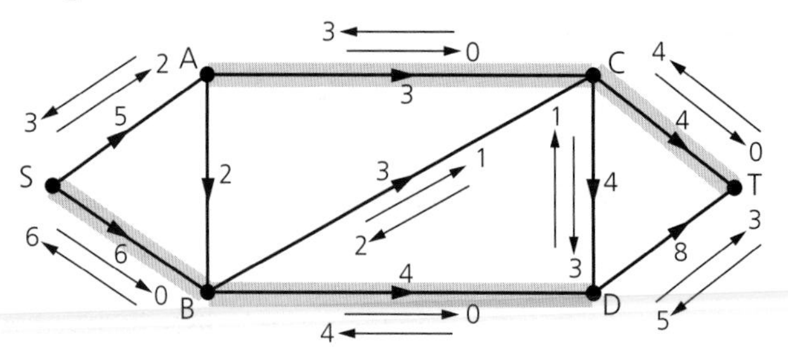

Path	Flow	Saturated arcs	
SACT	3	AC	
SBDT	4	BD	
SBCT	1	CT	
SBCDT	1	SB	Total flow 9

Then we take SABCDT which also has a flow of 1 available.

Step 4:

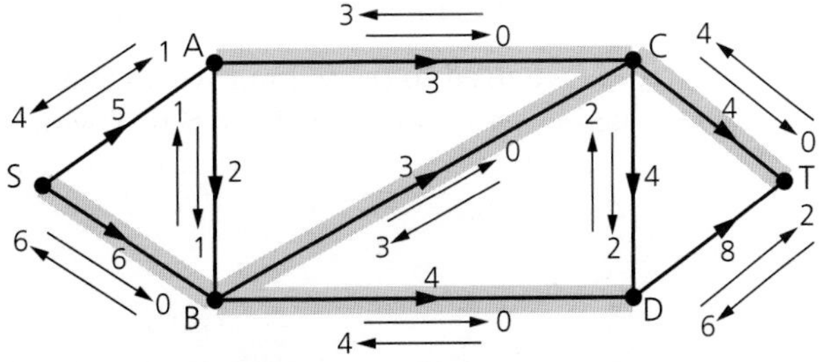

Path	Flow	Saturated arcs	
SACT	3	AC	
SBDT	4	BD	
SBCT	1	CT	
SBCDT	1	SB	
SABCDT	1	BC	Total flow 10

There are now no other paths through the network from S to T; every way we try is 'blocked' by saturated arcs, so our final solution is a flow of 10, as illustrated in this diagram.

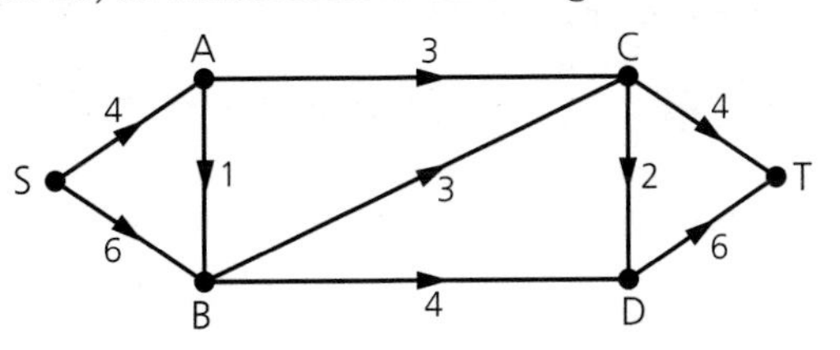

Hint: It is useful to use different colours to mark the labels on your network.

With practice, you do not need to draw a separate network at every stage. If you start with a large diagram and work neatly, you can show all your working on that diagram, listing the paths clearly underneath.

We can demonstrate this technique by finding a solution to the problem set in Exploration 12.1. First we need to represent the proposed one-way system as a directed network, where the arcs represent the roads and the vertices the road junctions, as shown in the diagram overleaf. Then we follow the steps as listed above, showing all the flow paths on one network, and listing the working underneath.

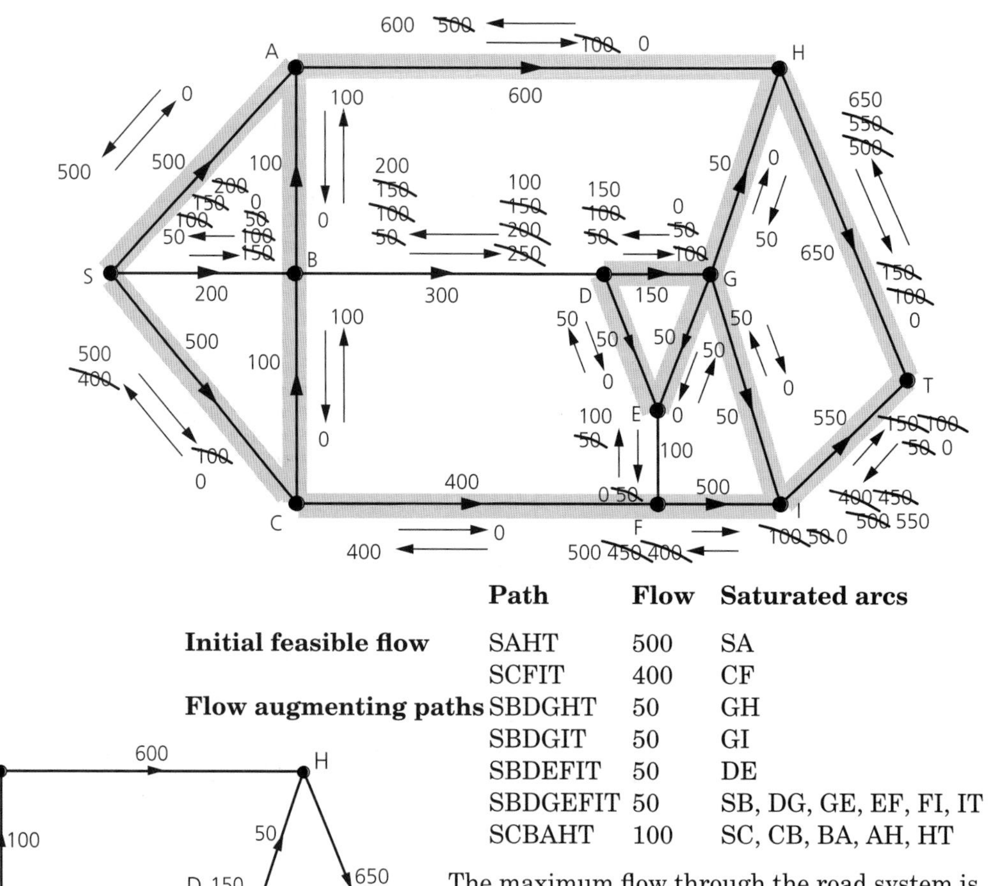

	Path	Flow	Saturated arcs
Initial feasible flow	SAHT	500	SA
	SCFIT	400	CF
Flow augmenting paths	SBDGHT	50	GH
	SBDGIT	50	GI
	SBDEFIT	50	DE
	SBDGEFIT	50	SB, DG, GE, EF, FI, IT
	SCBAHT	100	SC, CB, BA, AH, HT

The maximum flow through the road system is 1200 (so the planning officer was correct).

This diagram, which shows the final flows along each arc in our network, is a way of displaying the solution clearly on a diagram.

Exploration 12.2 *Networks with many sources and sinks*

Two reservoirs, R_1 and R_2, supply water to two towns, T_1 and T_2, via three pumping stations, A, B and C as shown in the network. The numbers represent the amount of water which can be pumped from station to station, in thousands of gallons per hour.

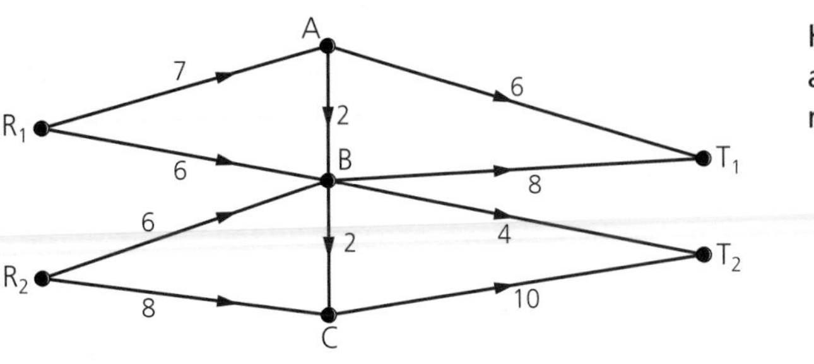

How can the labelling procedure be adapted to deal with a situation with more than one source and sink?

Supersource and supersink

The solution to the problem in Exploration 12.2 is to introduce a **supersource** S which feeds into R_1 and R_2. Since there are thirteen units of flow leaving R_1, the arc SR_1 must have a capacity of 13. Then the arc SR_2 will have a capacity of 14 units, to supply the arcs leaving R_2. This can be seen in the diagram.

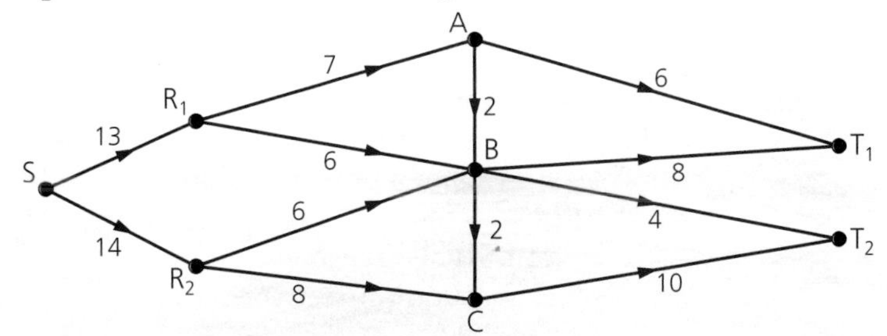

We deal with two sinks, T_1 and T_2, in a similar way by introducing a **supersink**, T, into which T_1 and T_2 feed. Again the arcs T_1T and T_2T must have sufficient capacity to allow all the flow entering T_1 and T_2 to leave.

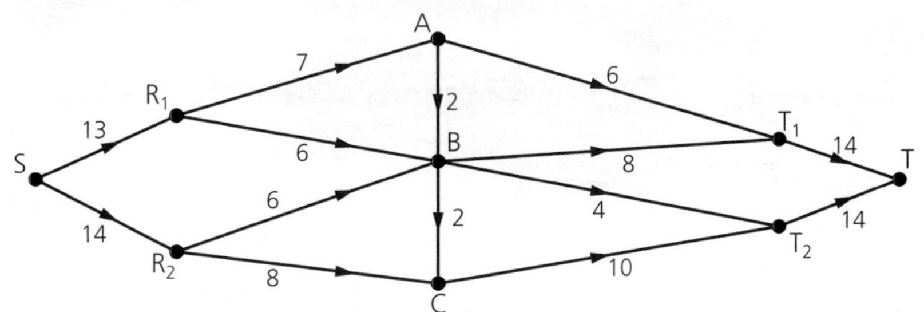

Once the supersource and supersink have been introduced, the flow through the network can be calculated using the labelling procedure.

EXERCISES

1 Use the labelling procedure to find the maximum flow from S to T in this network.

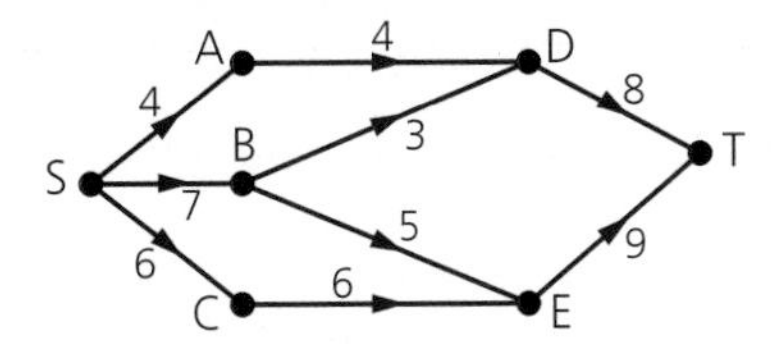

2 Six towns are connected by a local rail network.

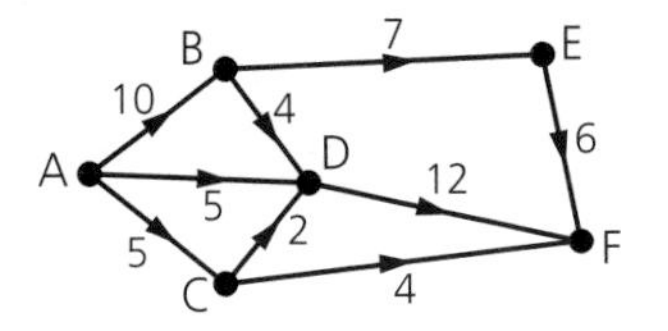

The network shows the maximum volumes of freight (in thousands of tonnes per week) that can be carried between the towns.

a) A large warehouse at A needs to supply 15 tonnes of goods each week to a retail outlet at F. Find a flow through the network that will make this possible.

b) High demand for the product means that the retailer wishes the supply to be as large as possible. Using your answer to part **a)** as an initial flow find the maximum tonnage of goods which the warehouse can supply each week.

c) The lines from A to C and C to D are upgraded and can now carry 8 thousand and 6 thousand tonnes per week respectively. How does this affect the maximum flow from A to F? Which one other line would you choose to upgrade to increase the flow still further?

3 The network shown in the diagram has a possible 26 units of flow leaving S and a possible 26 units of flow entering T. Explain why it is not possible to achieve a flow of 26 through the network from S to T.
Find the maximum flow through the network.

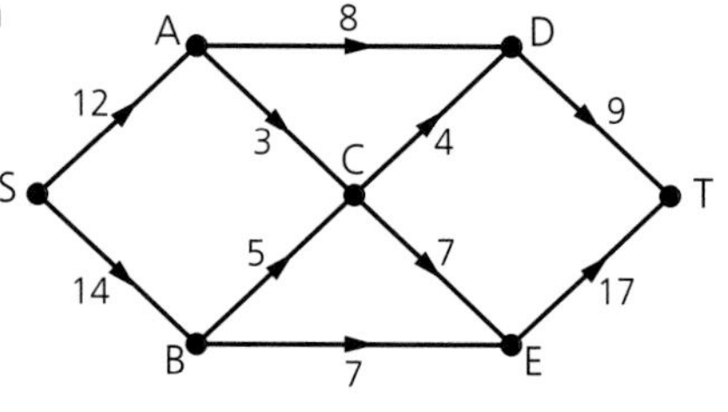

4 Calculate the maximum amount of water which can be supplied to the two villages in Exploration 12.2.

5 The network below shows the flow rates for traffic in a one-way system; the figures give maximum numbers of vehicles per hour (in hundreds). Find the maximum flow through this network.

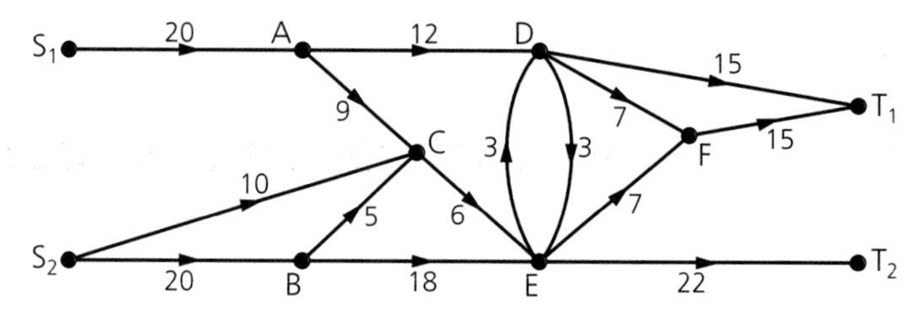

If roadworks at junction F restricted the flow from F to T_1 to 800 vehicles per hour, would this affect your maximum flow?

EXERCISES

12.2 HOMEWORK

1 Use the labelling procedure to find the maximum flow from S to T in the network illustrated in question 5 of Exercises 12.1 CLASSWORK. State the saturated arc in this case.

2 Repeat the procedure of the question above to find the maximum flow from S to T in the network you constructed in question 3 of Exercises 12.1 CLASSWORK, stating which arcs are saturated.

3 An exporter in Southampton wishes to ship containers of a product to an importer in Taiwan. There are several channels through which the containers may be sent, as shown in this network. The number on each arc represents the maximum load which each channel can handle. Determine the maximum number of containers that can be sent through the network.

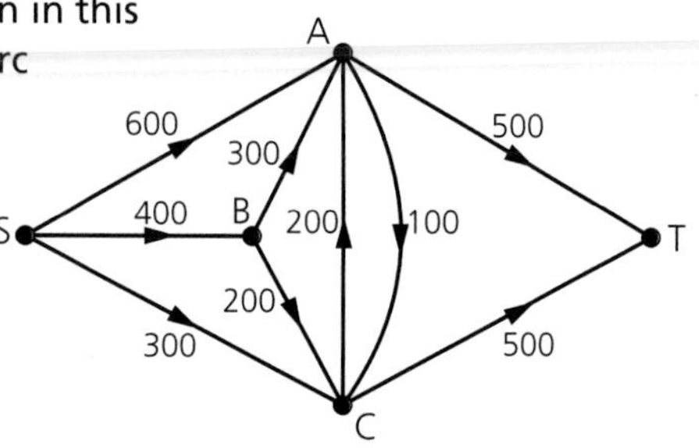

4 **a)** Natural gas is produced at two sources S_1 and S_2 and transported by a network of underwater pipelines to a refinery T. The maximum capacity of each pipe, in appropriate units, is given in the network. Determine the maximum flow in the network.

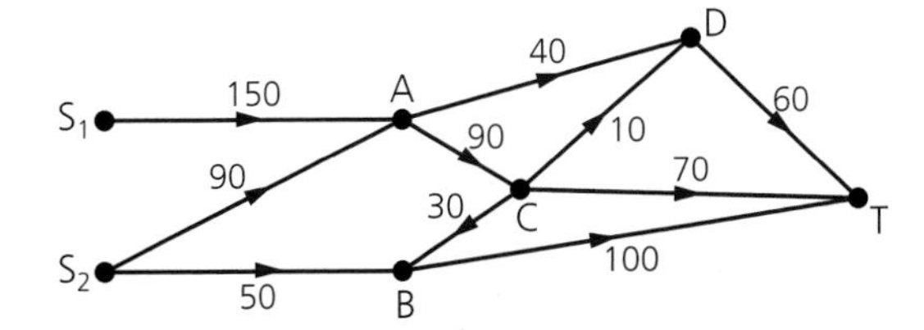

b) Due to the volatile nature of natural gas, there must be a minimum flow through each pipeline. If the figures above now represent the minimum capacity that each pipe must carry, devise a method by which the minimum flow may be found in a such a way that the given minimum capacities are satisfied.

5 Consider question 5 of Exercises 12.2 CLASSWORK. The roadworks now restrict the flow of traffic at junction E to 1500 vehicles, whereas the roadworks from F to T_1 have now been removed.

Redraw the network so that the vertex E is replaced by an arc of capacity 15 from E_1 to E_2 and all arcs directed towards E in the original network are directed towards E_1 and all arcs directed away from E in the original network are directed way from E_2 in the new network.

How is your maximum flow affected in this case?

THE MAXIMUM FLOW – MINIMUM CUT THEOREM

In the previous section we developed a method for finding the maximum flow through a directed network. Now we need to know how we can be sure that the flow we have found is the maximum possible, particularly if we have a large network. We can get some indication of the solution to this problem by considering question 3 in Exercises 12.2 CLASSWORK. You were asked to give a reason why the maximum flow in a network is not 26 even though it is possible for 26 units to flow from S and it is also possible for 26 units to flow into T.

We can start by considering a simple network with a bottleneck.

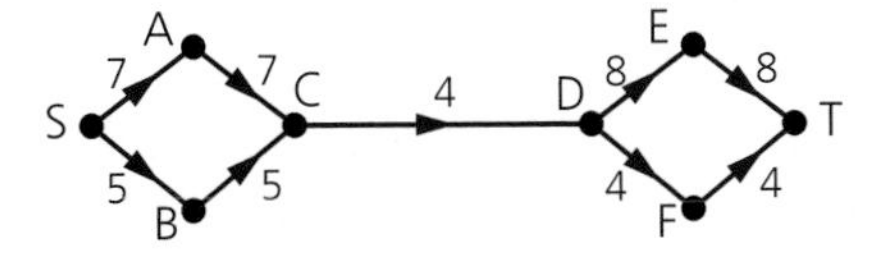

In this network we can clearly see that although twelve units can flow from S and twelve units can flow into T, the maximum flow through the network is only four because all flow must pass through arc CD.

We can make this idea more precise by introducing the idea of a **cut**, which is a line dividing the network into two parts, one part containing the source, S, and the other containing the sink T. The capacity of the cut is the sum of the capacities of the arcs it crosses, providing those arcs are directed from S to T.

Thus we can take a cut through SA and SB, marked C_1 on the diagram below. The capacity of this cut is $7 + 5 = 12$. We can certainly say that the maximum flow cannot exceed the capacity of this cut.

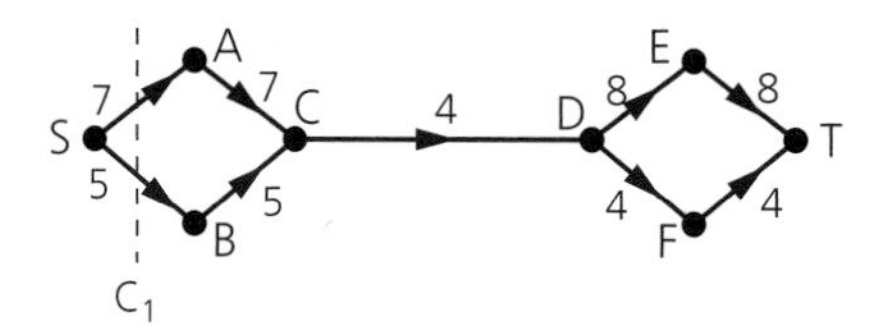

The table below describes some of the other possible cuts in this network.

Cut	Arcs in cut	Vertices to left of cut	Vertices to right of cut	Capacity of cut
C_1	SA, SB	S	A, B, C, D, E, F, T	7 + 5 = 12
C_2	SA, BC	S, B	A, C, D, E, F, T	7 + 5 = 12
C_3	AC, SB	S, A	B, C, D, E, F, T	7 + 5 = 12
C_4	AC, BC	S, A, B	C, D, E, F, T	7 + 5 = 12
C_5	CD	S, A, B, C	D, E, F, T	4 ← minimum
C_6	DE, EF	S, A, B, C, D	E, F, T	8 + 4 = 12
C_7	DE, FT	S, A, B, C, D, F	E, T	8 + 4 = 12
C_8	ET, DF	S, A, B, C, D, E	F, T	8 + 4 = 12
C_9	ET, FT	S, A, B, C, D, E, F	T	8 + 4 = 12

If we consider the other cuts through the network we can see that the cut through CD (C_5 in the table) has a capacity of 4. This is obviously the smallest capacity of any cut through this network and is called the **minimum cut**. We can also see that the maximum flow cannot exceed the value of this cut since it separates S from T and so all flow must go through the arcs in the cut. Hence it follows that:

maximum flow ≤ capacity of any cut

Since this is true for any cut, the maximum flow can never be greater than the capacity of the minimum cut.

This rule holds for any network. It is called the **Maximum flow – minimum cut theorem** and can be formally stated as:

maximum flow = value of minimum cut

If we return to the network in question 3 of Exercises 12.2 CLASSWORK, we can quickly deduce that, since the total flow out of vertices A and B is less than 26, this cannot be a feasible flow. Then we can use the Maximum flow – minimum cut theorem to find a value for the maximum flow and hence confirm the solution gained, using the labelling procedure.

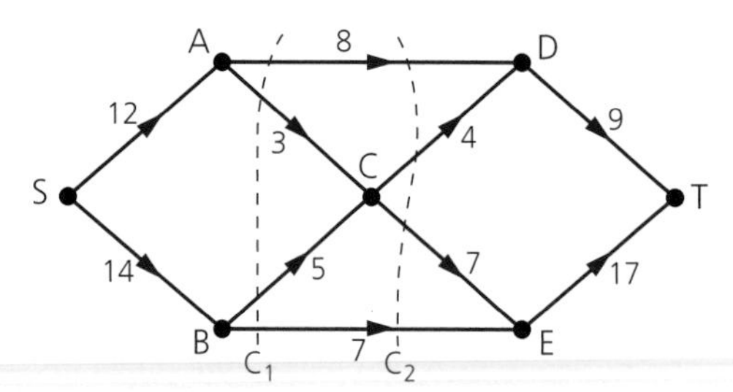

It is very tedious (and difficult) to list every cut through the network, so we use inspection to check the obvious choices for minimum value. Two such cuts are shown on the diagram.

C_1: arcs in cut: AD, AC, BC, BE capacity: 8 + 3 + 5 + 7 = 23

C_2: arcs in cut: AD, CD, CE, BE capacity: 8 + 4 + 7 + 7 = 26

Check: Was this the answer you obtained for question 3 in the exercises?

A quick inspection of the network will tell us that C_1 is the minimum cut with a capacity of 23, thus the maximum flow through the network is 23 units.

One final point to note is that in our original definition of a cut we specified that the arcs must be directed from S to T. We must therefore consider what happens in a case like the one here, where a cut crosses an arc which is directed from T to S.

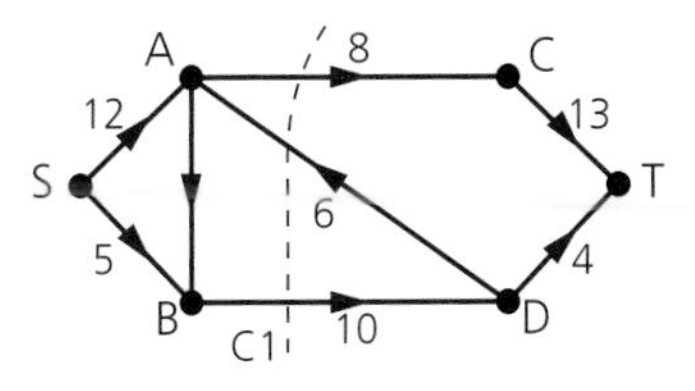

In this case the capacity of the cut through AC, AD and BD in the direction S to T is:

$8 + 0 + 10 = 18$

In other words, an arc directed from T to S does not contribute towards the value of the cut.

EXERCISES

12.3 CLASSWORK

1 – 4 Return to questions 1, 2, 4 and 5 in Exercises 12.2 CLASSWORK. Use the Maximum flow – minimum cut theorem to confirm that the flows you obtained were maximum flows.

5 Use the labelling procedure to find a maximum flow in this network.

Prove that the flow is a maximum by finding a cut of the same value.

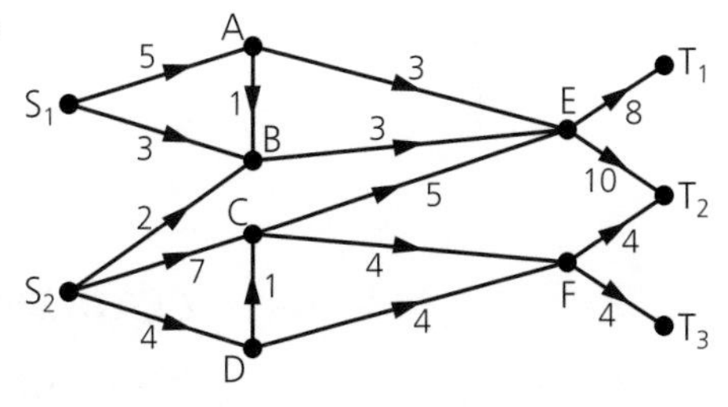

EXERCISES

12.3 HOMEWORK

1 **a)** In an analysis of a given basic network, a flow with value 10 is found. What can be deduced about the capacity of a minimum cut?

b) If a cut with capacity 10 is found in a basic network, what can be deduced about the value of the maximum flow?

c) If a cut with capacity 10 and a flow of value 7 are found in a basic network, what can be deduced about the value of the maximum flow?

d) If a cut with capacity 10 and a flow with value 10 is found in a basic network, what can be deduced about the value of the maximum flow?

e) If you calculate a flow with value 10 and a cut with capacity 7, what can you deduce?

2 The network below has eight cuts. Draw up a table listing the cuts and hence determine the minimum cut.

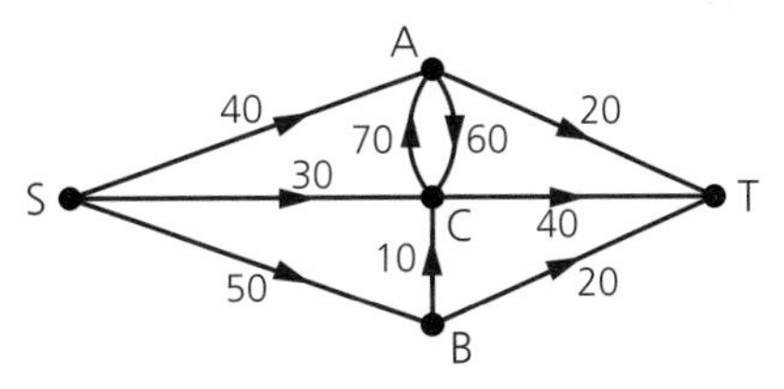

3 Use the Maximum flow – minimum cut theorem to verify that your answers to questions 1–4 in Exercises 12.2 HOMEWORK were correct.

4 A gas pipeline network is shown on the right. The left-hand figure on each arc is the minimum allowable flow and the maximum allowable flow is the right-hand figure. The problem is to determine, if it exists, the maximum flow such that the flow along each arc is not less than the lower capacity.

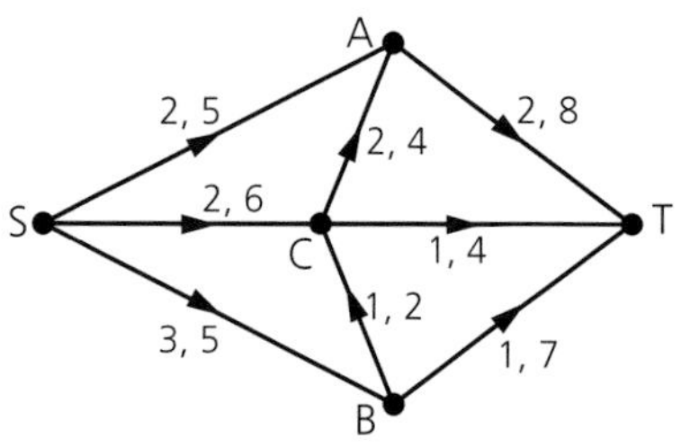

a) The first step is to find a flow which satisfies the maximum and minimum capacity conditions.
b) Find the maximum flow using the flow augmented method.
c) Find a cut through the network which proves your solution to **b)**.

5 Study these networks and decide whether a flow can exist in each of them.

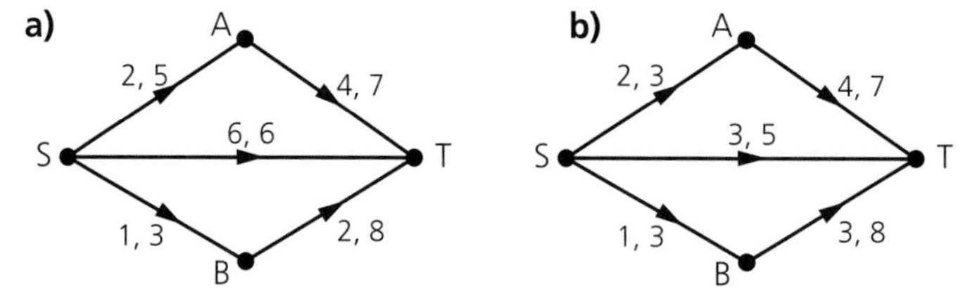

COURSEWORK INVESTIGATION

Investigation

Bottlenecks, a flow problem

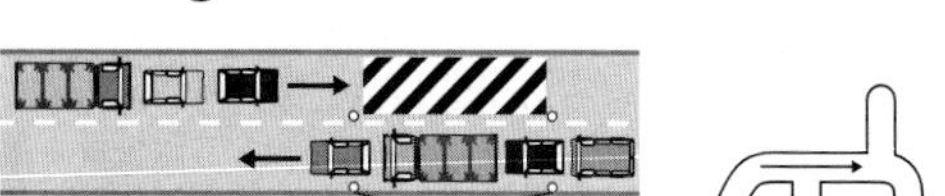

diversion

roadworks

Roadworks often cause bottlenecks by narrowing a road and restricting the flow of traffic. Use Maximum flow algorithms to model the effect of roadworks on roads in your area. Consider whether it may be better to divert some, or all, of the traffic. How could the introduction of temporary one-way systems alleviate the situation?

CONSOLIDATION EXERCISES FOR CHAPTER 12

1 **a)** If you can find a cut with capacity 9 in a network, what can you deduce about the maximum flow?
b) If you can find a flow of value 15 in a network, what can you deduce about the minimum cut?
c) If you have a flow of value 28 and a cut of capacity 35, what can you deduce?

2 **a)** Find the minimum cuts for these networks.

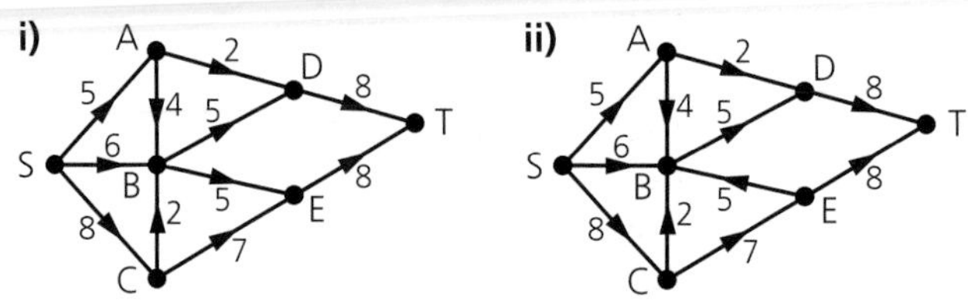

b) Confirm your cuts are minimum cuts by finding a flow of the same value in each network.

c) 'Every minimum cut in a network consists entirely of saturated arcs.' Is this statement correct? Give reasons for your answer.

3 A manufacturing company has factories F_1 and F_2 and wishes to transport its products to three warehouses W_1, W_2 and W_3.The capacities of the possible routes, in lorry loads per day, are shown in the figure.

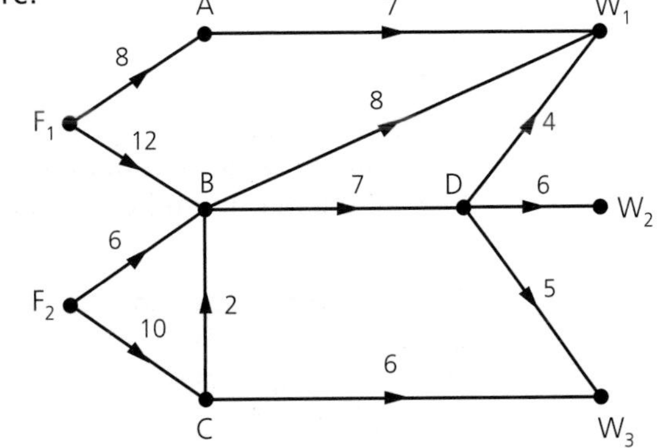

a) Add a supersource F and a supersink W to obtain a single-source, single-sink capicitated network. State the capacities of the arcs you have added.

b) Use the labelling procedure to obtain a maximal flow through the network.

c) Interpret your final flow pattern giving:

i) the number of lorry loads leaving F_1 and F_2,

ii) the number of lorry loads reaching W_1, W_2 and W_3,

iii) the number of lorry loads passing through B each day.

(Edexcel Decision Mathematics D1 Specimen Paper 2000)

4 The diagram shows a pipe network. The numbers on the arcs give the maximum capacities of the pipes.

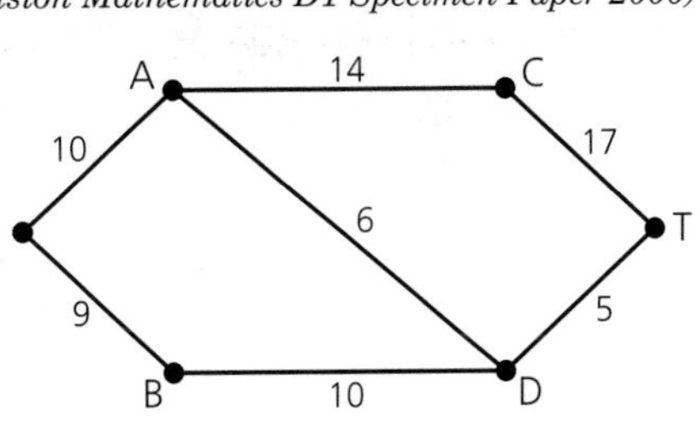

a) Use a labelling algorithm to find the maximum flow from the source S to the sink T. Show and describe the steps of your algorithm.

b) Find a cut to prove that your flow is maximal. Explain why it proves that the flow is maximal.

(UODLE Question 8, Paper 4, 1995)

5 The diagram shows a capacitated network representing traffic flow. The numbers on each of the arcs indicate the capacity in hundreds of cars per hour of the arcs (roads). The ringed numbers on the arcs indicate the flow along that arc for a particular flow *f*.

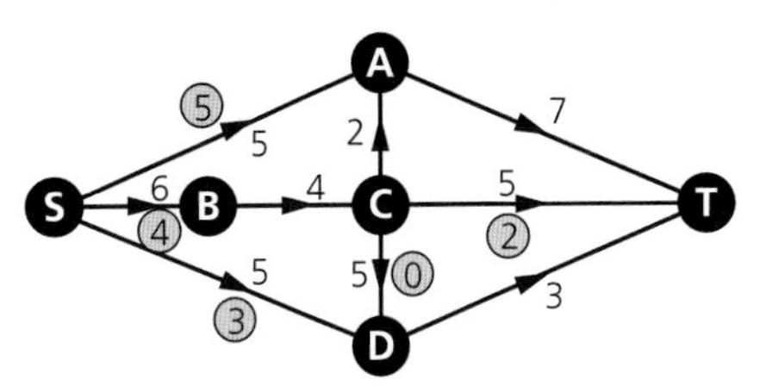

a) Copy and complete the figure by finding the flows along the remaining arcs.

b) Hence determine the value of the flow f.
c) Show that this flow is a maximum flow by finding a cut whose capacity has the same value.

(ULEAC Question 1, Specimen Paper, 1996)

6 The diagram shows a directed flow network with two numbers on each arc. The uncircled numbers show the capacity of each arc. The circled numbers are the flows currently passing through the network.

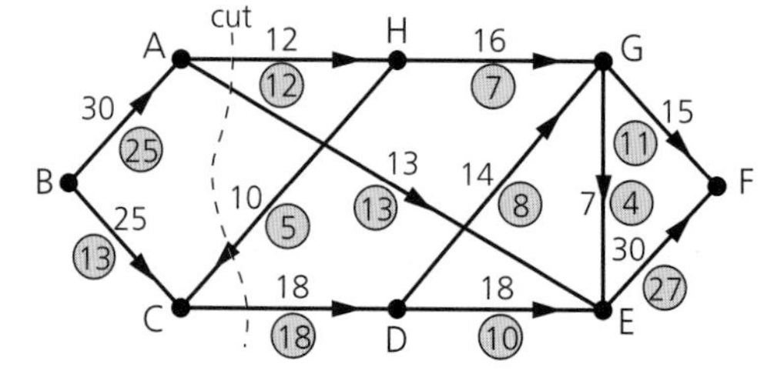

a) Identify the source and the sink.
b) What is meant by a cut? What is the flow across the given cut? Why is this not a maximum flow?
c) Find and indicate one flow-augmenting path, and use it to augment the flow as far as possible.
d) Find a maximal flow.
e) Explain how you know that the flow which you quoted in part **d)** is maximal.

(UODLE Question 10, Paper 4, 1990)

7 The network below shows the maximum rates of flow (in vehicles per hour) between towns S, A, B, C, D and T in the direction from S to T.

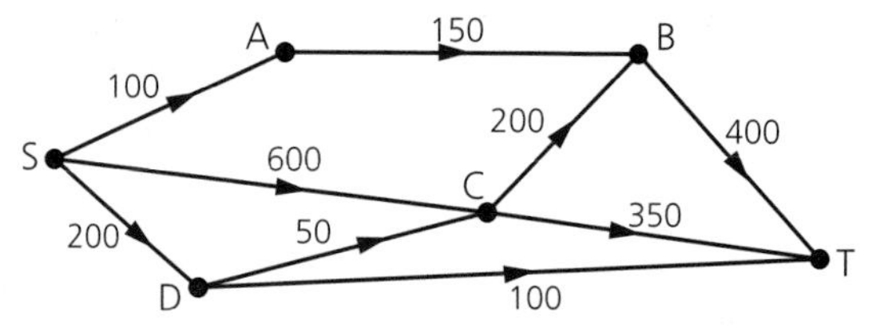

a) By choosing a minimum cut, or otherwise, find the maximum traffic flow from S to T. Give the actual rates of flow in each of the edges BT, CT and DT when this maximum flow occurs.
b) When a maximum flow occurs from S to T, how many of those vehicles per hour pass through C?
c) It is decided to reduce the traffic flow through C (in the direction from S to T) to a maximum of 480 vehicles per hour. In order to maintain the same maximum flow from S to T the capacity of a single edge is to be increased. Which edge should be chosen, and by how much must its capacity be increased?

(AEB Question 4, Specimen Paper, 1996)

8 The diagram shows a gas distribution network consisting of:

- three supply points A, B, C
- three intermediate pumping stations, P, Q, R
- two delivery points, X, Y
- connecting pipes.

The figures on the arcs are measures of the amounts of gas which may be passed through each pipe per day. The figures by A, B and C are measures of the daily availability of gas at the supply points.

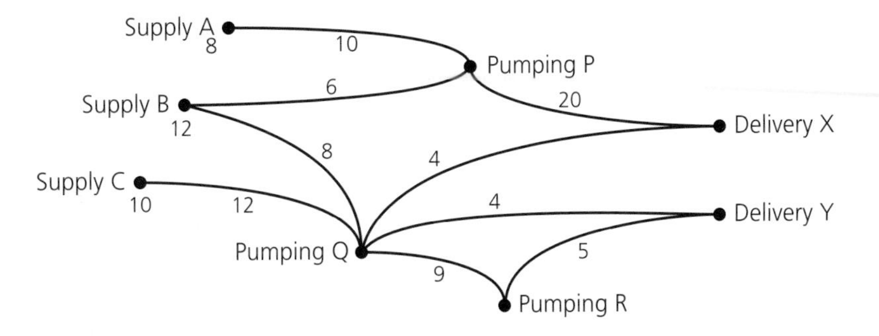

a) Copy the network, introducing a single source with links to A, B and C, the capacities on the links reflecting the supply availabilities.
b) Introduce a single sink linked to X and Y, the links having large capacities.
c) Use inspection to find the maximal daily flow through your network, making a list of the flows through each pipe.
d) Find a suitable cut to prove that your flow in part **c)** is maximal.
e) Interpret the flows in the new links, from the source and to the sink.
f) A new pipeline is proposed with capacity 5 units per day, connecting P and Q. Demonstrate the use of a labelling procedure to augment the flow, and thus find the new maximal flow.
g) R is now to become a delivery point. Explain how to adapt the approach of part **b)** and find the maximal daily flow of gas in total that can be delivered to R, X and Y. You do not need to calculate this flow.

(UODLE Question 19, Paper 4, 1993)

9 The two matrices represent a road network connecting seven towns. The first matrix gives the distances between the towns in kilometres. The second matrix gives the capacities of the roads – the maximum numbers of vehicles which can pass between towns in one hour (in thousands of vehicles).

	A	B	C	D	E	F	G
A	–	10	8	–	–	30	–
B	10	–	–	8	15	–	–
C	8	–	–	5	–	7	–
D	–	8	5	–	–	15	4
E	–	15	–	–	–	–	12
F	30	–	7	15	–	–	10
G	–	–	–	4	12	10	–

Distances

	A	B	C	D	E	F	G
A	–	3	3	–	–	2	–
B	3	–	–	6	2	–	–
C	3	–	–	2	–	4	–
D	–	6	2	–	–	2	1
E	–	2	–	–	–	–	2
F	2	–	4	2	–	–	2
G	–	–	–	1	2	2	–

Capacities

a) Draw the road network.
b) Find the maximum hourly flow of vehicles from B to F, showing how this may be achieved. *Prove* that this is a maximum.
c) Use, and demonstrate your use of, Dijkstra's algorithm to find the shortest route between B and F.
What percentage of maximum hourly flow of vehicles use the shortest route?

(UODLE Question 22, Paper 4, 1993)

10 The arcs in the flow network below represent oil pipelines with capacities in thousands of barrels of oil a day. Vertices 1, 2 and 3 are oil wells with production capacities of 150, 300 and 200 thousand barrels a day, respectively. Vertices 4, 5 and 6 are oil refineries which may process up to 200, 250, and 250 thousand barrels of oil a day, respectively. One barrel of crude oil processes to give one barrel of refined oil. Vertices 7, 8 and 9 are centres of demand which require 150, 100 and 300 thousand barrels of refined oil a day, respectively.

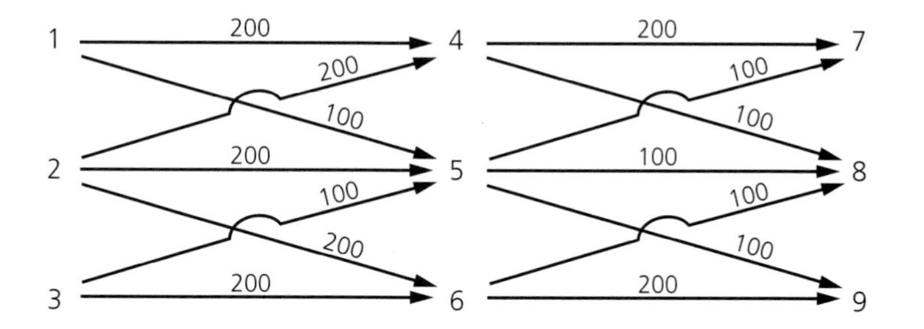

Modify the network, in a manner which you should explain, so that:
a) it has a single source
b) it has a single sink, and
c) capacities are associated only with arcs, not vertices.
Hence, by the application of a formal algorithm, find a feasible flow which satisfies the demands at 7, 8 and 9.

There is an accident at the refinery at 5 which reduces its production capacity to 50 thousand barrels a day. It is possible to increase, temporarily, production capacity at refineries at 4 and 6 to 300 thousand barrels a day. Show, by considering cut-sets, that there is no feasible flow through the network which satisfies all the demand, even with these temporary increases.

(UODLE Question 22, Paper 4, 1988)

Summary

After working through this chapter you should:

- understand that an arc is a directed edge in a network, a source is a starting point, a sink is a finishing point, the capacity is the maximum flow along the arc and an arc is saturated when it is carrying the maximum capacity
- know that a feasible flow through a network is an allowable path and flow-augmenting paths consist entirely of unsaturated arcs
- use the labelling procedure to find the maximum flow through a network
- know the Maximum flow – minimum cut theorem and its application.

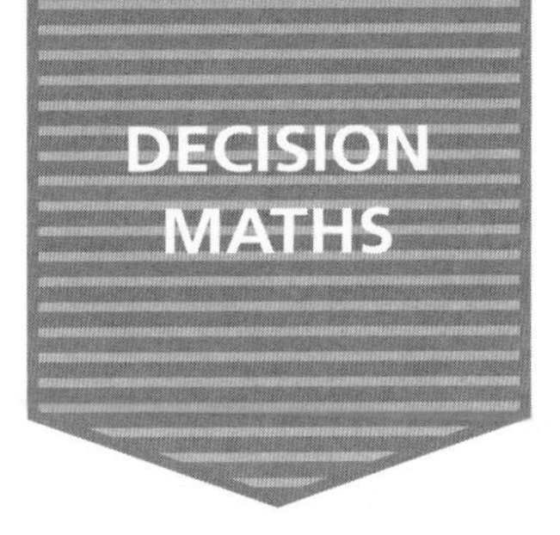

Answers

CHAPTER 1 Algorithms

Exercises 1.1 *Classwork*

1 $x = \frac{5}{4} \pm \frac{\sqrt{17}}{4}$

2 a)

a	105	75	30	15
b	180	105	75	30
R	1	1	2	2
S	75	30	15	0

HCF of 105 and 180 is 15.

b) 4 iterations

c)

a	180	105	1st iteration exchanges a and b
b	105	180	thus it gives the solution in 5
R	0	:	iterations
S	105	:	

3 The simplest (though not the most efficient) is to walk on the right-hand side of a passage (or the left) and turn left (or right) at every junction or dead end. This is a quicker route.

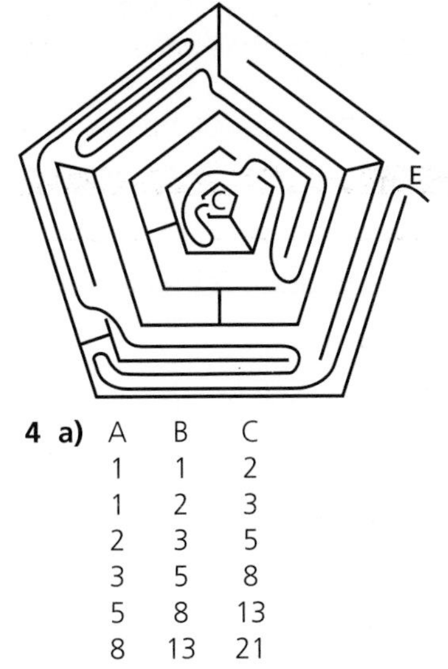

4 a)

A	B	C
1	1	2
1	2	3
2	3	5
3	5	8
5	8	13
8	13	21

b) The algorithm generates the first five terms of the Fibonnaci sequence.

Exercises 1.1 *Homework*

1 $x = -\frac{1}{5} \pm \frac{\sqrt{61}}{5}$

2

a	135	85	50	35	15	5
b	220	135	85	50	35	15
c	1	1	1	1	2	3
d	85	50	35	15	5	0

a) HCF of 135 and 220 is 5.
b) 6 iterations

3 a) $x = 0.441, 2.397$

Exercises 1.2 *Classwork*

1 Set 1 bubble sort

8	7	6	5	4	3	2	1
7	6	5	4	3	2	1	2
6	5	4	3	2	1	3	3
5	4	3	2	1	4	4	4
4	3	2	1	5	5	5	5
3	2	1	6	6	6	6	6
2	1	7	7	7	7	7	7
1	8	8	8	8	8	8	8

Set 1 quicksort

8	7	6	5	4	3	2	1
7	6	5	4	3	2	1	2
6	5	4	3	2	1	3	3
5	4	3	2	1	4	4	4
4	3	2	1	5	5	5	5
3	2	1	6	6	6	6	6
2	1	7	7	7	7	7	7
1	8	8	8	8	8	8	8

Set 1 Interchange sort

8	1	1	1	1	1
7	7	2	2	2	2
6	6	6	3	3	3
5	5	5	5	4	4
4	4	4	4	5	5
3	3	3	6	6	6
2	2	7	7	7	7
1	8	8	8	8	8

Set 1 Shuttle sort

8	7	6	5	4	3	2	1
7	8	7	6	5	4	3	2
6	6	8	7	6	5	4	3
5	5	5	8	7	6	5	4
4	4	4	4	8	7	6	5
3	3	3	3	3	8	7	6
2	2	2	2	2	2	8	7
1	1	1	1	1	1	1	8

Set 2 bubble sort

1	1	1	1
3	3	3	2
5	5	2	3
7	2	4	4
2	4	5	5
4	6	6	6
6	7	7	7
8	8	8	8

Set 2 quicksort

1	1	1	1	1
3	3	2	2	2
5	5	3	3	3
7	7	5	4	4
2	2	7	5	5
4	4	4	7	6
6	6	6	6	7
8	8	8	8	8

Set 2 Interchange sort

1	1	1	1	1	1	1	1
3	3	2	2	2	2	2	2
5	5	5	3	3	3	3	3
7	7	7	7	4	4	4	4
2	2	3	5	5	5	5	5
4	4	4	4	7	7	6	6
6	6	6	6	6	6	7	7
8	8	8	8	8	8	8	8

Set 2 Shuttle sort

1	1	1	1	1	1	1	1
3	3	3	3	2	2	2	2
5	5	5	5	3	3	3	3
7	7	7	7	5	4	4	4
2	2	2	2	7	5	5	5
4	4	4	4	4	7	6	6
6	6	6	6	6	6	7	7
8	8	8	8	8	8	8	8

Set 3 bubble sort

5	1	1	1	1
1	4	4	3	2
4	5	3	2	3
7	3	2	4	4
3	2	5	5	5
2	7	6	6	6
8	6	7	7	7
6	8	8	8	8

Set 3 quicksort

5	1	1	1	1
1	4	4	3	2
4	3	3	2	3
7	2	2	4	4
3	5	5	5	5
2	7	6	6	6
8	8	7	7	7
6	6	8	8	8

Set 3 Interchange sort

5 1 1 1 1 1 1 1
1 5 2 2 2 2 2 2
4 4 4 3 3 3 3 3
7 7 7 7 4 4 4 4
3 3 3 4 7 5 5 5
2 2 5 5 5 7 6 6
8 8 8 8 8 8 8 7
6 6 6 6 6 6 7 8

Set 2 Shuttle sort

5 1 1 1 1 1 1 1
1 5 4 4 3 2 2 2
4 4 5 5 4 3 3 3
7 7 7 7 5 4 4 4
3 3 3 3 7 5 5 5
2 2 2 2 2 7 7 6
8 8 8 8 8 8 8 7
6 6 6 6 6 6 6 8

Summary

Set	bubblesort		quicksort	
1	$c = 28$	$e = 28$	$c = 28$	$e = 28$
2	$c = 15$	$e = 6$	$c = 19$	$e = 3$
3	$c = 20$	$e = 11$	$c = 5$	$e = 9$

Set	interchange		shuffle	
1	$c = 25$	$e = 4$	$c = 28$	$e = 28$
2	$c = 28$	$e = 5$	$c = 13$	$e = 6$
3	$c = 28$	$e = 7$	$c = 17$	$e = 11$

2 Bubblesort

50 48 48 30 12 4 4 4 4
48 50 30 12 4 12 12 12 12
76 30 12 4 28 28 28 28 21
30 12 4 28 30 30 30 21 28
12 4 28 48 48 48 21 30 29
4 28 50 50 50 21 48 29 30
28 56 56 56 21 50 29 41 41
56 63 63 21 56 29 41 48 48
63 76 21 57 29 41 50 50 50
77 21 57 29 41 56 56 56 56
21 57 29 41 57 57 57 57 57
57 29 41 63 63 63 63 63 63
29 41 76 76 76 76 76 76 76
41 77 77 77 77 77 77 77 77

Quicksort

50 48 30 12 4 4 4
48 30 12 4 12 12 12
76 12 4 28 28 21 21
30 4 28 21 21 28 28
12 28 21 29 29 29 29
4 21 29 30 30 30 30
28 29 41 41 41 41 41
56 41 48 48 48 48 48
63 50 50 50 50 50 50
77 76 56 56 56 56 56
21 56 63 63 57 57 57
57 63 57 57 63 63 63
29 77 76 76 76 76 76
41 57 77 77 77 77 77

3 Solution by interchange sort

14 12 12 12 12 12 12 12
25 25 13 13 13 13 13 13
12 14 14 14 14 14 14 14
31 31 31 31 18 18 18 18
18 18 18 18 31 20 20 20
13 13 25 25 25 25 25 25
20 20 20 20 20 31 31 28
28 28 28 28 28 28 28 31

Solution by shuttle sort

14 14 12 12 12 12 12 12
25 25 14 14 14 13 13 13
12 12 25 25 18 14 14 14
31 31 31 31 25 18 18 18
18 18 18 18 31 25 20 20
13 13 13 13 13 31 25 25
20 20 20 20 20 20 31 25
28 28 28 28 28 28 28 31

4 a)

								c	e
28	20	13	18	31	12	25	14		
12	20	13	18	31	28	25	14	7	1
12	13	20	18	31	28	25	14	6	1
12	13	14	18	31	28	25	20	5	1
12	13	14	18	31	28	25	20	4	0
12	13	14	18	20	28	25	31	3	1
12	13	14	18	20	25	28	31	2	1
12	13	14	18	20	25	28	31	1	0

b) $c = 28$, $e = 5$

c) Comparisons – same as bubble sort but fewer exchanges therefore possibly more efficient on short list.

5 c) 9

Exercises 1.2 *homework*

1 Set 1

a)
4 5 6 7 8 9 10
5 6 7 8 9 10 9
6 7 8 9 10 8 8
7 8 9 10 7 7 7
8 9 10 6 6 6 6
9 10 5 5 5 5 5
10 4 4 4 4 4 4

b)
4 5 6 7 8 9 10
5 6 7 8 9 10 9
6 7 8 9 10 8 8
7 8 9 10 7 7 7
8 9 10 6 6 6 6
9 10 5 5 5 5 5
10 4 4 4 4 4 4

c)
4 10 10 10 10
5 5 9 9 9
6 6 6 8 8
7 7 7 7 7
8 8 8 6 6
9 9 5 5 5
10 4 4 4 4

d)
4 5 6 7 8 9 10
5 4 5 6 7 8 9
6 6 4 5 6 7 8
7 7 7 4 5 6 7
8 8 8 8 4 5 6
9 9 9 9 9 4 5
10 10 10 10 10 10 4

Set 2

a) Bubblesort
9 9 9 9
7 7 7 8
5 5 8 7
3 8 6 6
8 6 5 5
6 4 4 4
4 3 3 3
2 2 2 2

b) Quicksort
9 9 9 9 9
7 7 8 8 8
5 5 7 7 7
3 3 5 6 6
8 8 3 5 5
6 6 6 3 4
4 4 4 4 3
2 2 2 2 2

c) Interchange
9 9 9 9 9 9 9
7 7 8 8 8 8 8
5 5 5 7 7 7 7
3 3 3 3 6 6 6
8 8 7 5 5 5 5
6 6 6 6 3 3 4
4 4 4 4 4 4 3
2 2 2 2 2 2 2

c) Shuttle
9 9 9 9 9 9 9 9
7 7 7 7 8 8 8 8
5 5 5 5 7 7 7 7
3 3 3 3 5 6 6 6
8 8 8 8 3 5 5 5
6 6 6 6 6 3 4 4
4 4 4 4 4 4 3 3
2 2 2 2 2 2 2 2

Set 3

a) Bubblesort
6 8 8 8 8
8 6 6 7 7
2 3 7 6 6
3 7 4 4 5
7 4 3 5 4
4 2 5 3 3
1 5 2 2 2
5 1 1 1 1

b) Quicksort
6 8 8 8 8
8 7 7 7 7
2 6 6 6 6
3 2 3 4 5
7 3 4 5 4
4 4 5 3 3
1 1 2 2 2
5 5 1 1 1

c) Interchange
6 8 8 8 8 8 8 8
5 6 7 7 7 7 7 7
2 2 2 6 6 6 6 6
3 3 3 3 5 5 5 5
7 7 6 2 2 4 4 4
4 4 4 4 4 2 3 3
1 1 1 1 1 1 1 2
5 5 5 5 3 3 2 1

c) Shuttle

6 8 8 8 8 8 8 8
8 6 6 6 7 7 7 7
2 2 2 3 6 6 6 6
3 3 3 2 3 4 4 5
7 7 7 7 2 3 3 4
4 4 4 4 4 2 2 3
1 1 1 1 1 1 1 2
5 5 5 5 5 5 5 1

2 Bubblesort

41 41 57 57 77 77 77 77 77 77 77 77
29 57 41 77 63 63 63 63 63 63 76 76
57 29 77 63 57 57 57 57 57 76 63 63
21 77 63 56 56 56 56 56 76 57 57 57
77 63 56 41 41 41 41 76 56 56 54 54
63 56 29 29 29 30 76 48 48 48 48 48
56 28 28 28 30 76 48 50 50 50 50 50
28 21 21 30 76 48 50 41 41 41 41 41
4 12 30 76 48 50 30 30 30 30 30 30
12 30 76 48 50 29 29 29 29 29 29 29
30 76 48 50 28 28 28 28 28 28 28 28
76 48 50 21 21 21 21 21 21 21 21 21
48 50 12 12 12 12 12 12 12 12 12 12
50 4 4 4 4 4 4 4 4 4 4 4

Quicksort

57 77 77 77 77 77 77
77 63 76 76 76 76 76
63 76 63 63 63 63 63
53 57 57 57 57 57 57
76 53 53 53 53 53 53
48 48 48 50 50 50 50
50 50 50 48 48 48 48
41 41 41 41 41 41 41
29 29 29 29 30 30 30
21 21 21 21 29 29 29
28 28 28 28 21 28 28
4 4 4 4 28 21 21
12 12 12 12 4 4 12
30 30 30 30 12 12 4

3 Interchange sort

28 12 12 12 12 12 12 12
20 20 13 13 13 13 13 13
13 13 20 14 14 14 14 14
18 18 18 18 18 18 18 18
31 31 31 31 31 20 20 20
12 28 28 28 28 28 25 25
25 25 25 25 25 25 28 28
14 14 14 20 20 31 31 31

Shuttle sort

28 20 13 13 13 12 12 12
20 28 20 18 18 13 13 13
13 13 28 20 20 18 18 14
18 18 18 28 28 20 20 18
31 31 31 31 31 25 25 20
12 12 12 12 12 31 28 25
25 25 25 25 25 25 31 25
14 14 14 14 14 14 14 31

Exercises 1.3 *Classwork*

1 Full bin

$A + B =$ 1 m
$C + E + F =$ 1 m
$D + G =$ 90 cm
3 racks used

First fit

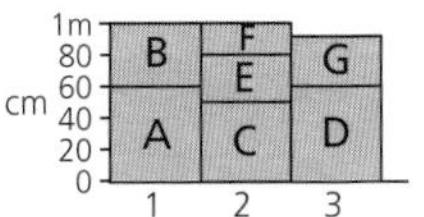

3 racks used
(same solution as full bin)

First fit decreasing

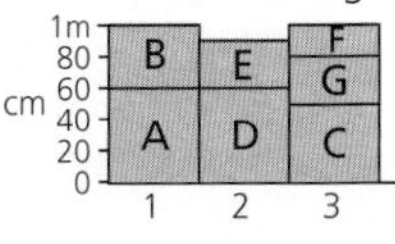

3 racks used
(variation of first two solutions)

2 Full bin

$I + L =$ 12 ft
$J + K =$ 12 ft
$A + B + C + D + E =$ 12 ft
$F + G + H =$ 11ft
4 lengths of pipe
One 1 ft length left over
First fit

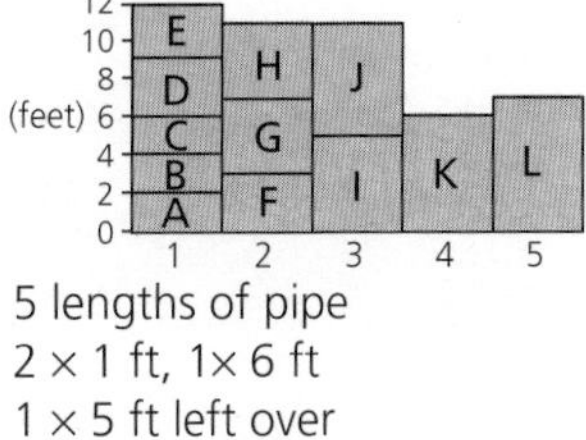

5 lengths of pipe
2 × 1 ft, 1× 6 ft
1 × 5 ft left over

First fit decreasing

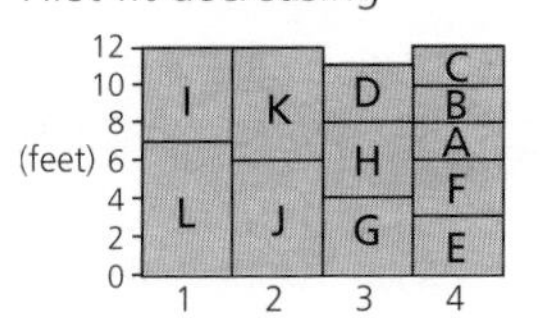

4 lengths of pipe (as full bin)

3 Full bin e.g.

$B + J + P =$ 800
$C + I + E + H + K =$ 800
$L + O + N + G + F =$ 750
$D + M + A =$ 490
4 discs
It is very hard to find full bin combinations in data such as this.

First fit

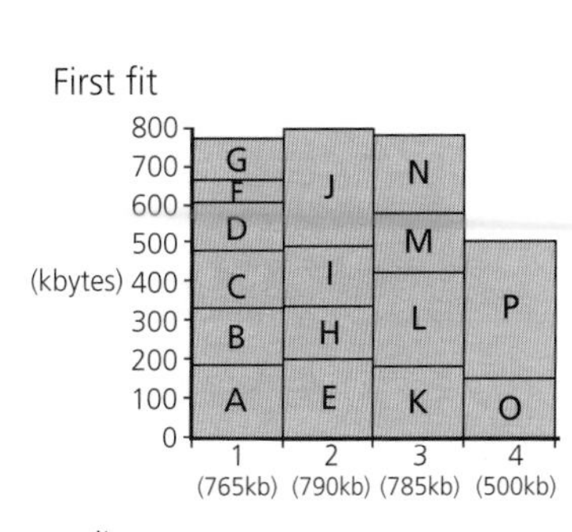

4 discs
Reasonably quick and fairly space efficient.

First decreasing order P J L E N A K M I B O D C H G F

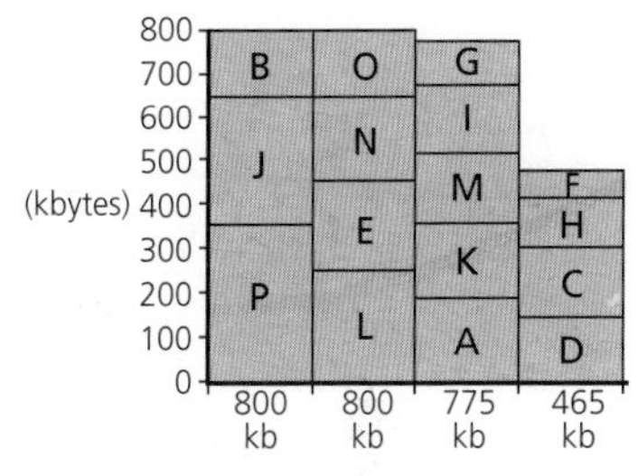

4 discs
Probably best since in terms of space usage, but it takes a long time.

4 Total breaks time =
180 × 3 = 540 seconds
Total Ads time = 540 seconds
Can be done in theory.
First fit

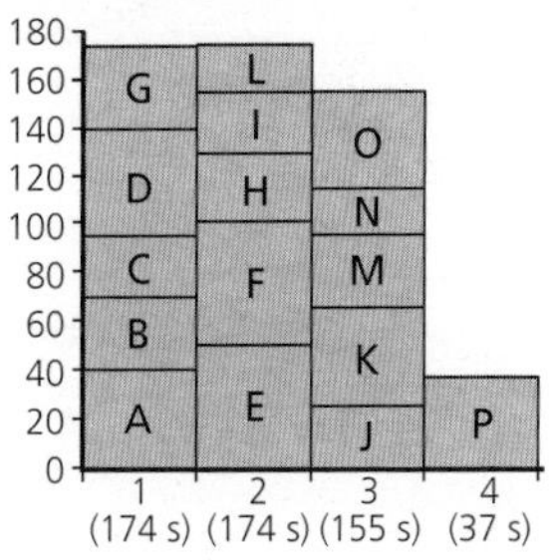

Comments: Full bin would take ages. Neither FF nor DFG will do it in 3 but DFF could manage if 3rd break was made 5 seconds longer.
First fit decreasing
Reorder: E52, D46, F46, A40, K40, O40, H38, P37, G33, B30, M30, C25, J25, I20, N20, L18

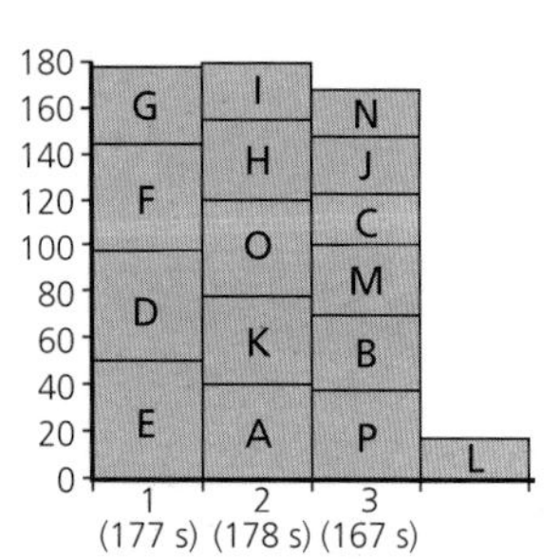

5 Sort by size

Activity	G	J	A	K	B	I	D	F	L	C	E	H
Duration	9	9	8	8	7	7	6	6	6	5	5	4

Using first fit decreasing, 8 workers are needed.
Using full-bin
A + H = 12
B + C = 12
D + F = 12
E + I = 12 8 workers
G
J
K
L

Exercise 1.3 *Homework*

1 60 cm : A B C
80 cm : D E F G
120 cm : H
First fit

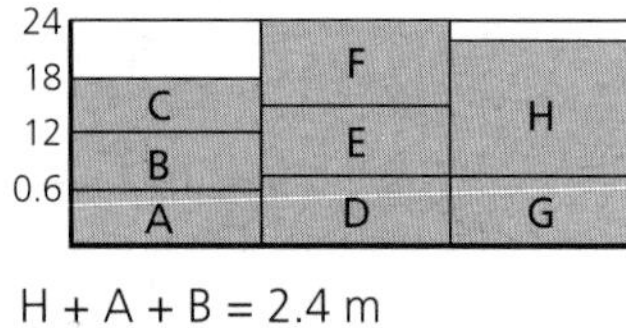

H + A + B = 2.4 m
D + G + F = 2.4 m
G + C = 1.4 m
First fit decreasing

3 racks

2 2 m : ABCD
2.5 m : EFG
3 m : HI
3.5 m : JK
First fit

6 lengths

Full bin
A + B + C = 6
H + I = 6
J + E = 6
F + K = 6
D + G = 4.5
First fit decreasing

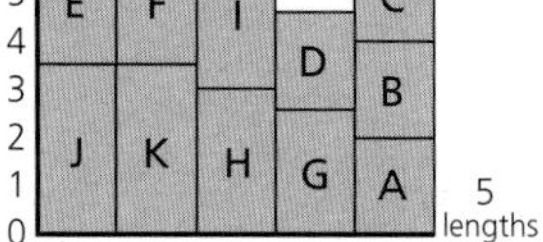

5 lengths

3 Too difficult to find full disks.
First fit
A + B + C = 1342
D + E + F + G + H + I = 1312
J + K + L + M + N + O = 947
P + Q = 1204
4 disks – quick and easy
First fit decreasing
B + Q = 1351
P + H + L = 1357
A + F + N + E + G = 1363
M + O + D + C + K + J + I
= 734
4 disks – more effective use but long time

4 Total show time = 90 minutes
Total act time = 87 minutes so therefore it is possible to show all acts.
Good use of time
Full bin

5 5 suitcases

Consolidation Exercises for Chapter 1

1 i)

A	2520	900	720	180
B	5940	2520	900	720
Q	2	2	1	4
R	900	720	180	0

B = 720

ii)

A	5940	2520
B	2520	5940
Q	0	
R	2520	

It reverses A and B and the algorithm needs one more iteration to obtain the solution.

iii) The worst case could be HCF = 1

2
5 7 2 8 6 9 11 17
2 5 7 5 8 9 11 17
2 5 6 7 8 9 11 17

3 $x = 3, y = 41$

a)

r	41	38	35	32	29	26	23	20	17	14	11	8	5	2
q	0	1	2	3	4	5	6	7	8	9	10	11	12	13

$q = 13$ $r = 2$
repetitions of box 3 13

b) $\frac{y}{x} = q$ remainder $r \Rightarrow y = qx + r$

c) $x = 3, y_1 = 4, y_2 = 1$

r	4	1	11	8	5	2
q_1	0	1	1	1	1	1
q_2			0	1	2	3

repetitions of box 3 1 $q_1 = 1$, $q_2 = 3$
repetitions of box 6 3 $r = 2$

d) First – easier to follow
Second – more efficient (less iterations needed)

4 a)

	i = 1	2	3	4	5	
	7	5	1	1	1	
	5	1	5	3	3	
	1	7	3	5	5	
	9	3	7	7	7	
	3	9	9	9	9	
	11	11	11	11	11	
comparisons	5	4	3	2	1	$c = 15$
swaps	3	3	2	1	0	swaps = 9

b)

7	7	5	1	1	1
9	5	7	11	3	3
5	9	9	3	11	11
comparisons	2	1		2	1
swaps	1	1		1	0

Comparisons 6, swaps 3

c)

	i = 1	2	3
5	5	5	1
7	7	1	3
9	1	3	5
1	3	7	7
3	9	9	9
11	11	11	11
comparisons	3	3	3
swaps	2	2	2

Total comparisons 15, total swaps 9
There is no advantage to splitting the list.

d) You don't need to check the first two pairs on first pass or first pair on second pass. By the third pass the list is guaranteed to be sorted.

5 (i) reorder 1.2, 1.1, 0.7, 0.4, 0.4, 0.3, 0.3, 0.2

Bin			
1.2	0.7		
1.1	0.4	0.4	
0.3	0.3	0.2	

1 m 2 m

(ii) Full bin 1.2 + 0.4 + 0.4 = 2 m
0.7 + 0.2 + 1.1 = 2 m
0.3 + 0.3 = 0.6 m

Bin		
1.2	0.4	0.4
1.1	0.7	0.2
0.3	0.3	

1 m 2 m

6 a)

Store	Q	R
0	7	9
9		
	3	18
27		
	1	36
63		

output 63 = 7 × 9

b)

Store	Q	R
0	4	8
	2	16
	1	32
32		

output 32

c) Algorithm stops when Q has been reduced to 1 by repeated division by 2.
10 will be so reduced more rapidly than 25.

7 Middle 6 FULLER (below)
Middle 9 LEECH (above)
7 GRANT
8 GREGORY

8 a)

Second digit (columns), First digit (rows)

	0	1	2	3	4	5	6	7	8	9
0	X	X			X		X		X	X
1	X		X		X	X	X		X	
2	X	X	X		X	X	X	X	X	
3	X		X	X	X	X	X		X	X
4	X		X		X	X	X		X	

Find the prime numbers up to 49.

b)

	0	1	2	3	4	5	6	7	8	9
0					2		2		2	3
1	2		2		2	3	2		2	
2	2	3	2		2	5	2	3	2	
3	2		2	3	2	5	2		2	3
4	2		2		2	3	2		2	7

9 a) (i) Choose the middle word each time.
(ii) 17
b) (i) K4 **(ii)** K6
(iii) Choice of either E or F column. **(iv)** 8
c) 9

10 i) *A, B, C, D* all fit onto one tape but *E* doesn't. Tapes hold *ABCD, EFG, HI, J, K, L*. So 6 tapes are needed.
ii) First tape contains *K* and *D* (or other '2 + 1' pair). Tapes hold *KD, LE, JF, HI, GCAB* (or equiv). So 5 tapes are needed.

CHAPTER 2
Graphs and Networks 1: Shortest path

Exercises 2.1 *Classwork*

1 a) 4 vertices, 5 edges, connected; matches played by football teams
b) 10 vertices, 11 edges, not connected; local bus routes
c) 7 vertices, 7 edges, connected; teachers and subjects which they teach

2 a)

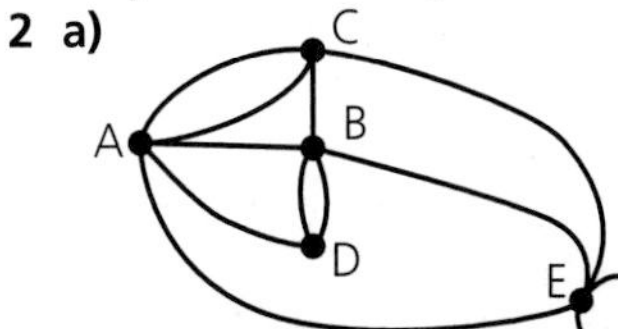

The 2 represents the loop at E which can be travelled in either direction.

b)

	P	Q	R	S	T
P	0	1	2	2	1
Q	1	0	2	0	0
R	2	2	0	0	1
S	2	0	0	2	1
T	1	0	1	1	2

3 a) AZ, ABZ, ABCZ, ACZ, ACBZ
b) ABDZ, ABDEZ, ABEZ, ABEDZ, ABCEZ, ABCDZ, ACEZ, ACEDZ, ACDZ, ACDEZ, ACBDZ, ACBEZ, ABCDEZ, ACBEDZ
c) ABEH, ABEFH, ABFH, ABCFH, ABCEH, ADGH, ADGFH, ADFH, ADCFH, ADCGH, ABCGH, ACFH, ACEH, ACEFH, ACGH, ACGFH, ADCEH

5 Number of arcs in K_n is $\frac{1}{2}n(n-1)$.

Exercises 2.1 *Homework*

1 a)

b) There are several: ABCEFDA is one example.

2

	A	B	C	D	E	F
A	0	1	0	2	1	0
B	1	0	2	1	0	0
C	0	2	0	2	1	0
D	2	1	2	0	2	1
E	1	0	1	2	0	2
F	0	0	0	1	2	0

3 a) ABCS, AS
b) ABCS, ABCIS, ABCIJS, AGHKEJS, AGHKEJIS, AGHKEJICS, AGHFEJS, AGHFEJICS, AGHFEGJIS, AFHKEJS, AFHKEJIS, AFHKEJICS, AFEJS, AFEJIS, AFEJICS.

4 a) 1 **b)** 4 **c)** $n - 1$

Exercises 2.2 *Classwork*

1 a) SBCT length 20
b) SBFT length 93
c) 3 possible routes: SADET, SCET, SCFIT all of length 22

2 Route: Bickleigh, Cadbury, Thorverton, Cowley, Exeter; length 24 km
Route: Bickleigh, Silverton, Cowley, Exeter; length 25 km

3 a) Ashford – Folkestone – Dover; £12.00 × 4 = £48.00
b) Deal – Dover, Canterbury, Maidstone; £21.50 × 4 = £86.00

4

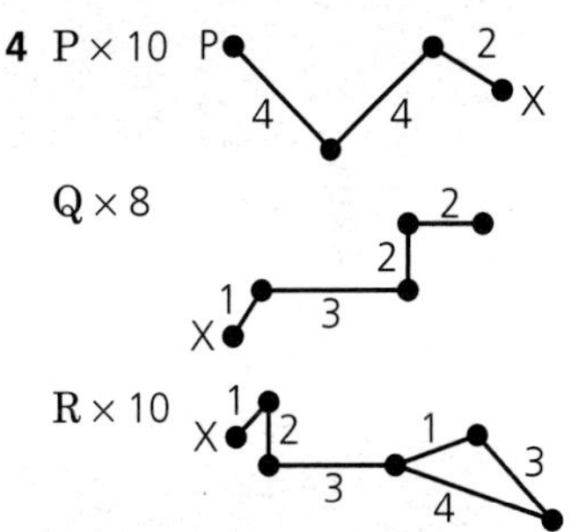

By considering three parts of the network: Use firestation at Q.

5 Table – shortest distances between pairs of towns.

	A	B	C	D	E	F	G	Totals
A	–	10	17	8	13	23	19	90
B	10	–	7	9	19	13	20	78
C	17	7	–	16	23	6	15	84
D	8	9	16	–	10	20	11	74
E	13	19	23	10	–	17	8	90
F	23	13	6	20	17	–	9	88
G	19	20	15	11	8	9	–	82

Build sports centre at D, assuming numbers using centre are roughly the same in all villages.

Exercises 2.2B *Homework*

1 a) SABT has length 13.
b) Three possible routes SBDT, SAEFT, SCFT; all of length 17.

2 SS–TS–B–L–Sp–W–WH 6.5 miles

3 a)

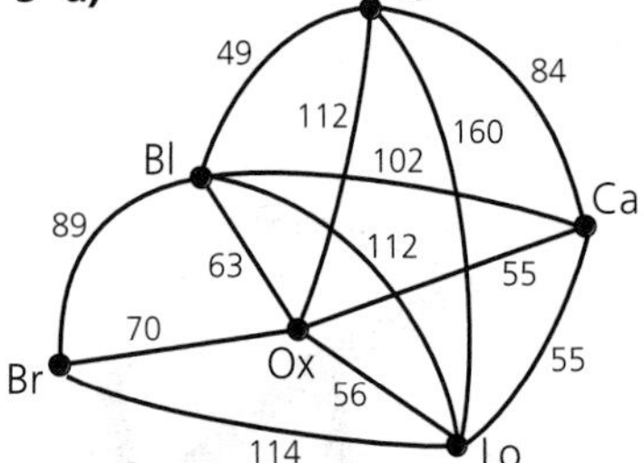

b) (i) Bristol – Oxford – Cambridge 125 miles
(ii) assuming an average speed of 40 mph then Bristol – London – Cambridge = 169 miles might be better assuming no hold-ups round London.

4 a) Helsinki to Pori: 255 FIM
b) Turku → 460 FIM
Tampere →
Mikkeli →
Varkaus →
Joensuu →

5 Lowest costs to:

Helsinki	£340
Amsterdam	£68
Copenhagen	£204
Stockholm	£136
Berlin	£272
Frankfurt	£102
Warsaw	£442

Consolidation exercises for Chapter 2

1 i) 3
ii) 6
iii) n – 1
iv) (n – 1) + (n – 2) + (n – 3) + … + 2 + 1 = $\frac{1}{2}n(n-1)$

2 BDEGF, total = 90

3 Lowest fare is £1000 along route SBCT.

4 a)

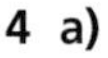

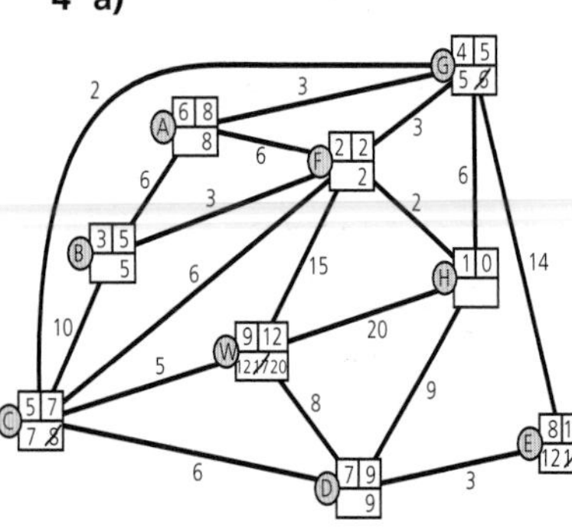

Tracing back:
W — C — G — F — H
Quickest route: HFGCW
Time: 12 mins

b) Add 10 to all routes through F and repeat Dijkstra, *or* add 10 to all routes through F – this means that shortest route cannot pass through F so delete F and repeat Dijkstra.

5 Label Q, R, S, T with direct distances from P, and select smallest (say Q).
Update distances to R, S, T if route from P via Q is shorter. The initial 3 possible updates require 3 additions so far.
Select the smallest of R, S, T and repeat the updating process; this stage requires another 2 additions.
The final stage similarly requires1 addition, so the total is 3 + 2 + 1 = 6.
Trace back to find the shortest path, including an arc XY whenever the distance to X plus the length of XY is the same as the distance to Y.
Worst case number is 4 + 3 + 2 + 1 i.e. 10
i) The number of comparisons
ii) The number of operations needed to complete the algorithm is (approximately) proportional to the square of the vertices.
So, for example, doubling the number of vertices will (roughly) quadruple the amount to be done.

6 a) ABEFG length 11
b)

n	(n – 2)!	$\frac{1}{2}(5n^2 - 3n) + 1$
7	120	113

large values of n make trial and improvement much less efficient

7 a) (i) Maidstone, Canterbury, Dover: £14 per kg, cost £56
(ii) Deal, Dover, Canterbury, Rochester: £19 per kg, cost £76

b)

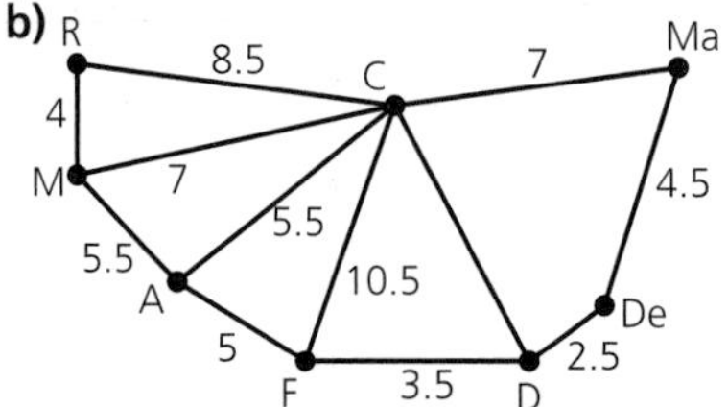

Maidstone, Ashford, Folkestone, Dover: £14 per kg, same price, different route

Deal, Dover, Canterbury, Rochester: £17.50 per kg, cheaper, same route

8 a)

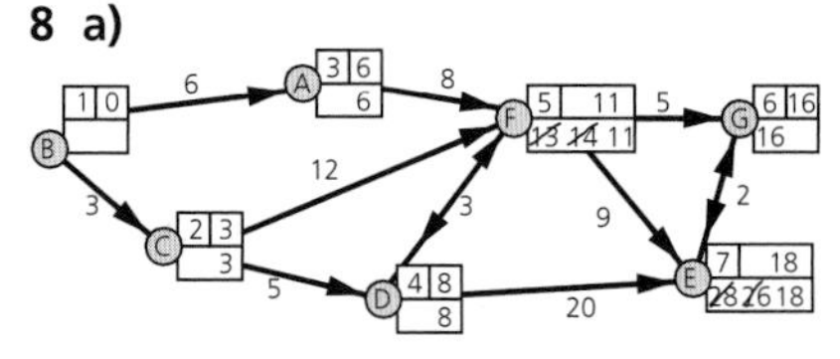

Lowest cost 18
Cheapest route BCDFGE
By tracing back: label – weight on arc = previous label

b)

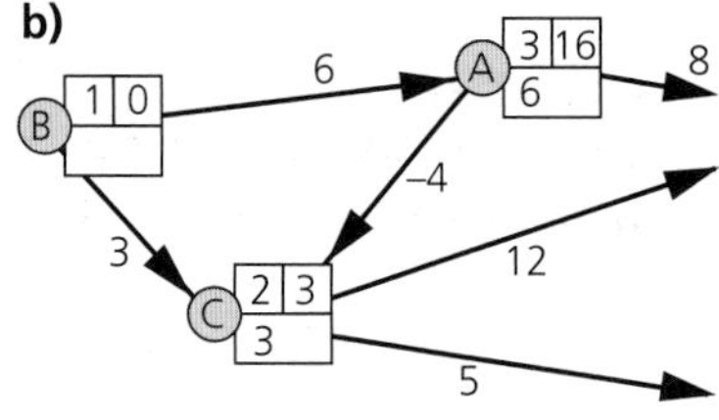

Dijkstra fails because C is labelled at stage 2 so the cheapest route to C (BAC cost 2) is not found.
The algorithm assumes that extra arcs always have extra cost.

9 a) (i)
1. label GO
2. T – labels all vertices directly connected to G.
3. P – label the smallest T – label.
4. Consider arcs directly connected to newly P – labelled vertex if weight on arc > P label, T label the vertex with the weight on previous P label.
5. Repeat 3 and 4 until all vertices have been P – labelled (or until B has a P – label).
6. Find minimum time by tracing back.

(ii)

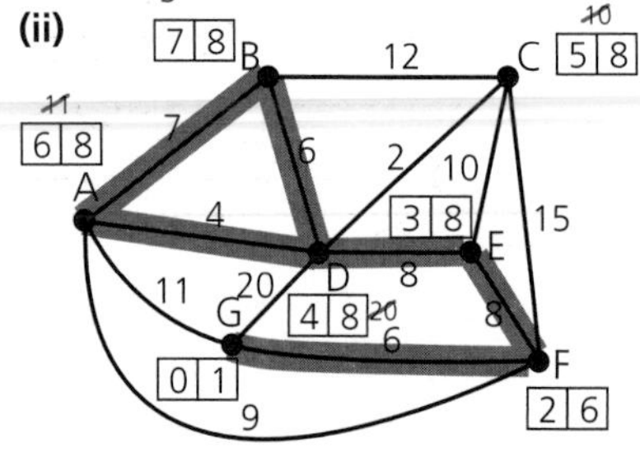

(iii) Shortest time 8, route GFEDB *or* GFEDAB

b) By selecting slowest edge to be in solution e.g. GDB; time 20

10 a) graph 1 not possible – cannot prove D from A. Solution represented by ?? needs 55 lines of proof.

b)

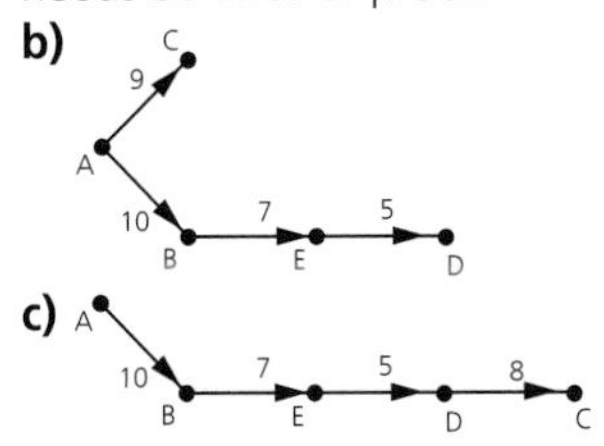

c)

algorithm gets to C first.

CHAPTER 3 Graphs and Networks 2: Minimum-connector

Exercises 3.1 *Classwork*

1

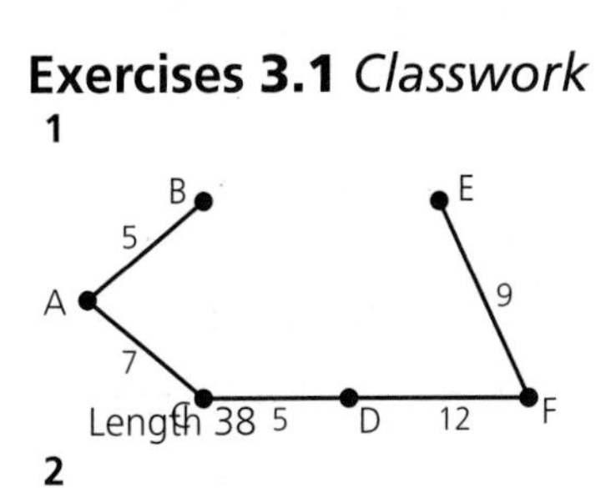

Length 38

2

	1	2	3	4	6	5
	A	B	C	D	E	F
A	∞	5	7	∞	∞	∞
B	(5)	∞	11	∞	13	∞
C	(7)	11	∞	5	18	∞
D	∞	∞	(5)	∞	15	12
E	∞	13	18	15	∞	(9)
F	∞	∞	∞	(12)	9	∞

Start at A order of connection ABCDFE length 38

3 Order of connection AIEGC Length 365 miles. Kruskal's appears quicker to apply once the network is drawn with more vertices. Prim's is quicker to apply as network may be large and difficult to use.

4

	1	6	3	4	5	2	3	7
	B	C	D	G	Li	Lon	S	W
B	∞	205	103	196	204	70	126	197
C	205	∞	154	122	(58)	281	200	73
D	103	154	∞	135	120	146	133	(96)
G	196	122	135	∞	64	176	(90)	141
Li	204	58	120	(64)	∞	231	147	77
Lon	(70)	281	146	176	231	∞	86	242
S	126	200	133	90	147	(86)	∞	176
W	197	(73)	96	141	77	242	176	∞

Order of connection Belfast – Londonderry – Sligo – Galway – Limerick – Cork – Waterford – Dublin; length 70 + 73 + 64 + 58 + 86 + 90 + 96 = 537 miles

5

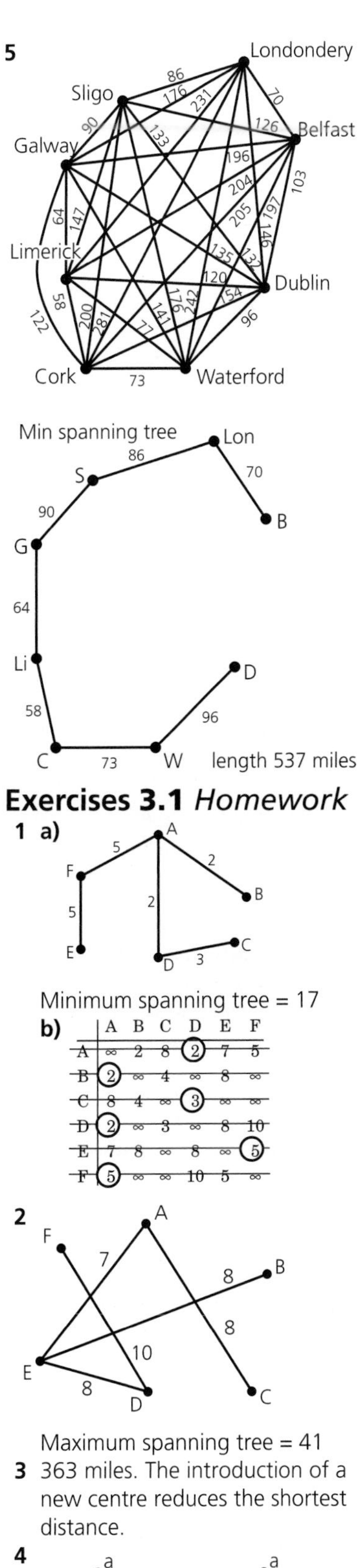

length 537 miles

Exercises 3.1 *Homework*

1 a)

Minimum spanning tree = 17

b)

	A	B	C	D	E	F
A	∞	2	8	(2)	7	5
B	(2)	∞	4	∞	8	∞
C	8	4	∞	(3)	∞	∞
D	(2)	∞	3	∞	8	10
E	7	8	∞	8	∞	(5)
F	(5)	∞	∞	10	5	∞

2

Maximum spanning tree = 41

3 363 miles. The introduction of a new centre reduces the shortest distance.

4

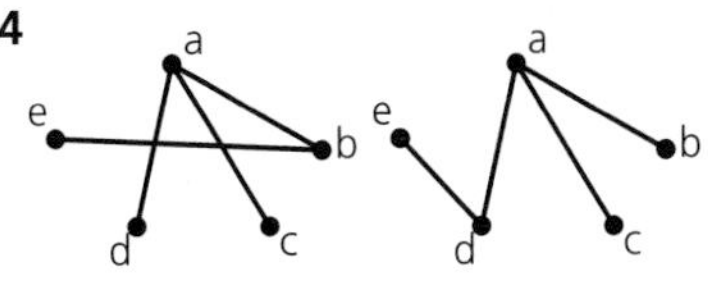

Minimum spanning tree = 94

5 a) i) 45 **ii)** $\frac{1}{2}n(n-1)$

b) 991 miles

Consolidation exercises for Chapter 3

1

Rank		
Sals – South	23	1
Sals – Bourn	27	2
Dorch – Bourne	28	3
Warm – Sals	28	4
South – Bas	31	5
Bourn – South	31	loop
Sals – Basing	36	loop
Warm – Swin	37	6
Sals – Dorch	40	
Sals – Swin	40	
Swin – Bas	41	

min spanning tree	length
23 + 27 + 28 + 28 + 31 + 37	= 174

2 a)

	1	5	7	6	4	3	2
	A	B	C	D	E	M	S
A	–	–	–	–	–	–	(50)
B	60	–	–	–	–	(50)	98.5
C	98.5	(50)	–	50	98.5	60	–
D	–	–	–	–	60	(50)	98.5
E	–	–	–	–	–	–	(50)
M	(50)	–	–	–	50	–	60
S	–	–	–	–	–	–	–

b)

Length 50 × 6 = 300 m
order S A M E B D C
Drains correctly, there is no house where water accumulates.

3 a)

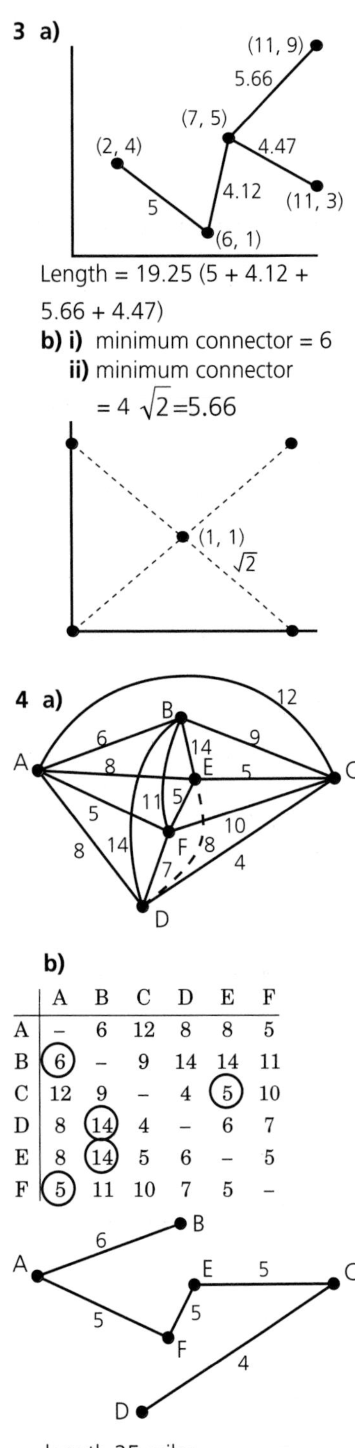

Length = 19.25 (5 + 4.12 + 5.66 + 4.47)

b) i) minimum connector = 6

ii) minimum connector $= 4\sqrt{2} = 5.66$

4 a)

b)

	A	B	C	D	E	F
A	–	6	12	8	8	5
B	(6)	–	9	14	14	11
C	12	9	–	4	(5)	10
D	8	(14)	4	–	6	7
E	8	(14)	5	6	–	5
F	(5)	11	10	7	5	–

length 25 miles

5 Edges: AD 13, AE 14, EF 14, FB 13, FC 15

total cost £69

6 Prims

1. Select a starting vertex.
2. Join to nearest vertex.
3. Select shortest edge joining vertex not in solution to one already in solution.
4. Repeat 3 for all vertices in tree.

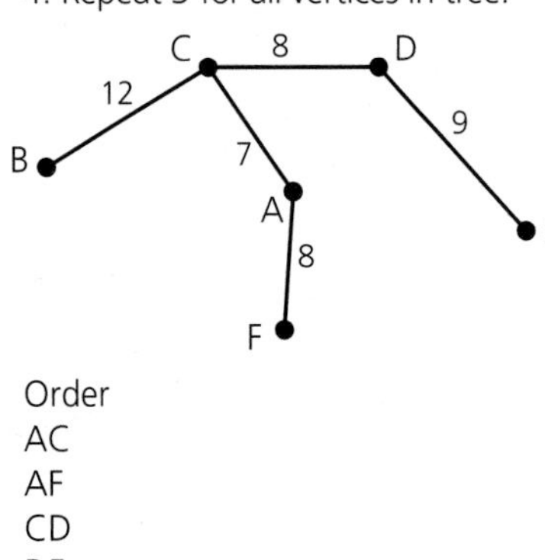

Order
AC
AF
CD
DE
CB

Minimum length of road is 44 km.

7 Edges: AE 9, EB 7, BC 10, AF 10, FD 9

Length 45

8 There are four possible solutions all of length 34.

AC, CB, BD, AE, AF
AC, CB, CD, AE, AF
AC, CB, BD, DE, AF
AC, CB, CD, DE, AF

9 Order of connection:

Crediton-Exeter	8
Crediton-Tiverton	12
Launceston-Okehampton	19
Launceston-Bude	19
Minehead-Taunton	21
Taunton-Tiverton	21
Okehampton-Exeter	23
Launceston-Plymouth	24
Length	147

10 Minimum connector

CE, 2; CD or ED, √5; BD, √5; AC√5

Length $2 + 3\sqrt{5}$

CHAPTER 4

Graphs and Networks 3: Travelling Salesman Problem

Exercises 4.1 *Classwork*

1 i) a) e.g. ABCA, ABDCA, ABDECA

b) e.g.ABEFCA, ABEGIHFA, ABDEHFCA

c) e.g. ABCDEFA, AGHIJKLFA, AGMKEFA

ii) There are several answers, here is one tree for each graph.

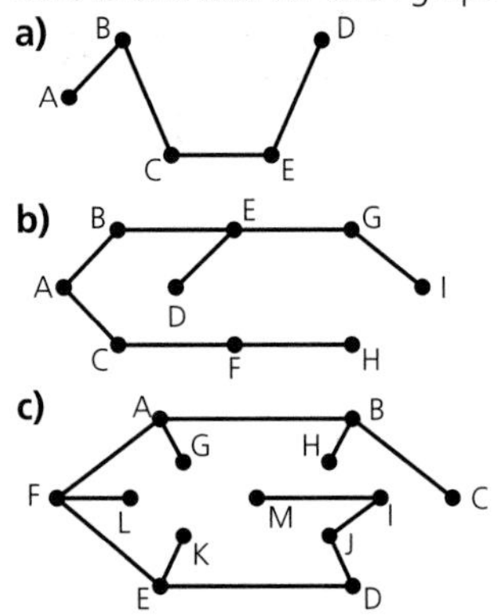

iii) The answer depends on your chosen cycles.

2 For k_1 and k_2 no cycles exist.

3 a) You have to go through vertex F twice.

b) Add edge DG or EI.

4

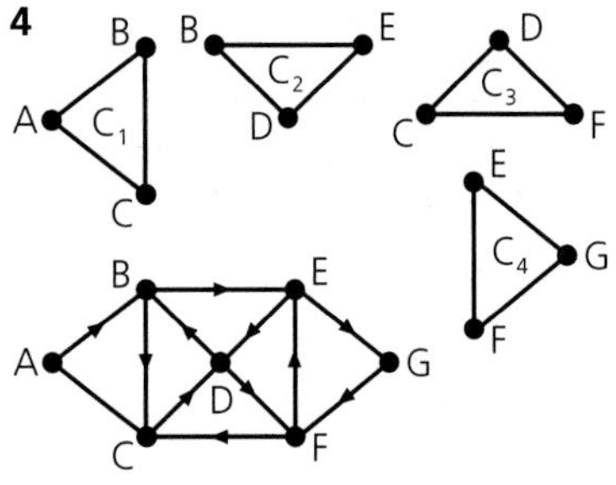

e.g. ABEGFEDFCDBCA

5 BGHJKFDMLTVWXZQRSNPCB
BGHJKFDCPQRSNMLTVWXZB

Exercises 4.1 *Homework*

1 a)

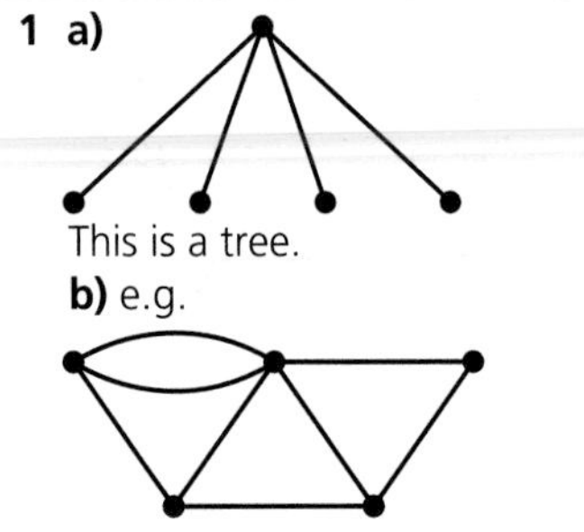

This is a tree.

b) e.g.

2 There are no cycles of length 7.

3

Graph	Hamiltonian	Eulerian
G_1	No	Yes
G_2	No	No
G_3	Yes (Several H-cycles)	Yes (Several E-cycles)

5 a) $n - 1$ **b)** $\frac{1}{2}n(n - 1)$

Exercises 4.2 *Classwork*

1 a)

	1	2	3	4	6	5
	A	B	C	D	E	F
A	∞	14	12	18	(39)	31
B	14	∞	(9)	26	28	28
C	(12)	9	∞	17	27	19
D	18	(26)	17	∞	40	25
E	39	28	27	40	∞	(15)
F	31	28	19	(25)	15	∞

Upper bound 12 + 9 + 26 + 25 + 15 + 39 = 126 km

A C B D F E A

b)

Vertex deleted	LB for vertex	LB for network	LB for TSP
A	26	60	86
B	23	63	86
C	21	72	93
D	35	55	90
E	42	57	99
F	34	65	99

Best lower bound by deleting E or F = 99

2

	1	5	3	4	2
	A	C	E	G	I
A	∞	(221)	125	145	105
C	221	∞	100	(95)	263
E	125	100	∞	40	(160)
G	145	95	(40)	∞	168
I	(105)	263	160	168	∞

Upper bound (beginning in Aberdeen) A – I – E – G – C – A; length 621 miles

Lower bound by deleting Aberdeen 270 + 295 = 565 miles

3 a)

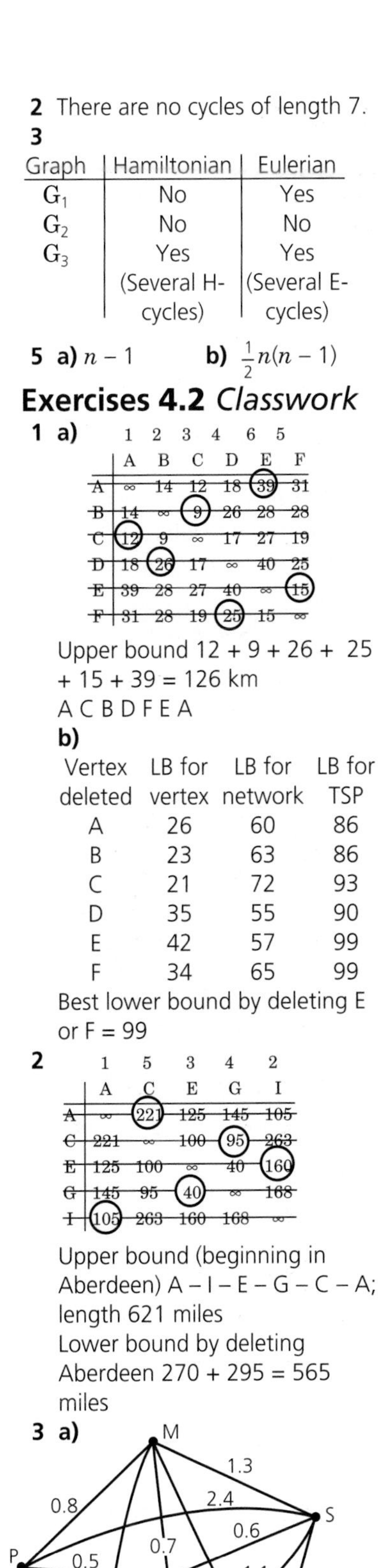

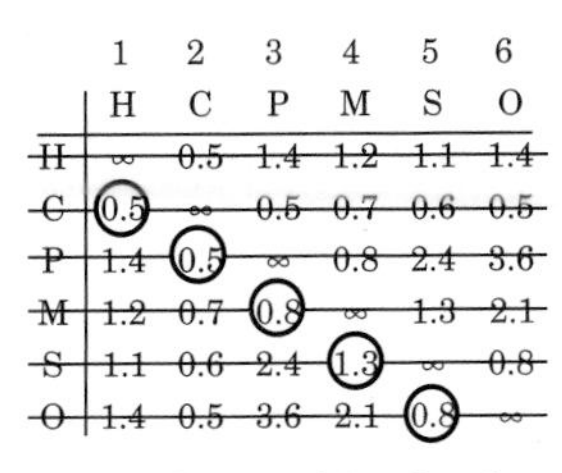

	1	2	3	4	5	6
	H	C	P	M	S	O
H	∞	0.5	1.4	1.2	1.1	1.4
C	(0.5)	∞	0.5	0.7	0.6	0.5
P	1.4	(0.5)	∞	0.8	2.4	3.6
M	1.2	0.7	(0.8)	∞	1.3	2.1
S	1.1	0.6	2.4	(1.3)	∞	0.8
O	1.4	0.5	3.6	2.1	(0.8)	∞

H – C – P – M – S – O – H; distance 5.3 miles

b) home – offices – pool – park – museum – shops – park – home; length 8 miles

4

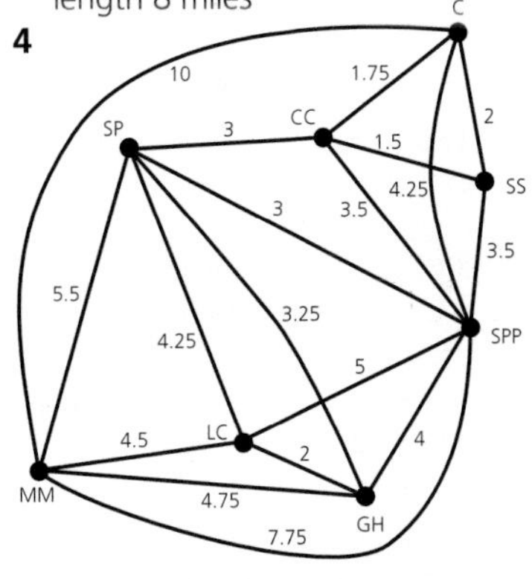

	1	4	5	3	2	8	7	6
	SSP	SS	C	CC	SP	MM	LC	GH
SSP	∞	3.5	4.25	3.5	3	[7.75]	5	4
SS	3.5	∞	2	(1.5)	4.5	10	8.5	7.5
C	4.25	(2)	∞	1.75	4.75	10	9	8
CC	3.5	1.5	1.75	∞	(3)	8.5	7.25	6.25
SP	(3)	4.5	4.75	3	∞	5.5	4.25	3.25
MM	7.75	10	10	8.5	5.5	∞	(4.5)	4.75
LC	5	8.5	9	7.25	4.25	4.5	∞	(2)
GH	4	7.5	(8)	6.25	3.25	4.75	2	∞

by nearest neighbour a possible tour is: St Peter Port → Saumarez Park → Craft Centre → St Sampsons → L'Ancresse common → German underground hospital → Little Chapel → Maritime Museum → St Peter Port; length 31.75 miles

Improved tour (heuristic)
SPP – SS – C – CC – SP – MM – LC – GH – SPP
length 26.25 miles

5 Use a directed network. Decide which to make last and work backwards.
Don't forget to add time to clean the machine before restarting the cycle.
You must follow on from point just labelled G – nearest neighbour algorithm.
b) butter – almond – ginger – chocolate – butter; upper bound for cleaning time = 71 minutes

Exercises 4.2 *Homework*

2 a – c – d – b – e – a; 110 seconds
3 Upper bound 622 miles; lower bound 467 miles
5 53, 75

Consolidation exercises for Chapter 4

1 a) AB, ACB, ADB, ACDB, ADCB
b) 16
c) Graph in b) has even vertices -Eulerian.
Graph in a) has odd vertices -not Eulerian

2 i) Travel time available is $30 - 10 - 4 \times 2\frac{1}{2} = 10$ mins.
Return time from last delivery is not relevant. This can be at least 3 mins (since any route can be followed in either direction).
Hence TSP solution of 10 + 3 = 13 or less suffices.
ii) The order is DABC
Total time is
$10 + 2 + 4 + 1 + 3 + 4 \times 2\frac{1}{2}$
= 30 minutes

3 a)

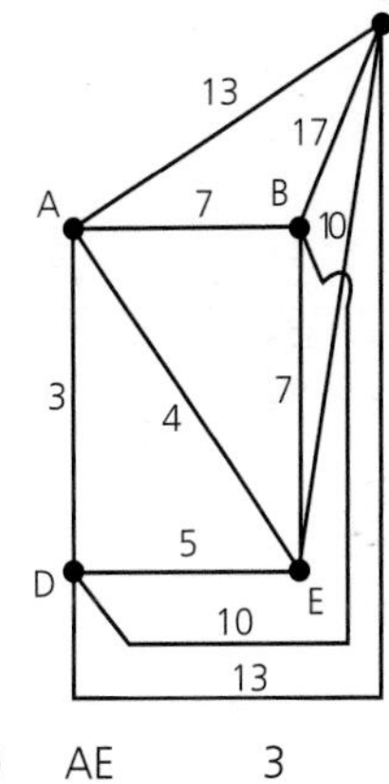

b)

AE	3
ED	5
DB	7
BD	17
CA	13
Length	45

c) tour of the park.
A→E→D→B→D→C
with length 45 km

4 a) The optional tour has length ≤440.
The nearest neighbour algorithm (starting from A)

ADBECFA	:340
ADFBECA	:330

b) The least edge from E is 40.

The minimum spanning tree for A, B, C, D, F has length 210.
So lower bound = 210 + 40 × 2 = 290

5 a) ABDECFA; 45 + 25 + 35 + 70 + 320 + 145 = 640, e.g. ABDEFCA costs 537

b) 5! = 120

c)

	A	B	C	D	E	F
A	–	65	80	78	110	165
B	75	–	97	55	113	130
C	80	90	–	70	90	340
D	90	65	90	–	75	250
E	110	90	80	45	–	82
F	165	130	320	195	100	–

d) B and E;
without tax BDE 25 + 35 = 60;
with tax BDE 55 + 75 = 130,
but BE = 113 with tax, 83 without tax

e) No difference – airport taxes are paid once at each town so they add 140 total cost regardless of which tour you use.

6 a)

	1 B	3 C	4 D	2 E	 F	 Arcs
B	0.5	10	31	7	23	BE
C	(10)	57	14	12	20	BC
D	31	14	11	(11)	9	ED
E	(7)	12	11	18	12	DF
F	23	20	(9)	12	43	

length 37

b) Add AE and AF, a total of 19, giving a lower bound <u>56</u>.

c) Hamiltonian cycle exists in this network.
If A is deleted (and arcs to it) we have a connector for other vertices which is less than or equal to a cycle through other vertices. Since AE + AF are minimum lengths for connecting A back into the network, then AE + AF + min connector ≤ length of min Hamiltonian cycle.

d) A – E – B – C – D – F – A
9 7 10 14 9 10
= 59 > 56

7 a) Minimum spanning tree AD, DE, EC, EB, CF, BG with length 298.

b) Since network satifies triangle inequality, initial upper bound can be found by doubling the minimum connector.
Upper bound ≤2 × 298 = 596km

c) Using nearest neighbour gives ADECFGBA of length 420km

d) Min connector for A
42 + 54 = 96
Min connector for rest of network BE, EC, ED, CF, BG. 256
Lower bound 256 + 96 = 352km

8 a) AC, CE, ED, DB, BA with cost 19

b) BC with cost 7 must be included AD, DB, BC, CE, EA with cost 27

9 a) London – Cambridge – Bristol – Exeter – Manchester – Liverpool – Leeds – Sheffield – Birmingham – Oxford – London; 938 miles

b) 9 stores ⇒ 27 hours
938 miles ⇒ 23 hours 27 minutes
total time of 50 hours 27 minutes

c)

Day 1: 7.00 – 8.30
London – Cambridge
8.30 – 11.30
Cambridge store
11.30 – 16.00
Cambridge – Bristol
Overnight in Bristol

Day 2: 8.30 – 11.30
Bristol store
11.30 – 13.36
Bristol – Exeter
13.36 – 16.36
Exeter store
16.36 – 21.00
Exeter to stop point N. of Birmngham
→ to Manchester

Day 3: 8.51 – 11.51
Manchester store
11.51 – 12.44
Man – Liv
12.44 – 15.44
Liverpool store
15.44 – 17.39
Liv – Leeds
Overnight in Leeds

Day 4: 8.30 – 11.30
Leeds store
11.30 – 12.24
Leeds – Sheff
12.24 – 15.24
Sheffield store
15.24 – 17.33
Sheff – Birm
Birmingham overnight

Day 5: 8.30 – 11 30
Birmingham store
11.30 – 13.12
Birm – Oxford
13.12 – 16.12
Oxford store
16.12 – 17.36
Oxford – London and home

CHAPTER 5
Networks 4: Route inspection problems

Exercises 5.1 *Classwork*

1 a) Planar

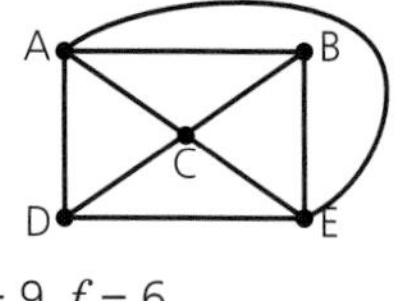

$v = 5, e = 9, f = 6$
$v - e + f = 2$

b) Planar

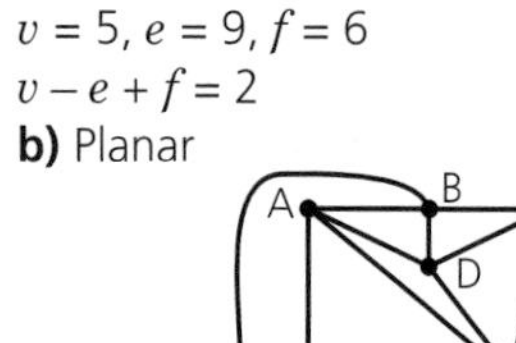

$v = 6, e = 12, f = 8$
$v - e + f = 2$

c) Not planar
$v = 6, e = 15$
For Euler to hold
$6 - 15 + f = 2 \Rightarrow f = 11$
$2e \geq 3f \Rightarrow 30 \geq 3f$
$\Rightarrow 10 \geq f$
∴ cannot be planar.

3 a) Not Eulerian, (B and D odd)

b) Eulerian, all vertices even
ADBAFDCFECBEA

c) Not Eulerian, all vertices odd

4 Graph 2 and graph 3 have the same sequence of vertex degrees and so are isomorphic.

5 Traversable graphs are not necessarily Eulerian – they can have two odd vertices, provided you do not end at the same vertex as you began.

Exercises 5.1 *Homework*

1 **b)** and **c)** are planar graphs, **a)** $K_{3,3}$ and **d)** are non-planar graphs.

2 **b)** is Eulerian.

3 There are several edge-disjoint cycles. Here are two for vertex a.

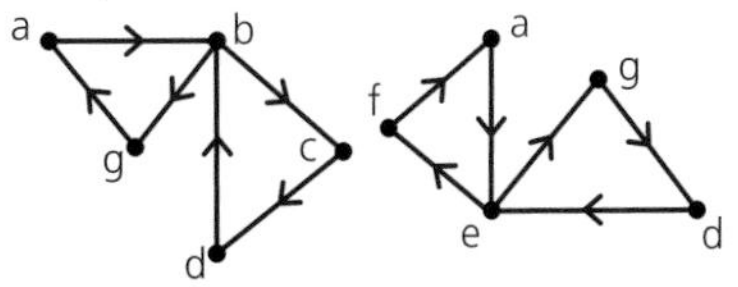

They give the Eulerian cycle abcdbgaegdefa.

5 **a)** **b)**

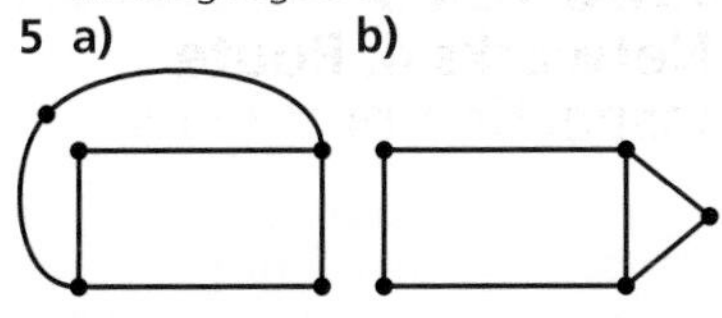

Exercises 5.2 *Classwork*

1

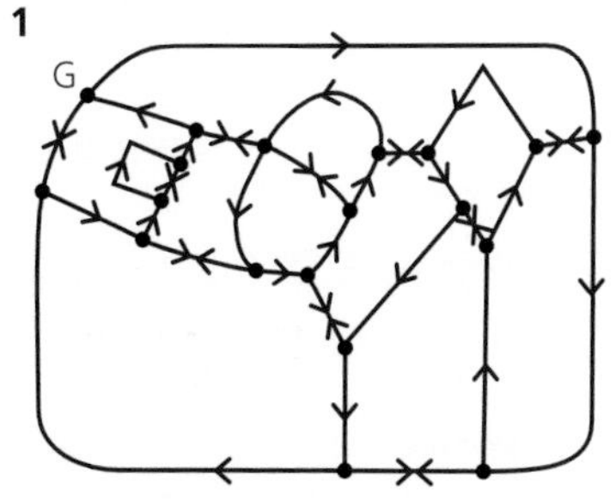

It is not possible without repeating paths since there are lots of odd vertices.

2

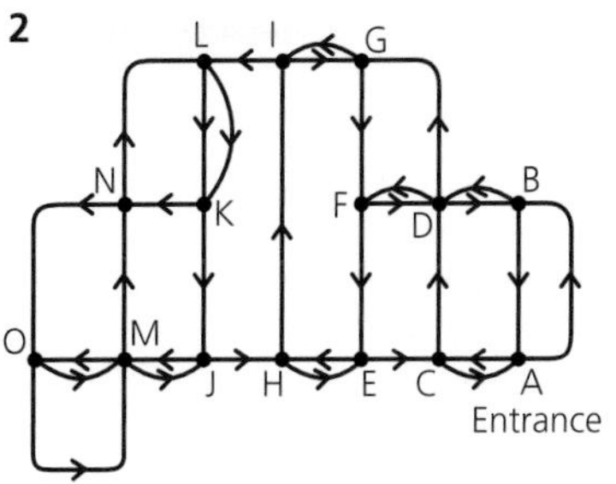

Graph vertices are at junctions of paths.
12 odd vertices – repeat 6 paths e.g. AC,GI, EH, FB(FDB), OJ (OMJ), KJ
Possible route:
ABACDBDGFDFEHIGILKJMNLKN OMOMJHECA

3 Odd vertices S, T ⇒ repeat ST length 6; distance along all edges 111 km; total to travel 111 + 6 = 117 km.

4 Odd vertices A, C, D, G; distance along all edges 148 km; total travel 148 + 35 = 183 km by repeating ABEG and CD.

5 A route of minimum length, 98 miles, is ABEGFEBACBFGDCDA
If road between E and F is closed then the route ABEGFBCADGDCA is 87 miles (note that these answers are not unique).

Exercises 5.2 *Homework*

1

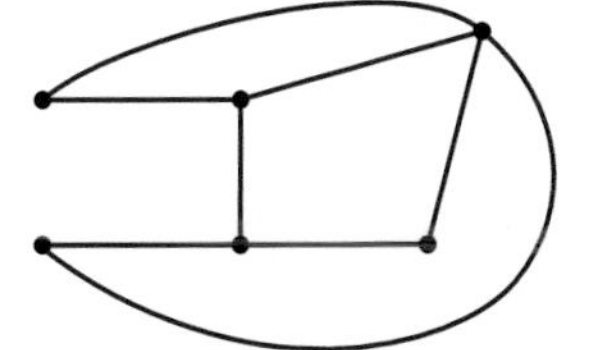

No Eulerian cycle is possible. Each door may be used at least once but you do not end where you started.

2 Graph is K_7 and each vertex is order 6.
No change by adding the doubles, as each vertex then has a loop.

3 61 miles

4 **a)** 909 miles
b) KF length 101

5 **a)** 460 mm **b)** 548 mm

Consolidation exercises for Chapter 5

1 **a)** 18
b)

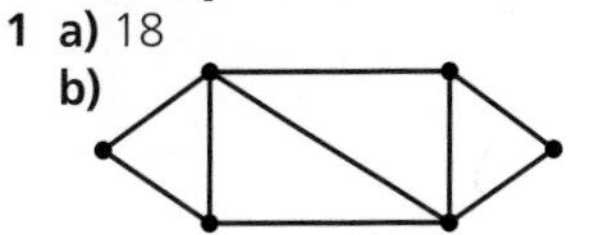

degrees of vertices 4, 4, 3, 3, 2, 2
c) i) planar – yes
ii) Eulerian – no. It has two odd vertices.

2 **a)** 0, 2, 4 (Σdegv must be even)
b) 2, 4 **c)** 2, 4 **d)** 2

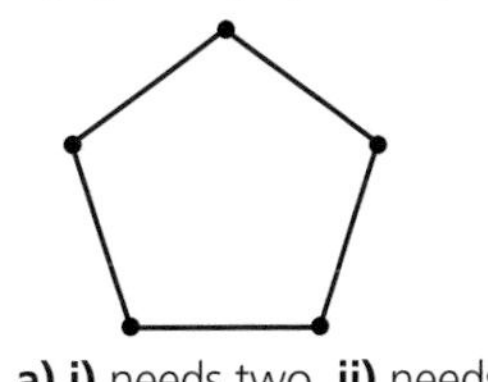

3 **a) i)** needs two, **ii)** needs three
b) $\frac{1}{2}k$

4 Minimum length is 54 repeating SV.

5 Route S A B S D E G F B E F S C G C B D A S with cost £18.48

6 **a)** Minimum length is 48 cm repeating AF and CG.
b) Upper bound is 28 cm (AEFDBCGFA)
1. A E F D B C G A 28
2. A E F D C G B A 24
3. A E F D G C B A 23

7 **i) Orders are**

a)

A	B	C	D	E
5	5	4	5	5

b)

A	B	C	D	E
5	4	4	4	3

c)

A	B	C	D	E
4	4	4	4	4

Hence a) is neither, since there are 4 odd nodes, b) is semi-Eulerian, since there are 2 odd nodes, c) is Eulerian, since all the nodes are even.
ii) No such route exists for graph **(a)**
There are routes for graph **(b)**, but they have to start and finish at odd nodes, i.e. A and E.
There are routes for graph **(c)**, where all routes start and finish at the same node, which can be any of the five.

8 **a) i)** odd vertices A, C, F, I; CPP route must duplicate an edge at every odd vertex since it must enter and leave the vertex.
ii) The minimum distance is 2620 m by repeating ABC and FI.
iii)

B	C	D	E	F	G	H	I
3	2	2	2	3	2	2	2

b) i) Redraw the network with two edges for every road.
ii) All vertices are now even ⇒ traversable distance = 2300 × 2 = 4600 m.
c) He can only go one way along each side of the road, so edges will need to be directed.

9 **a)**

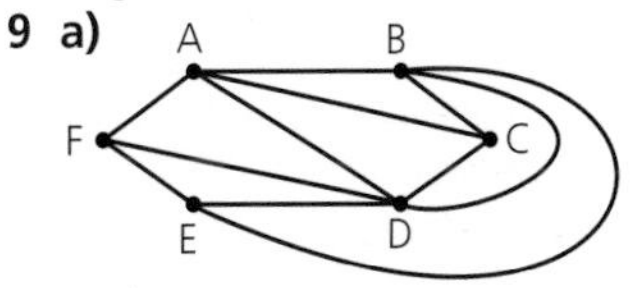

ABCDEF is a Hamiltonian cycle.
AC 'in' BE, BD 'out'
⇒ AD, FD 'in' ⇒ no way to connect F and C
b) i) One example is

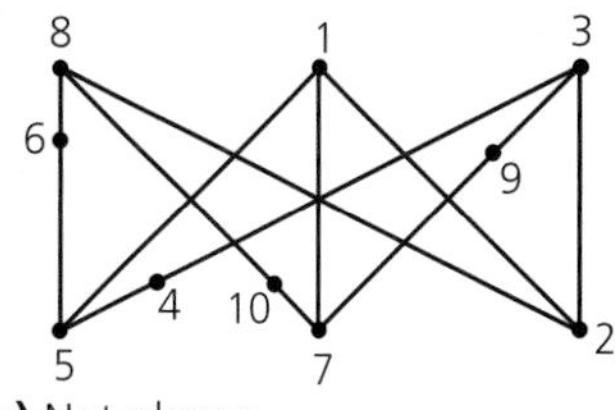

c) Not planar

10 **a)** 4 odd vertices ⇒ network not transversable.

b) By Dijkstra, shortest path A H K J or A H I J , 230 mm order of labelling vertices A G H B K I F C J D E.

c) Shortest distance is 2220 mm by repeating AB and EJ. Route
A B A H B C H G K H I C D I J K F J D E J E F G A
(no icing)

d) 14 odd vertices; no of pairings $13 \times 11 \times 9 \times 7 \times 5 \times 3 \times 1 =$ 135 135

CHAPTER 6
Critical path analysis

Exercises 6.1 *Classwork*

1

Activity	Duration	Preceding
A	2	–
B	4	A
C	5	A
D	2	C
E	3	B
F	8	C
G	4	D, E
H	1	G, F

2 Activity on vertex

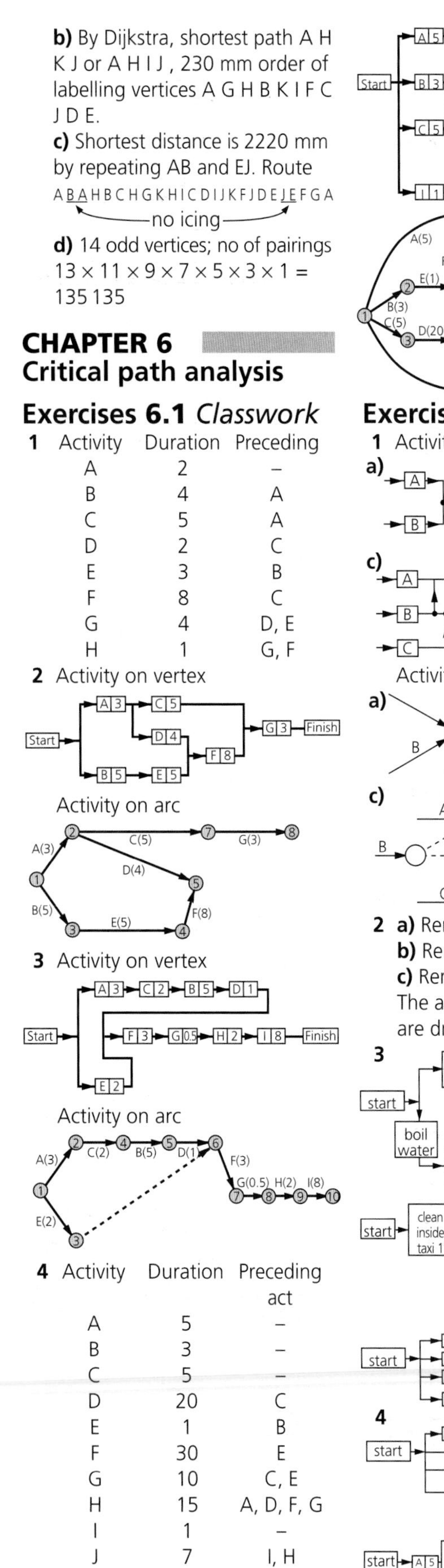

Activity on arc

3 Activity on vertex

Activity on arc

4

Activity	Duration	Preceding act
A	5	–
B	3	–
C	5	–
D	20	C
E	1	B
F	30	E
G	10	C, E
H	15	A, D, F, G
I	1	–
J	7	I, H
L	8	J

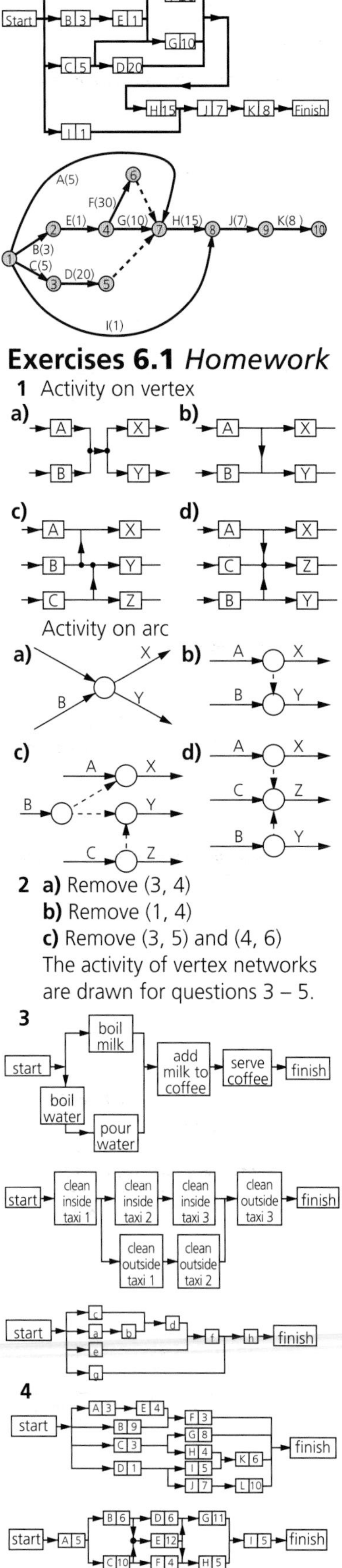

Exercises 6.1 *Homework*

1 Activity on vertex

a) **b)** **c)** **d)**

Activity on arc

a) **b)** **c)** **d)**

2 a) Remove (3, 4)

b) Remove (1, 4)

c) Remove (3, 5) and (4, 6)

The activity of vertex networks are drawn for questions 3 – 5.

3

4

5

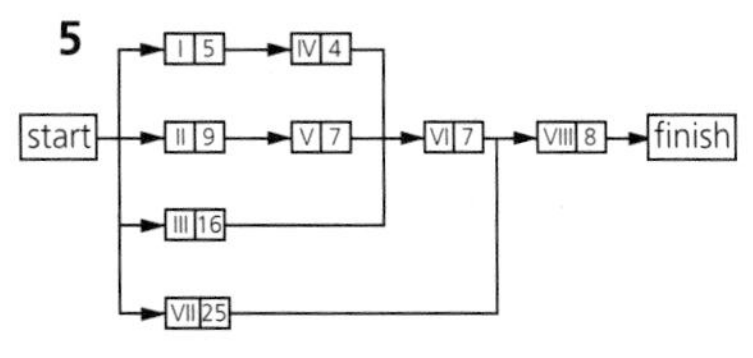

Exercises 6.2 *Classwork*

1. Critical path ACFH
2. Critical path BEFG
3. Critical path ACBDFGHI
4. Critical path BEFHJK

Exercises 6.2 *Homework*

1. Critical path CGIL; minimum project time 47.
2. Critical path KFGM; minimum completion time 16.
3. Exercise 6.1B question 4:
   a) Critical path DJ L, time 18 weeks
   b) Critical path ACG EI, time 43
   Exercise 6.1B question 5:
   Make test rig, test and dispatch; 33 weeks.

Exercises 6.3 *Classwork*

1

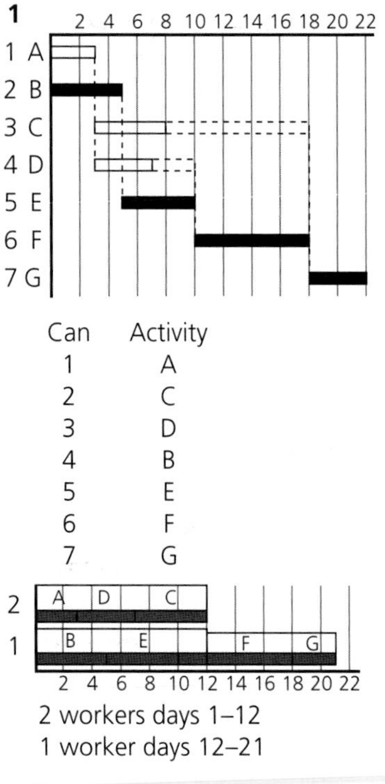

Can	Activity
1	A
2	C
3	D
4	B
5	E
6	F
7	G

2

2 workers days 1–12
1 worker days 12–21

2

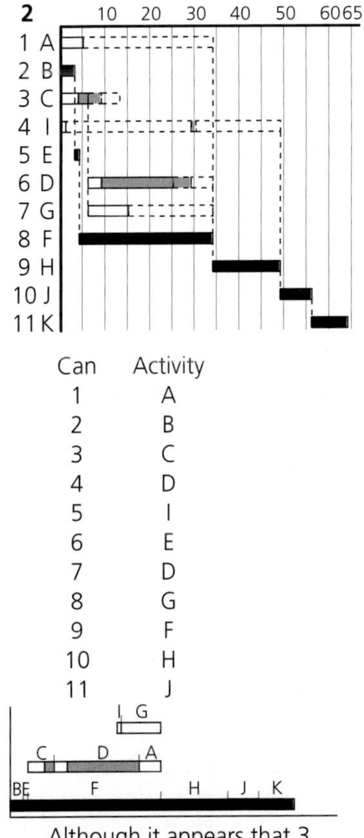

Can	Activity
1	A
2	B
3	C
4	D
5	I
6	E
7	D
8	G
9	F
10	H
11	J

Although it appears that 3 people are needed for a minimum time, if Marc times his activities correctly, he can do it alone since he doesn't need to work while the pie is cooking for example, so he can do another task. There is no time saved if his girlfriend helps.

3

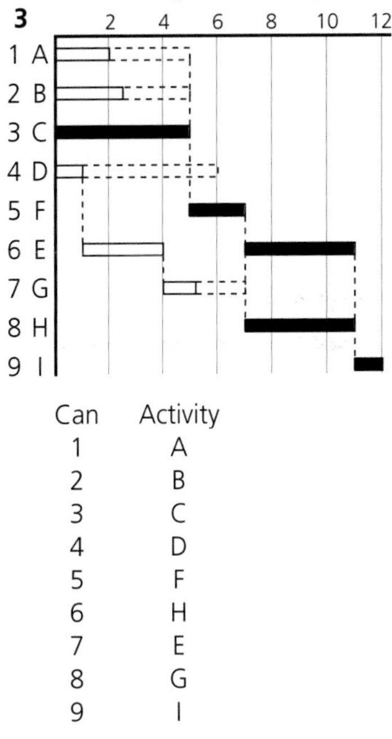

Can	Activity
1	A
2	B
3	C
4	D
5	F
6	H
7	E
8	G
9	I

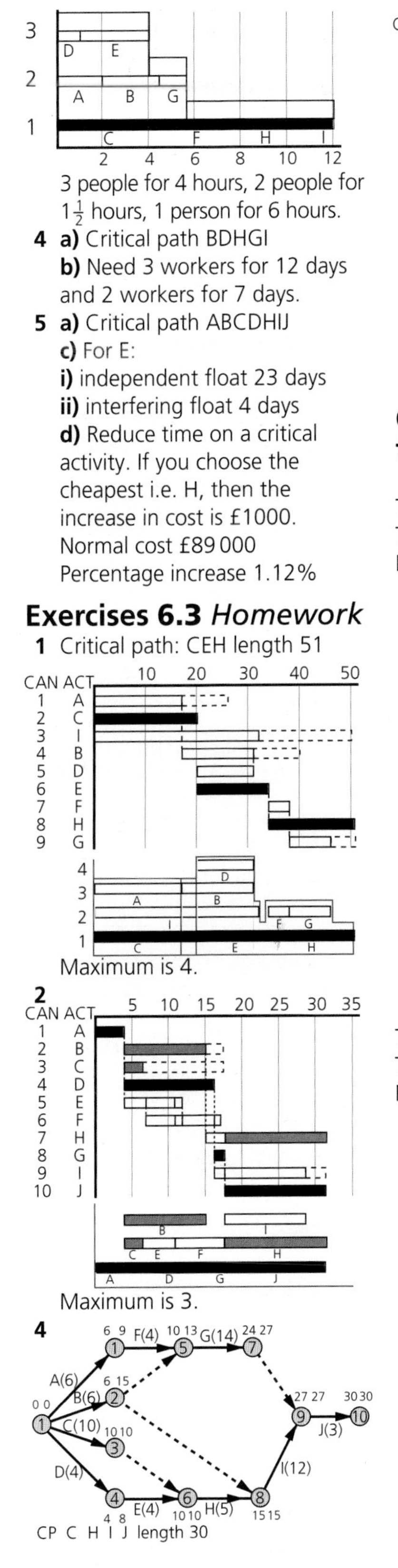

3 people for 4 hours, 2 people for $1\frac{1}{2}$ hours, 1 person for 6 hours.

4 a) Critical path BDHGI

b) Need 3 workers for 12 days and 2 workers for 7 days.

5 a) Critical path ABCDHIJ

c) For E:

i) independent float 23 days

ii) interfering float 4 days

d) Reduce time on a critical activity. If you choose the cheapest i.e. H, then the increase in cost is £1000. Normal cost £89 000 Percentage increase 1.12%

Exercises 6.3 *Homework*

1 Critical path: CEH length 51

Maximum is 4.

2

Maximum is 3.

4

CP C H I J length 30

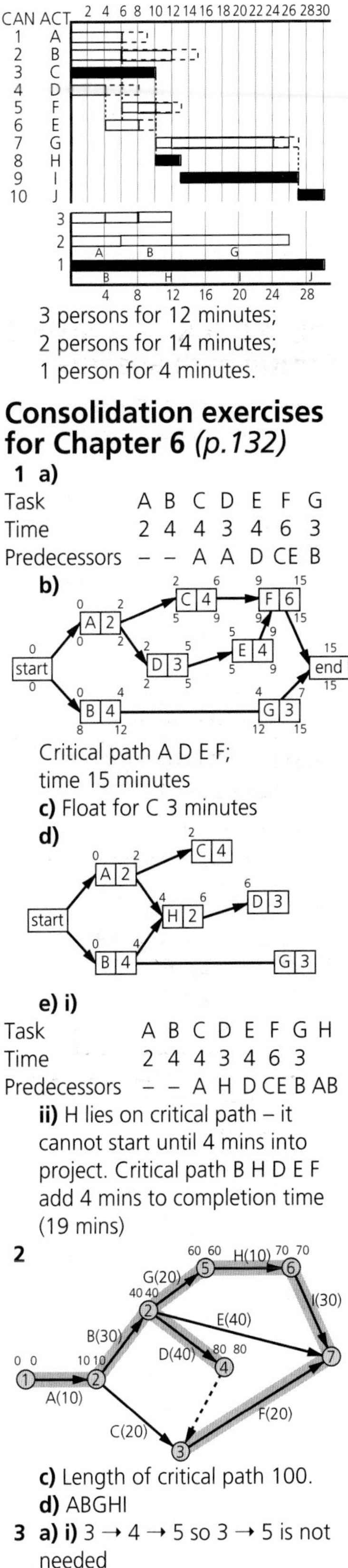

3 persons for 12 minutes;
2 persons for 14 minutes;
1 person for 4 minutes.

Consolidation exercises for Chapter 6 *(p.132)*

1 a)

Task	A	B	C	D	E	F	G
Time	2	4	4	3	4	6	3
Predecessors	–	–	A	A	D	CE	B

b)

Critical path A D E F; time 15 minutes

c) Float for C 3 minutes

d)

e) i)

Task	A	B	C	D	E	F	G	H
Time	2	4	4	3	4	6	3	
Predecessors	–	–	A	H	D	CE	B	AB

ii) H lies on critical path – it cannot start until 4 mins into project. Critical path B H D E F add 4 mins to completion time (19 mins)

2

c) Length of critical path 100.

d) ABGHI

3 a) i) 3 → 4 → 5 so 3 → 5 is not needed

ii)

Activity	A	B	C	D	E	F	G
Predecessors	–	A	–	A	BDC	CD	E
Event	1	2	3	4	5	6	
Early	0	2	6	6	8	9	
Late	0	2	6	6	8	9	

iv)

Activity	A	B	C	D	E	F	G
Float	0	1	3	0	0	1	0

v) Minimum completion time 9
Critical activities A D E G

b) Interfering float will affect the scheduling of other activities. Independent float in amount C can be changed without affecting the scheduling of other activities.

c) Interfering float = 3
Independent float = 0

4 i)

4 4 | 8 9 | 2 | C(4) | 4 | A(4) | F(1) | 0 0 | 1 | D(3) | dummy | 6 | 10 10 | B(2) | G(6) | 3 | E(1) | 5 | 2 4 | 4 4

ii) Critical activities A, G
iii) Float for C is 1

5 a) Critical activites A, D, E, G. Length of critical path is 17

b) Floats

B	1
C	6
F	6
H	9

c)

2 4 6 8 10 12 14 16 18 20
1 A
2 B
3 C
4 D
5 E
6 F
7 H
G

d)

2 4 6 8 10 12 14 16 18 20
1 A
2 B
3 C
4 D
5 E
6 F
7 H
G

Minimum number of workers = 2

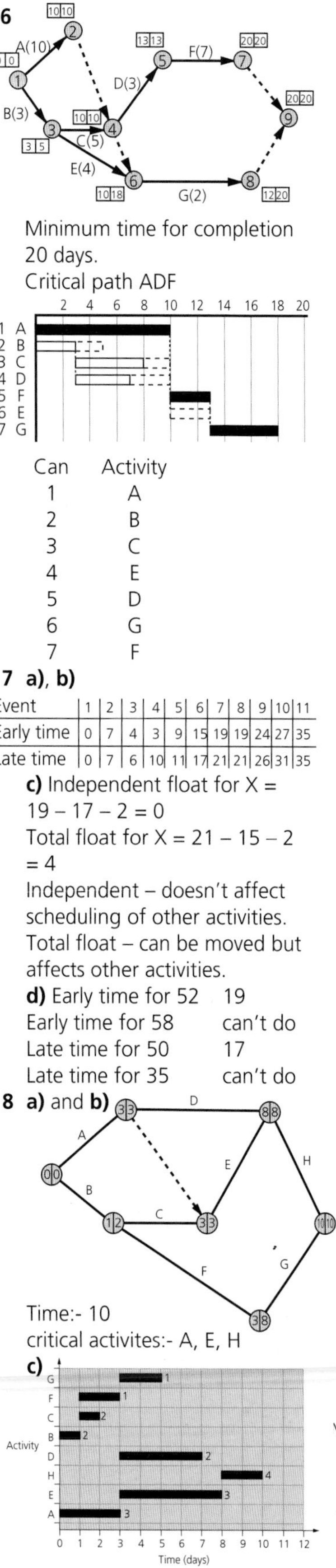

6

Minimum time for completion 20 days.
Critical path ADF

Can	Activity
1	A
2	B
3	C
4	E
5	D
6	G
7	F

7 a), b)

Event	1	2	3	4	5	6	7	8	9	10	11
Early time	0	7	4	3	9	15	19	19	24	27	35
Late time	0	7	6	10	11	17	21	21	26	31	35

c) Independent float for X = 19 – 17 – 2 = 0
Total float for X = 21 – 15 – 2 = 4
Independent – doesn't affect scheduling of other activities.
Total float – can be moved but affects other activities.

d)

Early time for 52	19
Early time for 58	can't do
Late time for 50	17
Late time for 35	can't do

8 a) and **b)**

Time:- 10
critical activites:- A, E, H

c)

d) F – 2; G – 8; D – 4

9 Minimum completion time 9 days. Critical path BEG
For C – at time = 3
For F – at time = 4

10 Critical activities BEGCF
Time to completion 13 days
13 day cost £30 200
In 12 days – choose to cut one day of cheaper activity on critical path
C (on C, F path) £700
E (on B E G path) £600
So extra cost is £1300.
% increase is 4.3%.

CHAPTER 7 Matchings

Exercises 7.1 *Classwork*

1 a) i)

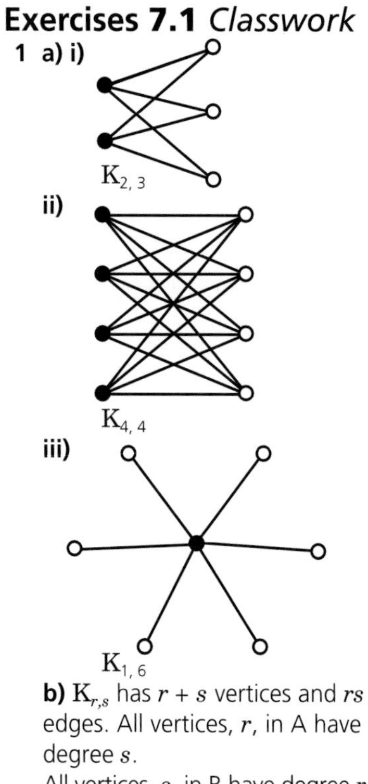

ii)

iii)

b) $K_{r,s}$ has $r + s$ vertices and rs edges. All vertices, r, in A have degree s.
All vertices, s, in B have degree r.

2 a) non-planar – subdivision K_5
b) non-planar – contains $K_{3,3}$
c) planar

3
A 1, 3
B 1, 3, 4
C 1, 2, 5
D 3, 5
E 4

4 Girls Boys

Vi BR G BR
L BU P BU
T BA E BA
K F J F

5
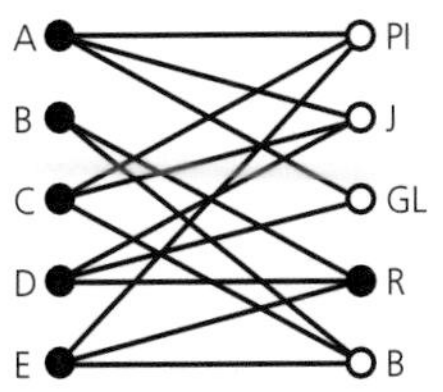

Exercises 7.1 *Homework*

1 **a)** and **c)** are planar
b) is non planar

2
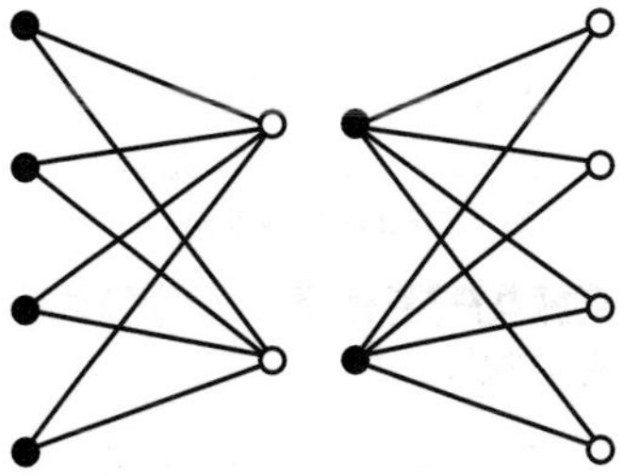

$K_{4,2}$ $K_{2,4}$

3 $2n$, $n(2n - 1)$, $n = 1, 2$

4
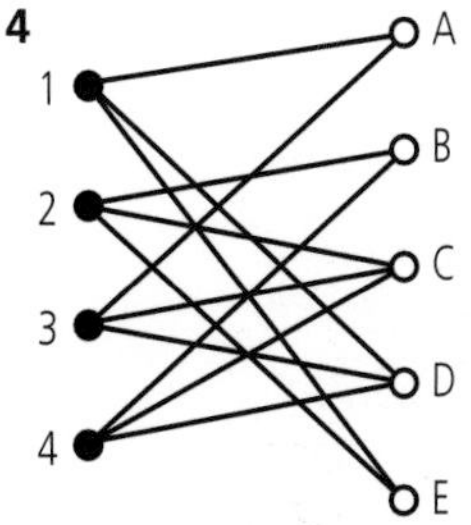

Exercises 7.2 *Classwork*

1 Miss Abraham – Environmental studies
Mr Bowen – Science
Mrs Carter – Geography
Ms Delgardo – Maths
Mr Eggham – Graphics

2 A – 1, B – 3, C – 2, D – 5, E – 4 (**or** A – 3, B – 1, C – 2, D – 5, E – 4 etc)

3 There are several solutions.
i) Meena – backstroke
Lauren – breaststroke
Tara – freestyle
Kerry – butterfly
ii) Meena – breaststroke
Lauren – freestyle
Tara – backstroke
Kerry – butterfly
iii) Meena – backstroke
Lauren – freestyle
Tara – breaststroke
Kerry – butterfly
i) Gary– breaststroke
Paul – backstroke
Eddie – butterfly
James – freestyle
ii) Gary– breaststroke
Paul – freestyle
Eddie – backstroke
James – backstroke
iii) Gary – butterfly
Paul– freestyle
Eddie – butterfly
James – breaststroke
iv) Gary – freestyle
Paul – backstroke
Eddie – butterfly
James – breastroke
v) Gary – freestyle
Paul – butterfly
Eddie – butterfly
James – breastroke

4 Anne – plastering
Brian – bricklaying
Carl – joinery
Daljit – glazing
Errol – roofing

5 Annabel – ladies fashions
Bettina – furnishing
Chris – electrical goods
Doreen – hardware
Imran – menswear
Answer not unique

Exercises 7.2 *Homework*

1 $x_1 - y_3$, $x_2 - y_2$, $x_3 - y_4$, $x_4 - y_1$

2 $1 - b$, $2 - a$, $3 - d$, $4 - c$

3 $a - E$, $b - G$, $c - D$, $d - C$, $e - F$, $f - B$, $g - A$; This is a unique solution

5 $1a$ – Doing, $16a$ – Going, $20a$ – Frame, $8di$– draft, $12d$ – Grant

Consolidation exercises for Chapter 7

1
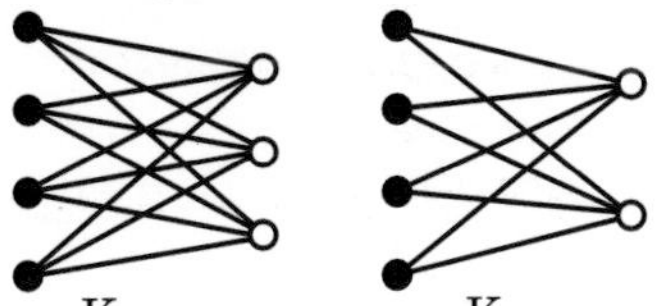

$K_{4,3}$ $K_{5,2}$

2 **i)** Non-planar (contains $K_{3,3}$)
ii) Planar $v = 8$, $e = 15$, $f = 9$
$v - e + f = 2$

3 **a)**
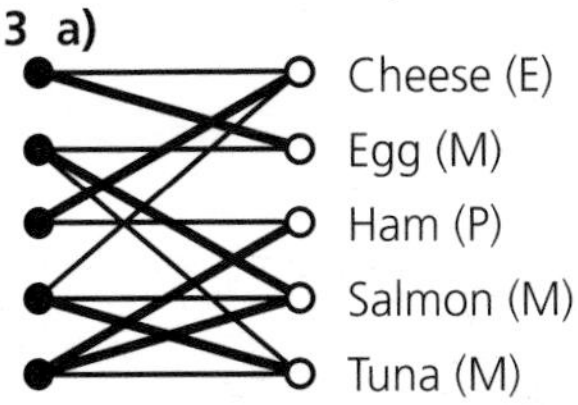

Initial matching L-E, N-C, O-T, P-S
Alternating path H-P-S-M

c) Solution

Mr Large	Egg
Ms Nice	Cheese
Mr Oliver	Tuna
Miss Patel	Ham
Mrs Moore	Salmon

4 (i)
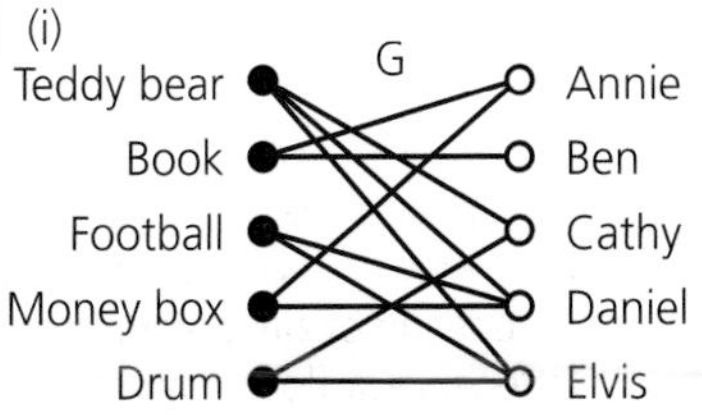

(ii) initial matching (M)
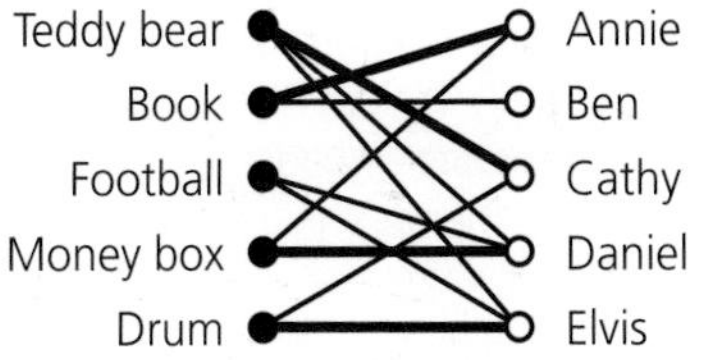

(iii)
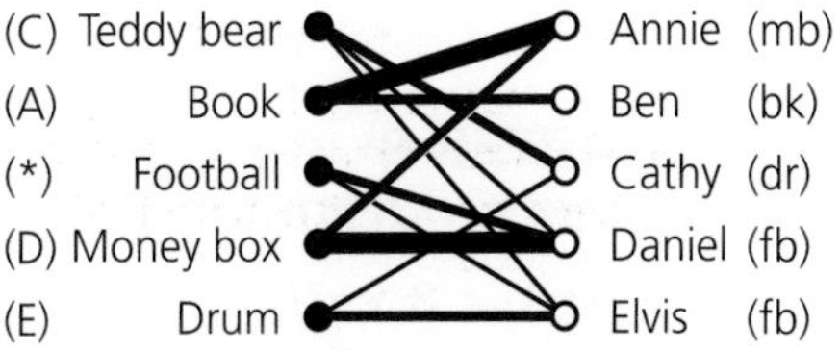

Alternating path is B-bk-A-mb-D-fb, and so a maximal mating is (Annie, Money Box), (Ben, Book), (Cathy, Teddy Bear), (Daniel, Football), (Elvis, Drum)

5 **a)**
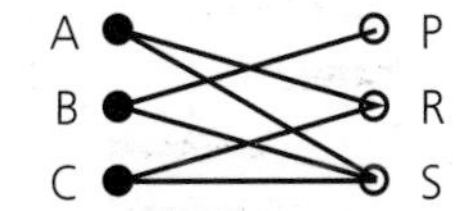

b)
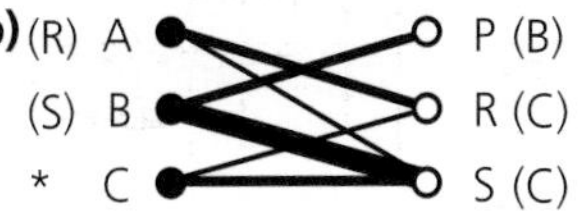

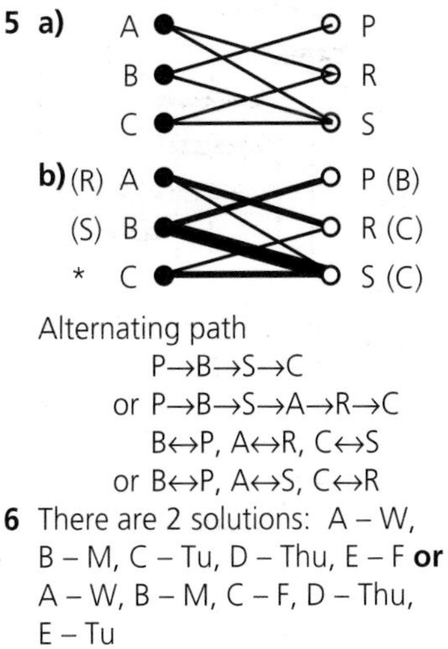

Alternating path
P→B→S→C
or P→B→S→A→R→C
B↔P, A↔R, C↔S
or B↔P, A↔S, C↔R

6 There are 2 solutions: A – W, B – M, C – Tu, D – Thu, E – F **or** A – W, B – M, C – F, D – Thu, E – Tu

7 a) (S) A — E (B)
* B — F (B)
(F) C — S (B)
(E) D — W (C)

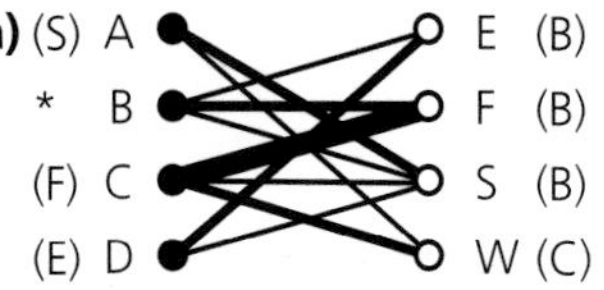

Ahmed – stationary
Bridget – food
Chris – watches
Diane – electrical

8 Mrs Frawt – publicity, Miss Delgardo – tickets, Miss Perry – costumes, Mr Quince – lights, Ms Rouse – make-up, Mrs Standish – scenery, Mr Taylor – producer, Mr Unwin – sound

9 Couple – twin room
Family 4 – family room
Family 5 – appartment 6
Family 7 – flat 8
3 friends – appartment 4
Five friends will be disappointed.

10 a) 4 ways
b) Mike – butterfly
Ned – freestyle
Oliver – breaststroke
Pete – backstroke

CHAPTER 8
Linear programming 1

Exercises 8.1 *Classwork*

1 Maximise $x + y$
Where $2x + 3y \leq 36$ (eggs)
$250x + 150y \geq 3000$ (margarine)
$x \geq 0, y \geq 0$
and x is the number of chocolate cakes and y is the number of coffee cakes.

2 Maximise $p = 15x + 11y$
Where $x \leq 3$ (moulder)
$3x + y \leq 12$ (lathe)
$x + y \leq 7$ (assembler)
$x \geq 0, y \geq 0$
and x is the number of bicycles and y is the number of trucks.

3 Minimise $x + y$
Where $6x + 18y \geq 54$ (distance covered)
$4x + 3y \leq 12$ (aerobic points)
$y \geq 2x$
$x \geq 0, y \geq 0$
and x is time spent jogging and y is time spent cycling.

4 maximise $P = 8000x + 11500y$
Where $150x + 200y \leq 24000$ (labour)
$x + 2y \leq 150$ (plots)
$40x + 105y \leq 9600$ (capital)
and x is the number of 'one plot houses' and y is the number of 'two plot houses'.

5 Maximise $P = 1.55X - (0.7x_1 + 1.05x_2 + 1.05x_3 + 1.3x_4 + 1.1x_5)$
where $X = x_1 + x_2 + x_3 + x_4 + x_5$ = amount of Alma margarine.
$5.6X < 1.2x_1 + 3.4x_2 + 8x_3 + 10.8x_4 + 8.3x_5 < 7.4X$
$x_1 + x_2 + x_3 \leq 40\,000$
$x_4 + x_5 \leq 32000$
and x_1, x_2, x_3, x_4, x_5 are the weights of groundnut, soya bean, palm, lard and fish respectively.

Exercises 8.1 *Homework*

1 $\frac{5}{x} + \frac{3}{y} \leq \frac{4}{3}, x > 0, y > 0$

2 Minimise £C where
$C = 0.9x + 0.25y$
subject to $5x + 20y \geq 2$, $0.5x + 0.2y \geq 0.3$, $0.04x + 0.08y \leq 0.04$
$x \geq 0, y \geq 0$ where x kg of meat and y kg of biscuits are purchased.

3 If x denotes the number of thousand kg of XTRA and y the number of thousand kg of YTER produced per week, $x \geq 0, y \geq 0$, $5 \leq 0.9x + 0.5y \geq 20$, $2 \geq 0.1x + 0.5y \geq 7$, $x + y \geq 25$ and profit $P = 1.5x + 2.1y$ (in thousand pounds).

4 Minimise £C where $C = 20x + 16y$ subject to $x \geq 0, y \geq 0$, $6x + 2y \geq 12$, $2x + 2y \geq 8$, $4x + 12y \geq 24$, $x + y \leq 8$
where x hours is spent in Orchard A and y hours in Orchard B.
New cost function $C = 34 + 8x$.

5 If x, y in 1000 litres are the capacities of CAH and BBM respectively $x, y \geq 0$, $\frac{1}{3}x \leq y \leq 4x$, $x + y \leq 5$, $y \geq 2 - \frac{x}{2}$
Cost (\$) C = $300x + 220y$, profit (\$) P = $1700x + 1780y$

Exercises 8.2 *Classwork*

1 a) $f = 132$ at (12, 36)
b) $f = 190$ at (10, 90)
c) $f = 4$ at (2, 2)
d) $f = 47$ at (7, 6)
e) $f = 38\frac{1}{3}$ at (20, 25)

2 a) $c = 12360$ at (1080, 180)
b) $f = 20$ at (5, 10)
c) $f = 18$ on line $2x + y = 36$ for $9 \leq x \leq 16.5$ and $3 \leq y \leq 18$

3 a) $f = 1550$ at (90, 65)
b) $c = 60$ at (60, 0)

4 1) $14\frac{2}{3}$ at $(8, \frac{20}{3})$
2) 89 at (3, 4)
3) $\frac{11}{3}$ at $\left(1, \frac{8}{3}\right)$
4) Build 150 'one-plot houses'
5) Too many variables.

Exercises 8.2 *Homework*

1 (0, 0), (0, –1), (2, 2)
2 (–1, –1) and $z = 5$
3 (15, 0), –30
4 a) –2 at (1, 2) **b)** 10 at (–1, 2)
5 2) $x = 0.5, y = 0.25$; $C = 0.5125$
3) $x = 13.75, y = 11.25$; $P = 44250$
4) $x = 1, y = 3, C = 68$; $x = 0, 6 \leq y \leq 8, C = 44$
5) $x = \frac{2}{9}, y = \frac{8}{9}, C = 262.22$; $x = 1, y = 4, P = 8820$.

Consolidation exercises for Chapter 8

1 i) $a + 4b + 3c \leq 50$
$a + 5b + 2c \leq 20$
ii) $3a + 8b + 6c$
iii) Maximise $P = 3a + 8b + 6c$
Subject to $a + 4b + 3c \leq 50$
and $a + 5b + 2c \leq 20$
Together with $a \geq 0, b \geq 0, c \geq 0$
iv) The variables a, b, c must be integers

2 a) $3x + y \geq 9$
$x + y \geq 7$
$x + 2y \geq 8$
b)

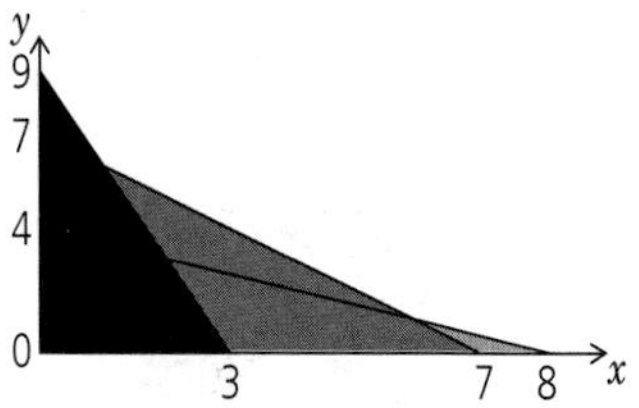

c) 1 Xtravit tablets and 6 Yeastalife tablets

3 a) Maximise P = $2.5x + 3y$
subject to $3x + 6y \leq 90$
$2x + y \leq 35$
$x + y \leq 20$
b)

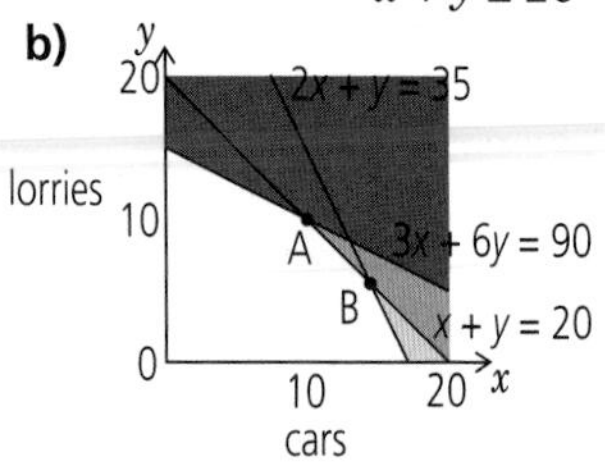

c) At A (10, 10), I = £55
At B (15, 8), I = £52.50
Make 10 cars and 10 lorries
d) Department B, 5 hours

4 b) $P = 7x + 8y$
c) 3 cushions and 4 ragdolls

5 a) $4x + 3y \leq 5100$
b) $4x + 5y \leq 5500$
$2x + 3y \leq 3000$
c) $P = 5x + 7y$
d) Maximise $P = 5x + 7y$ subject to $4x + 3y \leq 5100$, $4x + 5y \leq 5500$, $2x + 3y \leq 3000$, $x, y \geq 0$
Bake 750 biscuits and 500 buns.
Income £72.50

6 a) Maximise $f = 3a + 4b + 1600$ subject to $a + b \leq 200$, $5a + 7b \leq 1200$, $a, b \geq 0$
b) Produce 100 m of plan A and 100 m of plan B
Income £23.00

7 a), b) and **c)** $3x + y \geq 240$
$3x + y \leq 480$
d) Cheapest 100 ml gin, 100 ml martini
Most expensive 140 ml gin, 60 ml martini
e) $\frac{1}{2} \leq p \leq \frac{7}{10}$ corresponding to $100 \leq x \leq 140$

8 Produce 4 Alpha and 4 Beta spaceships.

9 a) i) demand must be satisfied in month 1.
ii) demand must be satisfied in months 1 and 2.
iii) No more than total demond may be produced in months 1 and 2.
b)

P_2
(5, 4)
(9, 0)
P_1

c) $p_1 + 5p_2 + 2(15 - p_1 - p_2) + 3(p_1 - 5) + 3(p_1 + p_2 + - 9) = 5p_1 + 6p_2 - 12$
d) Best plan: $p_1 = 9$, $p_2 = 0$ ($p_3 = 6$). Cost = 33

10 a) $4x \geq y$
b) Minimise c = $0.3x + 0.2y$
Subject to
$y \leq 4x, x \geq 0$
$x + y \geq 60\,000$
$y \geq 40\,000$
$0.1y \geq 0.05(x + y) \Rightarrow y \geq x$

c)

y
$y = 4x$
feasible region
$y = x$
A
$y = 40000$
B
$x + y = 60000$
x

d) Minimum c = £13 200 at (12 000, 48 000)

CHAPTER 9 Linear programming 2

Exercises 9.1 *Classwork*

1 a) (2, 3), $r = 3$
b) $\left(\frac{31}{5}, -\frac{9}{5}\right)$, $r = 4$
c) $\left(\frac{94}{11}, \frac{23}{11}\right)$, $r = \frac{1}{2}$
d) $\left(\frac{111}{41}, \frac{1}{41}\right)$, $r = \frac{4}{3}$
e) (1, 1), $r = \frac{3}{2}$
f) (0.92, 3.2), $r = \frac{1}{5}$

2 a) (–3, 0, 1)
b) (0.8, –1.8, 1.8)
c) $\left(-\frac{8}{7}, -\frac{9}{7}, -\frac{3}{7}\right)$
d) (1, 2, 3)

3a) (5, 2, 1)
b) (1, –1, 2, 3)

4 a) $\left(-\frac{23}{3}, \frac{25}{3}\right)$ **b)** No solution, parallel lines

5 $h = 2 + 30t - 5t^2$
Launched at h = 2 metres

Exercises 9.1 *Homework*

1 a) $x = 1, y = 2$
b) $x = 2, y = 0$
c) $x = 1, y = 3$
d) $x = 2, y = -1$
e) $x = 1, y = -1$
f) $x = 2a + b, y = a$

2 a) Infinity of solutions
b) No solutions

3 a) $x = 1.7, y = -0.2$
b) No solutions

4 a) $x = 1, y = 1, z = -1$
b) $x_1 = -2, x_2 = 0, x_3 = -1$

5 $d = 8, a = 2, b = -5, c = 1$

Exercises 9.2 *Homework*

3 a) Unique solution $z = 2, y = 4, x = 2$
b) No solutions
c) Infinity of solutions, $z = t$, $y = 2 - 3t$, $x = 1 + 2t$

4 a) i) $\lambda \neq -1, \lambda \neq 1, x = \dfrac{\lambda}{\lambda + 1}$, $y = \dfrac{1}{\lambda + 1}$
ii) $\lambda = 1, y = t, x = 1 - t$, for any t
iii) $\lambda = -1$
b) i) $\lambda \neq 0, \lambda \neq -2, z = \dfrac{2}{\lambda}$, $y = \dfrac{4 - \lambda}{\lambda}$, $x = \dfrac{4\lambda - 6}{\lambda}$
ii) $\lambda = -2, z = t, y = 2t - 1$, $4 - 3t$, for any t
iii) $\lambda = 0$

Exercises 9.3 *Classwork*

1 a) $f = \frac{2502}{19}$ at $\left(\frac{222}{19}, \frac{686}{19}\right)$
b) $c = 15$ at (15, 0)
c) $f = 1457.65$ at (88.46, 57.69)
d) $g = 65$ at (15, 5)
e) $M = 806$ at (81.5, 241.5)

2 a) $f = 150$ at (0, 30)
b) $f = 4.5$ at (1.5, 3)
c) $c = 12$ on line $2x + y = 24$ between (9, 6) and (10, 2)
d) $c = 10\,560$ at (1080, 180)

3 3 bicycles and 3 trucks (assuming whole number of toys; if not, $2\frac{1}{2}$ bicycles and $4\frac{1}{2}$ trucks.

4 8 chocolate cakes and 6 coffee cakes (assuming integer number of cakes.

Exercises 9.3 *Homework*

1 a) $x = 10, y = 1, f = 230$
b) $x = 2, y = 0, g = 2$
c) $x = 0, y = \frac{35}{3}, h = \frac{245}{3}$
d) $x = 5, y = 15, M = 440$
e) $x = 10, y = 10, N = 200$

2 a) $x = 0, y = 20, f = 60$
b) $x = 80, y = 0, g = 640$
c) $x = 2, y = 32, h = 182$
d) $x = 8, y = 14, m = 46$

3 40 kg of A, 470 kg of B, income = £3060

4 Maximum $f = 4x + 3y$, subject to $15x + 4y \leq 1640$, $2x = y \leq 50$ and $y \leq 300$

5 150 16-bit, 250 32-bit, profit = £145 000

Consolidation exercises for Chapter 9

1 i)

	P	x	y	z	s_1	s_2	
	1	–3	–4	–5	0	0	0
	0	2	8	5	1	0	3
	0	9	3	6	0	1	2
$+5R_3$	1	$4\frac{1}{2}$	$-1\frac{1}{2}$	0	0	$\frac{5}{6}$	$\frac{5}{3}$
$-5R_3$	0	$-5\frac{1}{2}$	$5\frac{1}{2}$	0	1	$-\frac{5}{6}$	$1\frac{1}{3}$
$R_3 \div 6$	0	$\frac{3}{2}$	$\frac{1}{2}$	1	0	$\frac{1}{6}$	$\frac{1}{3}$
$+1\frac{1}{2}R_2$	1	3	0	0	$\frac{3}{11}$	$\frac{20}{33}$	$\frac{67}{33}$
$R_2 \div 5\frac{1}{2}$	0	–1	1	0	$\frac{2}{11}$	$-\frac{5}{33}$	$\frac{8}{33}$
$-\frac{1}{2}R_2$	0	2	0	1	$-\frac{1}{11}$	$\frac{8}{33}$	$\frac{7}{33}$

ii) After 1 iteration $x = 0$, $y = 0$, $z = \frac{1}{3}$, $P = 1\frac{2}{3}$
After 2 iterations $x = 0$, $y = \frac{8}{33}$, $z = \frac{7}{33}$, $P = \frac{67}{33}$
iii) No negative numbers in top row, therefore solution is optimal.

2 i) Let coaches moving AD be x_1
AE be x_2
AF be x_3
BD be y_1
BE be y_2
BF be y_3
CD be z_1
CE be z_2
CF be z_3

ii) Minimise $P = 40x_1 + 70x_2 + 25x_3 + 20y_1 + 40y_2 + 10y_3 + 35z_1 + 85z_2 + 15z_3$

iii) subject to $x_1 + x_2 + x_3 \le 8$
$y_1 + y_2 + y_3 \le 5$
$z_1 + z_2 + z_3 \le 7$
$x_1 + y_1 + z_1 \ge 4$
$x_2 + y_2 + z_2 \ge 10$
$x_3 + y_3 + z_3 \ge 6$

$x_1 + x_2 + x_3 + y_1 + y_2 + y_3 + z_1 + z_2 + z_3 \le 20$

3 Minimise $z = x_1 + 2x_2 + 3x_3$
subject to $x_1 + 3x_2 + 2x_3 > 300$
$2x_1 + x_2 + 3x_3 > 150$
$2x_1 + 2x_2 + 5x_3 > 200$
$x_1, x_2, x_3 > 0.$

4 a) $c = 3.8x + 4.8y$
b) $3x + 0.2y > 32$
$300x + 40y > 5500$
$x + 10y > 30$
Use integer solutions, the parents should serve 19 measures of milk and 2 measures of juice.

5 Ben should spend 6 hours per week mowing lawns and 9 hours per week cleaning cars.

6 Build 150 'one plot houses'.

8 a) $2x + y \le 10$, $x + y \le 6$, $x + 3y \le 15$, $x, y \ge 0$
b) $P = 160x + 120y$
c) 4 tonnes of product X and 2 tonnes of product Y gives a profit of £880.

9 Maximise $D = m + c + l$
subject to $2m + 3c + 6l \le 300$
$6m + 3c + 4l \le 360$
$m \le 55$
$c + l \le 55$
$m, c, l \ge 0$

b)

	m	c	l	u	v	w	x	P		
R_1	2	3	6	1	0	0	0	0	300	
R_2	6	3	4	0	1	0	0	0	360	Increase m R_3
R_3	1	0	0	0	0	1	0	0	55	
R_4	0	1	1	0	0	0	1	0	55	
R_5	−1	−1	−1	0	0	0	0	1	0	
$R_1A = R_1 - 2R_3$	0	3	6	1	0	−2	0	0	190	
$R_2A = R_2 - 6R_3$	0	3	4	0	1	−6	0	0	30	Increase c R_2
$R_3A = R_3$	1	0	0	0	0	1	0	0	55	
$R_4A = R_4$	0	1	1	0	0	0	1	0	55	
$R_5A = R_5 + R_3$	0	−1	−1	0	0	1	0	1	55	
$R_1B = R_1A - 3R_2B$	0	0	2	1	−1	4	0	0	160	
$R_2B = 3R_2A$	0	1	$\frac{4}{3}$	0	$\frac{1}{3}$	−2	0	0	10	
$R_3B = R_3A$	1	0	0	0	0	1	0	0	55	
$R_4B = R_4A - R_2A$	0	0	$-\frac{1}{3}$	0	$-\frac{1}{3}$	2	1	0	45	
$R_5B = R_5A - R_2B$	0	0	$\frac{1}{3}$	0	$\frac{1}{3}$	−1	0	1	65	

d) Not optimal (objective function line not all positive.) Travel 10 miles by car and 55 miles by moped; a total distance of 65 miles
All time is used and $1\frac{1}{3}$ gallons of fuel left.

10 b) Violates trivial constraints
c) i) $x = 0$, $y = 8$, $c = 8$
ii) $s_1 = 0$, $s_2 = 2$, $s_3 = 10$ which satisfies trivial constraints.
d) $x = 4$, $y = 1$, $s_1 = 1$, $s_2 = 0$, $s_3 = 0$ and $c = 5$

CHAPTER 10 Simulation

Exercises 10.1 *Classwork*

1 a) 331
b) 2791
c) $3t^2 + 3t + 1$

2 Using a dice to simulate spread to a tree rather than all at once.
e.g.

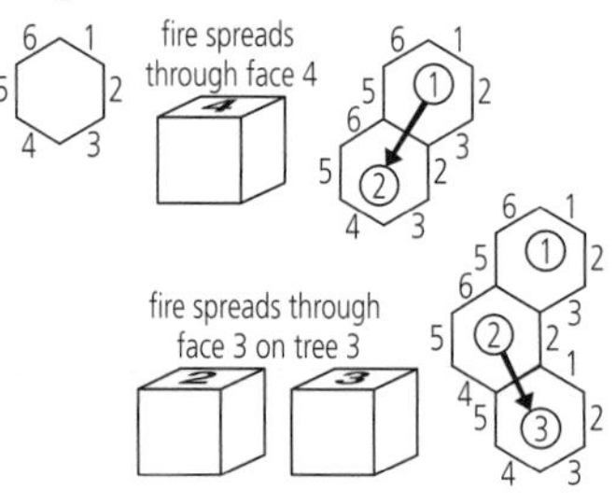

4 Use a dice and coin and the following table.

	H	T
1	TOY 1	TOY 7
2	TOY 2	TOY 8
3	TOY 3	TOY 9
4	TOY 4	TOY 10
5	TOY 5	TOY 11
6	TOY 6	TOY 12

Toy 6 would correspond to a 6 and a head.

5 a) Use numbers on dice from (1, 1) →.
Ignore (2, 6) to (6, 6) inclusive if they come up.
b) Use a dice and coin to generate 12 possibilities (or 2 dice: 1–6 and even/odd)
c) If you generate a non-existent date, ignore it and continue.

Exercises 10.1 *Homework*

3 $\frac{1}{16}, \frac{1}{4}, \frac{3}{8}, \frac{1}{4}, \frac{1}{16}$

Exercises 10.2 *Classwork*

1 Queueing — Use counters to record.
Situation 1 – Each new customer chooses to join shorter queue
Situation 2 – Record which service point is free hence where next customer goes.

2 Rules:

Interarrival time	1	2	3	4	5
RND	0–54	55–84	85–94	95–98	99

Service time	1	1.5	2	2.5	3
RND	0–19	20–43	44–71	72–89	90–99

e.g. Average wait – 4 minutes,
Maximum wait – 6.5 minutes
Average queue – 2.5
Maximum queue 4
Not very fast – would be fast food with 2 service points]

3 a)

Time to serve	1	2	5
RND	0–24	25–74	75–99

b)

Customer	1	2	3	4	5	6	7	8	9	10	11	12
Service time	5	2	2	2	1	2	1	2	5	5	1	1

c)

Time between customers	1	2	3	4
RND	1–15	16–45	46–75	76–90

Using 01–90 ($\frac{1}{6}$ is 15 numbers, $\frac{1}{3}$ is 30 numbers)
Ignore 91–99.

d)

Customer	1	2	3	4	5	6	7	8	9	10	11	12
Time (between)	2	1	1	2	1	3	3	4	4	4	1	1

e)

Time (minutes)	Customer arriving	Time of next arrival	Customer Start service	Time end service	Length of queue
0		2			0
1					0
2	1st	3	1st	7	0
3	2nd	4			1
4	3rd	6			2
5					2
6	4th	7			3
7	5th	10	2nd	9	3
8					3
9			3rd	11	2
10	6th	13			3
11			4th	13	2
12					2
13	7th	17	5th	14	2
14			6th	16	1
15					1
16			7th	17	0
17	8th	21	8th	19	0
18					0
19					0
20					0
21	9th	25	9th	26	0
22					0
23					0
24					0
25	10th	26			1
26	11th	27	10th	31	1
27	12th				2
28					2
29					2
30					2
31			11th	32	1
32			12th	33	0

f) Average queue length

$= \frac{3+3+3+2+3+2+2+2+1+1+0}{15}$ (minutes)

= 1.47 (people each minute).

Neglects zeros at the beginning and any build up at the end.

g) Not occupied for 2 minutes (19 + 20 minutes).

Percentage occupation 87%

h) Too small – needs a longer run *and* several repeats in order to be confident of reliability.

4 a) Present 0–84 Absent 85–99

1–4	PPPPP	PPPPP	PPPPP	APPPP
5–6	PPPPP	PPAPP	PAPAP	APPPP
9–12	PPAPP	PAAPA	PAPPP	PPAPP
13–16	PPPPP	PPPPA	PPPPA	PPPPP
17–20	PPPAP	PPAPP	PPPPP	PPPPP

b)

Company	Present
1	5
2	5
3	5
4	4
5	5
6	4
7	3
8	4
9	4
10	2
11	4
12	4
13	5
14	4
15	4
16	5
17	4
18	4
19	5
20	5

c) i) P(all P) $= \frac{8}{20} = \frac{2}{5} = 0.4$

ii) mean = 4.25 spaces occupied

d) i) $p = 0.4437$

ii) $p = 5 \times 0.85^4 \times 0.15$

$= 0.3915$

5 a)

Eggs	4	5	6	7	8	9	10
RND	0–7	8–17	18–33	34–58	59–88	89–95	96–99

Example simulation:

Pair	RND	Eggs re clutch	Survivors[1]	Surviving adults[2]	Surviving chicks[3]
1	90	9	7	2	2
2	46	7	3	0	1
3	10	5	3	2	1
4	43	7	3	2	2
5	10	5	2	1	0
6	42	7	5	0	1
7	69	8	6	2	3
8	63	8	8	0	6
9	38	7	5	2	2
10	33	6	2	2	0
11	38	7	2	2	0
12	25	6	3	1	2
13	59	8	5	1	1
14	55	7	4	1	2
15	27	6	1	1	0
16	70	8	5	2	1
17	42	7	5	2	2
18	49	7	5	1	2
19	57	7	4	2	1
20	27	6	4	0	1

1 Survivors calculated from 0–39 die, 40–99 survive for each individual.

2 Surviving adults (40 adults), 4 die, 7 survive.

3 Surviving chicks calculated from 0–69 die, 70–99 survive, for each individual with this simulation after one year 56 idividuals alive.

b) Start with a new lot of adults as those alive in spring, assume equal numbers of male and female (so breeding pairs). If an odd number, drop one from breeding totals.

Consolidation exercises for Chapter 10

1 i) Pink 00–19
Blue 20–49
White 50–99

ii) Pink 00–13
Blue 14–34
White 35–97

Ignore 98 and 99.

2 a) You need 3 numbers to simulate progress of each ball. At each junction, 0–4 left, 5–9 right.

Ball	RND	Path	Endpoint	Score
1	858	RRR	D	5
2	834	RLL	B	10
3	844	RLL	B	10
4	574	RRL	C	10
5	683	RRL	C	10
6			C	10
7			C	10
8			D	5
9			C	10
10			C	10

Ball	Distribution	Score
11	B	10
12	C	10
13	B	10
14	B	10
15	C	10
16	A	5
17	B	10
18	D	5
19	B	10
20	C	10
21	D	5
22	B	10
23	B	10
24	C	10

b) i) Mean score 8.958

ii) Expected mean $\frac{70}{8} = 8.75$

3 a) 00–09→1
10–49→2
50–69→3
70–89→4
90–99→5

b)

G1 times	4	2	1	2	3	2
G2 times	5	1	3	2	4	4
Totals	9	3	4	4	7	6

Mean waiting time = 5.5

c) i)

G1 times	2	2	4	3	2	2
G2 times	2	4	2	4	2	5
G3 times	3	4	4	5	1	2
Totals	7	10	10	12	5	9

ii)

Min time	7	3	4	4	5	6

Mean = 4.83

d) more runs
accounting for group size
considering other starting conditions
shorter time slices

4 a) Inter-arrival times 3, 2, 2, 4, 2
Rule:
2 minutes 0–24,
3 minutes 25–74,
4 minutes 75–99

b) Service times 2, 5, 2, 1 – 1, 11, 1
Rule:
1 minute 0–31,
2 minutes 32–79,
5 minutes 80–95
Ignore 96–99

c)

Customer	Arrival time	Inter-arrival	Start of service	Service time	End of service	Queue
1	0	3	0	2	2	0
2	3	2	3	5	8	0
3	5	2	8	2	10	1
4	7	4	10	1	11	2
5	11	2	11	1	12	0
6	13	–	13	1	14	0

d) Mean time in queue $\frac{6}{6}$ = 1 minute
Mean queue length (per minute) $\frac{6}{14}$ = 0.43 people
Server utilisation $\frac{12}{14} \times 100 = 86\%$

e) Needs a much longer run.

5 a) i) 30–49, 50–59, 60–69, 70–99

ii)

customer	1	2	3	4	5
time interval	90	10	10	30	30
arrival time	90	100	110	140	70

b) i) 48–59, 60–95
ii) 15, 80, 80, 80, 45

c)

Customer number	arrival time	till number	service starts	service ends	total time service and queuing
1	90	1	90	105	15
2	100	2	100	180	80
3	110	1	110	190	80
4	140	1	190	270	130
5	170	2	150	225	5

i) 72 seconds
ii) 55.6%

d)

Customer number	arrival time	till number	service starts	service ends	total time service and queuing
1	90	1	90	105	15
2	100	2	100	180	80
3	110	1	110	190	80
4	140	2	180	260	120
5	170	1	190	235	65

i) 72 seconds
ii) 57.7%

e) i) more efficient - better service utilisation
ii) individuals are not kept waiting too long

6 a) i) nut centres 0–14,
not nut 15–99

ii)

Carton	1	2	3	4	5	6	7	8	9	10
Nut centres	3	5	3	2	1	4	1	3	1	4

iii) p(3 nuts) = 0.3
p(< 3 nuts) = 0.4

iv) nut centre 0–13
not nut 14–97
ignore 98, 99

b) i) p is the number of nut centres in a carton.
ii) $f(15) = (1 - 15/100)^{20} + 20(15/100)(1 - 15/100)^{19} + \ldots$
= 0.405
simulation may be inaccurate increase the number of runs.

7 a) If today is wet
dry tomorrow – 0–69
wet tomorrow – 70–97
ignore 98,99

b) If today is Dry, then the following days are D, D, D, W, D, D, W, D, D, D, W, D, D, D

c) proportion of wet days 3/14 = 0.214

d) $p = \frac{7}{14}$

e) Longer run, more runs.

8 a) i) Winning centres
1.5-2.5, 5.5–7.5, 10.5–12.5, 15.5–17.5, 20.5–22.5, 25.5–27.5, 30.5–32.5, 35.5–37.5, 40.6–42.5, 45.5–47.5, 50.5–51.5

ii) Rules 00000 – 1.5 cm up to 98 000 for 51.5 cm
2000 to 1 cm

x-values:		34.118	41.5385	42.7905
Win		✗	✓	✗
		22.0425	45.1365	4.7835
		✗	✓	✗
	34.052	38.4715	2.337	19.7075
	✗	✗	✓	✗

b) p(win) = 0.3774

c) e = 6 cm

9 a) i) 00–09, 10–49, 50–69, 70–89, 90–99
ii) 2.0, 2.5, 1.0

b) i) 00–31, 32–79, 80–95
ii) 0.5, 0.5, 1.0

c) i)

2	12:00	12:02	12:02	12:02.5
1	12:01.5	12:04	12:04	12:04.5
2	12:04	12:05	12:05	12:06

ii)

2	12:00	12:02	12:02	12:02.5
2	12:02	12:04.5	12:04.5	12:05
1	12:02	12:03	12:04.5	12:05

d) Service of all three cars completed earlier … at expense of car 2

10 Break the time interval for 10–20 minutes into 1-minute intervals.
For extractions + fillings assume that time of treatment are equally distributed.

Time (mins)	Check up	Fillings	Extractions
10	33%	5%	
11		5%	
12		5%	
13		5%	
14		5%	
15		5%	
16		5%	2%
17		5%	2%
18		5%	2%
19		5%	2%
20		5%	2%
Totals	33%	55%	12%

Treatment time	% frequency	RND
10	38	0–37
11	5	38–42
12	5	43–47
13	5	48–52
14	5	53–57
15	7	58–64
16	7	65–71
17	7	72–78
18	7	79–85
19	7	86–92
20	7	93–99

Other assumptions – assume all patients arrive at the surgery when the appointment is due to begin. (Ignore waiting time before appointments)

Set out the table as follows:

Appointment time	Treatment starts	RND	Length	Treatment ends	Waiting
8.30	8.30	39	11	8.41	0
8.45	8.45	56	14	8.59	0
9.00	9.00	98	20	9.20	0
9.15	9.20	17	10	9.30	5
9.30	9.30	62	15	9.45	0
9.45	9.45	83	18	10.03	0
10.00	10.03	40	12	10.15	3
10.15	10.15	57	14	10.29	0
10.30	10.30	32	10	10.40	0
10.45	10.45	30	10	10.55	0
11.00	11.00	14	10	11.10	0
11.15	11.15	78	17	11.32	0
11.30	11.32	64	15	11.47	2
11.45	11.47	63	15	12.04	2
12.00	12.04	5	10	12.14	4
12.15	12.15	13	10	12.15	0

The system seems to work OK.

CHAPTER 11
Boolean Algebra

Exercises 11.1 *Classwork*

1 a) Ali studies Maths and German.
b) Ali studies Maths and not Art.
c) Ali studies Maths and German and not Art.
d) Ali studies Maths and not German or Art.

2

p	q	$p \Rightarrow \sim q$
T	T	F
T	F	T
F	T	T
F	F	F

3

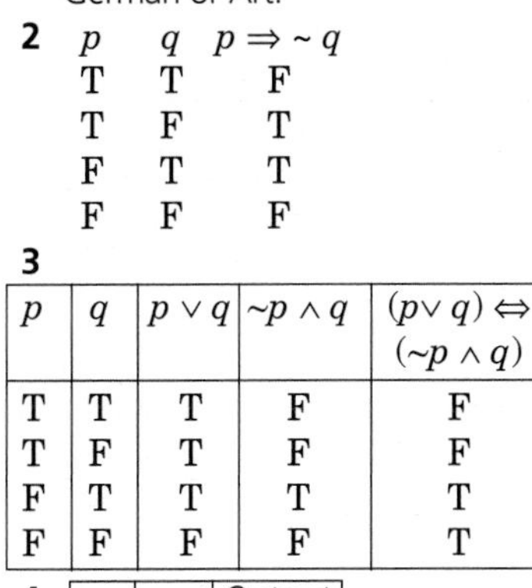

p	q	$p \vee q$	$\sim p \wedge q$	$(p \vee q) \Leftrightarrow (\sim p \wedge q)$
T	T	T	F	F
T	F	T	F	F
F	T	T	T	T
F	F	F	F	T

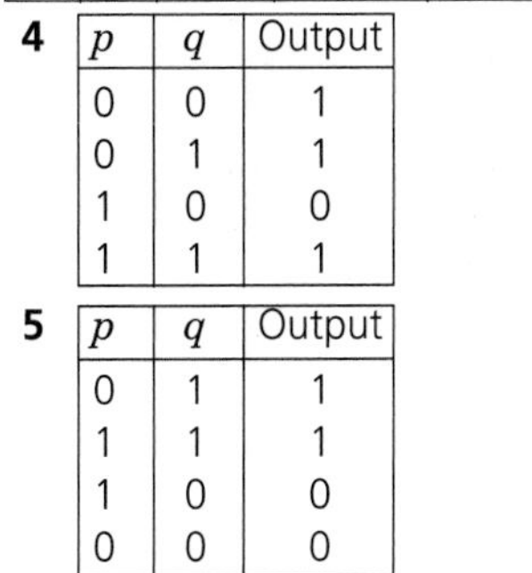

4

p	q	Output
0	0	1
0	1	1
1	0	0
1	1	1

5

p	q	Output
0	1	1
1	1	1
1	0	0
0	0	0

Exercises 11.1 *Homework*

1 a) Only r is sufficient for s to be true.
b) Neither statement is true.

3 b) $b \wedge (c \vee b') \wedge ((c' \wedge a') \vee (a \vee (b \wedge c)))$

4 a) $c = x_1 \wedge x_2$ $s = (x_1 \vee x_2) \wedge (x_1 \wedge x_2)'$
b) s is the right hand digit on the sum of two single digits, c is a single carry digit.

Consolidation exercises for Chapter 11

1

n	s	n⇒s	~s	~n	~s⇒~n	(n⇒s)⇔(~s⇒~n)
0	0	1	1	1	1	1
0	1	1	0	1	1	1
1	0	0	1	0	0	1
1	1	1	0	0	1	1

2 a) (a∧~b)∨(~a∧b)
b)
= [(a∧~b)∨~a]∧[(a∧~b)∨b]
= [(a∨~a)∧(~b∨~a)]∧[(a∨b)∧(~b∨b)]
= (~b∨~a)∧(a∨b)
= (a∨b)∧(~a∨~b)

c) a, b ; a, b

3 a)

a	b	carry	answer
0	0	0	0
0	1	0	1
1	0	0	1
1	1	1	0

b)

a	b	c	carry	answer
0	0	0	0	0
0	0	1	0	1
0	1	0	0	1
0	1	1	1	0
1	0	0	0	1
1	0	1	1	0
1	1	0	1	0
1	1	1	1	1

4 i)

ii) (a∨~b)∧(b∨~a)

0 1 1 0 **1** 0 1 1 0
0 0 0 1 **0** 1 1 1 0
1 1 1 0 **0** 0 0 0 1
1 1 0 1 **1** 1 1 0 1

a = 0 represents first switch down (or up!) etc.
(a∧b)∨(~a∧~b)

iii)

A	B	C	current
down	down	down	no
down	down	up	yes
down	up	down	yes
down	up	up	no
up	down	down	yes
up	down	up	no
up	up	down	no
up	up	up	yes

Starts flowing if not flowing; stops if was flowing.

5 (A∧B)∨(A∨B)

A	B	(A∧B)	(A∨B)	output
0	0	0	0	0
0	1	0	1	1
1	0	0	1	1
1	1	1	1	1

6 a) a — $\sim b$; $\sim a$ — b

b) a — a ; a — b ; a — c ; b — c

7 a) a, b ; c, $\sim a$; b — $\sim c$, b

b) [(a∨b)∧(c∨~a)]∨[b∧(~c∨b)]
= a∧(c∨~a)∨b∧(c∨~a)∨b∧(~c∨b)
(distributive law)
= (a∧c)∨(a∧~a)∨b∧(c∨~a∨~c∨b)
(distributive law)
= (a∧c)∨o∨b∧(c∨~c∨~a∨b)
= (a∧c)∨b∧(1∨~a∨b)
= (a∧c)∨(b∧1)∨(b∧~a)∨(b∧b)
= (a∧c)∨b∨(b∧~a)∨b
= (a∧c)∨(b∨b)∧(b∨~a)∨b
= (a∧c)∨b∧(b∨~a)∨b (absorption law
= (a∧c)∨b b∧(b∨~a)=b)
= b∨(a∧c)

8 a) $p \Rightarrow q$
b) I don't work hard and I pass my exams.
c) I work hard and I don't take regular exercise so I will not pass my exams.

9 a) $\sim (p \Rightarrow (q \Rightarrow p))$

p	q	$p \Rightarrow q$	$p \Rightarrow (q \Rightarrow p)$	$\sim (p \Rightarrow (q \Rightarrow p))$
T	T	T	T	F
T	F	T	T	F
F	T	F	T	F
F	F	T	T	F

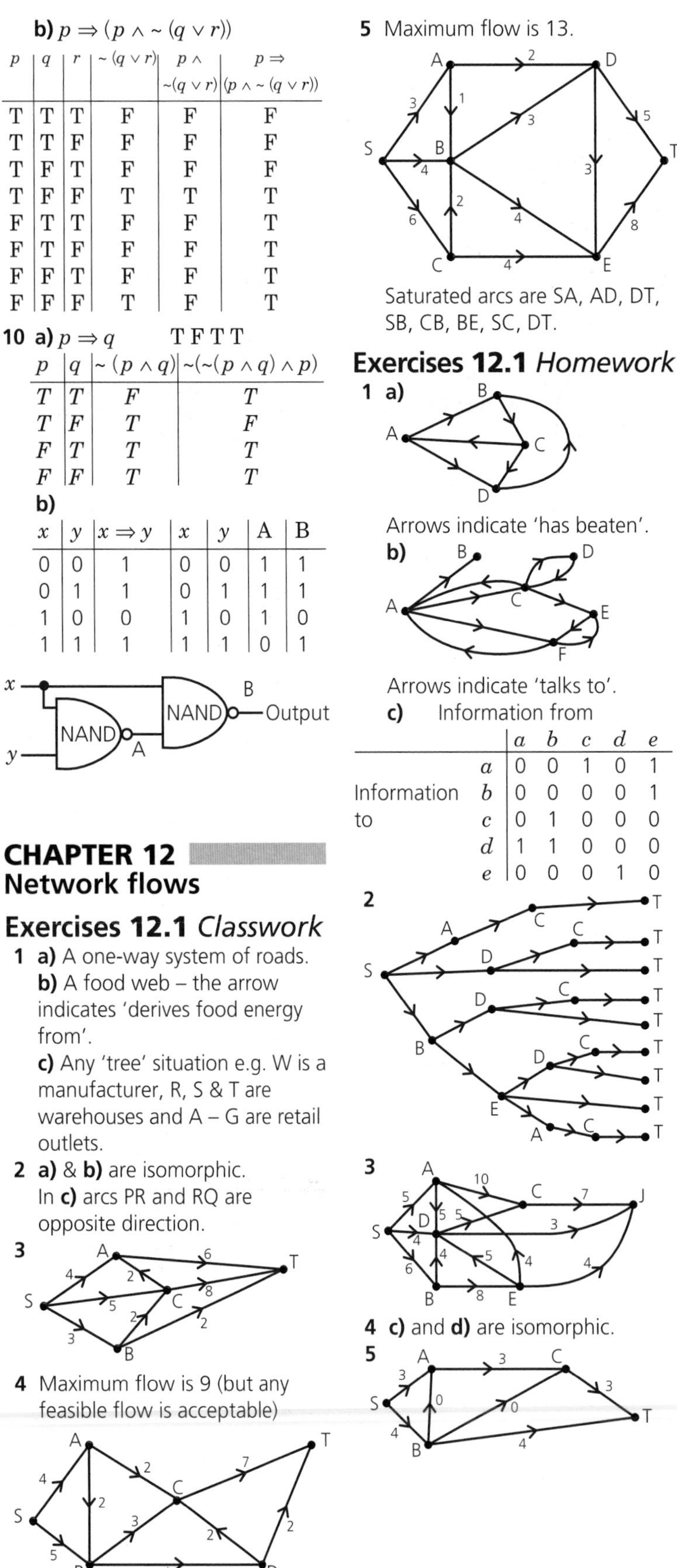

b) $p \Rightarrow (p \wedge \sim(q \vee r))$

p	q	r	$\sim(q \vee r)$	$p \wedge \sim(q \vee r)$	$p \Rightarrow (p \wedge \sim(q \vee r))$
T	T	T	F	F	F
T	T	F	F	F	F
T	F	T	F	F	F
T	F	F	T	T	T
F	T	T	F	F	T
F	T	F	F	F	T
F	F	T	F	F	T
F	F	F	T	F	T

10 a) $p \Rightarrow q$ T F T T

p	q	$\sim(p \wedge q)$	$\sim(\sim(p \wedge q) \wedge p)$
T	T	F	T
T	F	T	F
F	T	T	T
F	F	T	T

b)

x	y	$x \Rightarrow y$	x	y	A	B
0	0	1	0	0	1	1
0	1	1	0	1	1	1
1	0	0	1	0	1	0
1	1	1	1	1	0	1

CHAPTER 12 Network flows

Exercises 12.1 *Classwork*

1 a) A one-way system of roads.

b) A food web – the arrow indicates 'derives food energy from'.

c) Any 'tree' situation e.g. W is a manufacturer, R, S & T are warehouses and A – G are retail outlets.

2 a) & **b)** are isomorphic. In **c)** arcs PR and RQ are opposite direction.

3

4 Maximum flow is 9 (but any feasible flow is acceptable)

This saturates **all** arcs.

5 Maximum flow is 13.

Saturated arcs are SA, AD, DT, SB, CB, BE, SC, DT.

Exercises 12.1 *Homework*

1 a)

Arrows indicate 'has beaten'.

b)

Arrows indicate 'talks to'.

c)

		Information from				
		a	b	c	d	e
Information to	a	0	0	1	0	1
	b	0	0	0	0	1
	c	0	1	0	0	0
	d	1	1	0	0	0
	e	0	0	0	1	0

2

3

4 c) and **d)** are isomorphic.

5

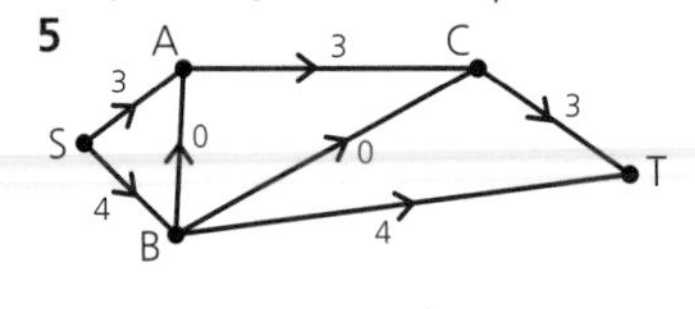

Exercises 12.2 *Classwork*

1

Path	Flow	Saturated
SADT	4	SA, AD
SCET	6	SC, CE
SBDT	3	BD
SBET	3	ET

Maximum flow 16.

2 a) 15 000 tonnes with flows ABEF 6 000 tonnes, ADF 5000 tonnes, ACF 4000 tonnes

b)

Path	Flow	Saturated
ABEF	6	EF
ACF	4	CF
ADF	5	AD

Flow augmenting paths		
ABDF	4	AB, BD
ACDF	1	AC

Maximum flow 20 000 tonnes

c) If AC and CD are upgraded to 8 and 6 the flow increases by 2 000 tonnes since 2 units are available in DF (CF is saturated). Improve either DF or CF to increase flow by a further 2000 tonnes per week.

3 12 to A but only 11 can leave, 14 to B but only 12 can leave, maximum flow 23 units.

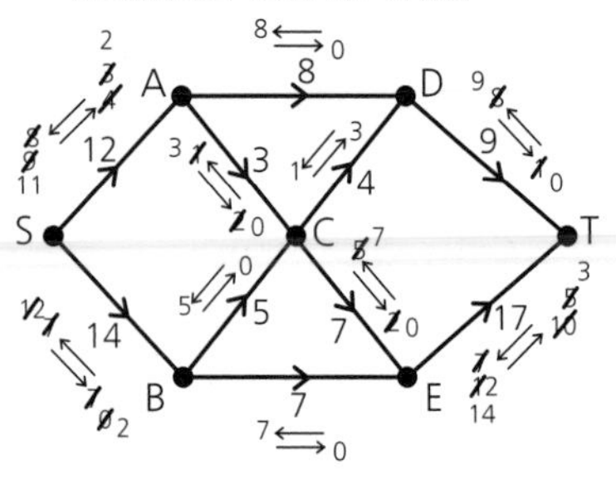

Path	Flow	Saturated
SADT	8	AD
SBET	7	BC
SACDT	1	DT
SBCET	5	BC
SACET	2	AC, CT

Maximum flow 23 units.

4 Maximum flow 27 000 gallons per hour.

5

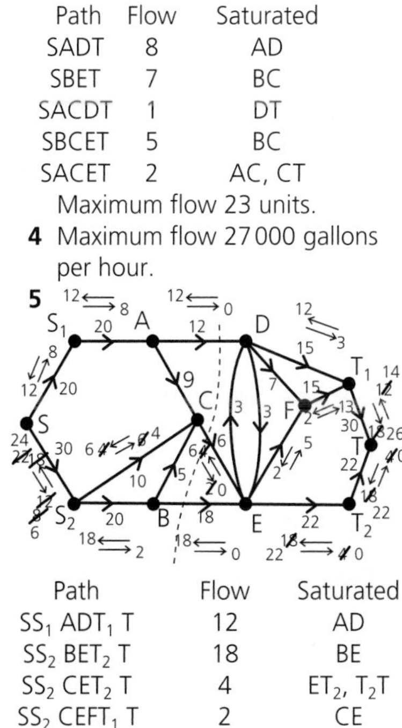

Path	Flow	Saturated
SS_1 ADT_1 T	12	AD
SS_2 BET_2 T	18	BE
SS_2 CET_2 T	4	ET_2, T_2T
SS_2 $CEFT_1$ T	2	CE

Maximum flow is 3600 vehicles per hour.
The problem at F has no effect as the solution has less than 8 units of flow passing through F.

Exercises 12.2 *Homework*

1

Path	Flow	Saturated
SADT	2	AD
SBET	4	BE, SB
SCET	4	CE
SCBDT	2	CB, SC
SABDT	1	AS, DT

Maximum flow of 13.

2

Path	Flow	Saturated
SAT	4	SA
SCT	5	SC
SBCT	2	
SABDT	1	SB

Maximum flow of 12.

3 Maximum flow of 1000 containers.

4 a) maximum flow of 180 units

5 Reduces maximum flow by 900 vehicles per hour.
Maximum flow is now 2700 vehicles per hour.

Exercises 12.3 *Classwork*

(Answers in thousands)

1 Cut AD, BD, ET
Capacity $4 + 3 + 9 = 16$

2 Cut AB, AD, AC
Capacity $10 + 5 + 5 = 20$

3 Cut R_1A, R_1B, R_2B, R_2C
Capacity $7 + 6 + 6 + 8 = 27$
OR RR_1, RR_2

4 Cut AD, CE, BE
Capacity $12 + 6 + 18 = 36$

5

Path	Flow	Saturated
SS_1 AET_1 T	3	AE
SS_1 BET_1 T	3	S_1B
SS_1 $ABET_1$ T	1	AB
SS_2 DFT_3 T	4	S_2D, DF, FT_3, T_3T
SS_2 CET_2 T	5	CE
SS_2 CET_2 T	2	S_2C
SS_2 BET_1 T	1	BE, ET, T_1T

Maximum flow 19 (other flows through paths also give 19).
Cut AE, BE, S_2C, S_2D
Capacity $3 + 5 + 7 + 4 = 19$

Exercises 12.3 *Homework*

1 a) Value of minimum cut ≥ 10
b) Maximum flow ≤ 10
c) $7 \leq$ maximum flow ≤ 10
d) Maximum flow = 10
e) You have made a mistake! Either the flow is too large or the cut is too small.

2

Cut	Arcs in cut	Capacity of cut
c_1	SA, SC, SB	120
c_2	SA, SC, BC, BT	100
c_3	SA, AC, CA, CT, BT	170
c_4	AT, CT, BT	80
c_5	AT, AC, CA, SC, SB	160
c_6	AT, AC, CA, SC, BC, BT	140
c_7	AT, CT, BC, SB	110
c_8	SA, CA, AC, CT, BC, SB	200

4 a) The following flow is an example of one that satisfies the maximum and minimum conditions.

Path	Flow
SAT	2
SCAT	2
SBCT	1
SBT	1

b) Maximum flow is 16.

5 a) Yes conditions can be satisfied.
b) No, cannot achieve a minimum flow through AT.

Consolidation Exercises for Chapter 12

1 a) Maximum flow ≤ 9
b) Minimum cut ≥ 11
c) Maximum flow and minimum cut most lie between 28 and 35 inclusive.

2 a) Minimum cut of 15 is through AD, BD and ET.
b) Minimum cut of 14 is through AD, BD, BE and CE.
c) Statement false. It should be: a minimum cut will consist of saturated arcs in direction S to T and arcs with zero flow in direction T to S.

3 a) and **b)**

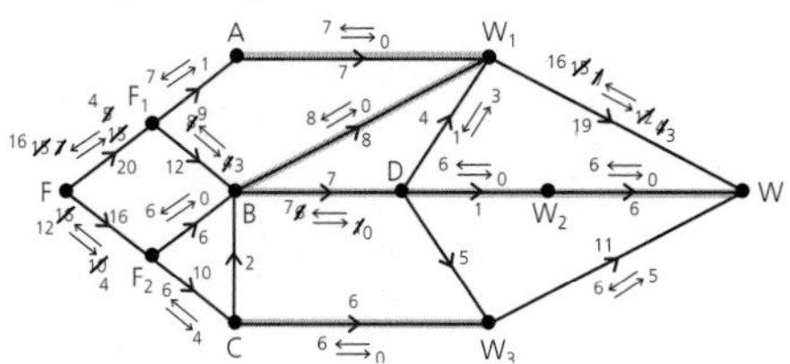

c) i) Lorries leaving F_1 is 20 and F_2 is 16
ii) Lorries reaching W_1 is 16, W_2 is 6 and W_3 is 6
iii) Lorries passing through B is 15

4 a) Note that in this problem the network is undirected so you can send flow in either direction.

Path	Flow
SACT	10
SBDT	5
SBDACT	4

This gives a maximum flow of 19 units.
b) The cut SA, SB has a capacity of 19 units.

5 a)

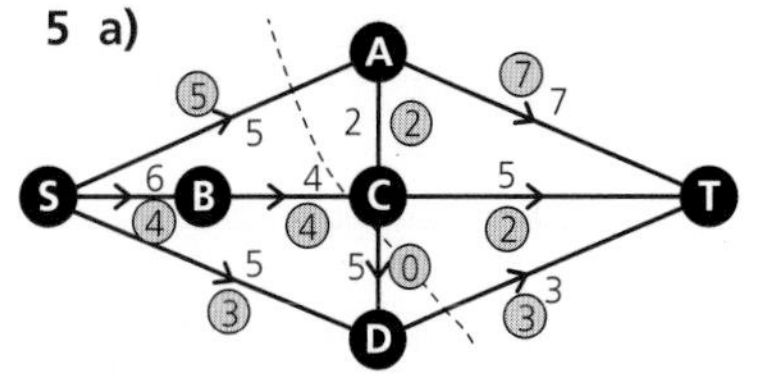

b) Value of f is 1200 cars per hour.
c) Cuts SA, BC, CD, DT has a capacity of 1200.

6 a) Source B, sink F.
b) Cut partitions network into 2 parts, one containing source and one containing sink.
Flow across cut
$12 + 13 + 18 - 5 = 38$
Not max because capacity of cut $12 + 13 + 18 + 0 = 43$

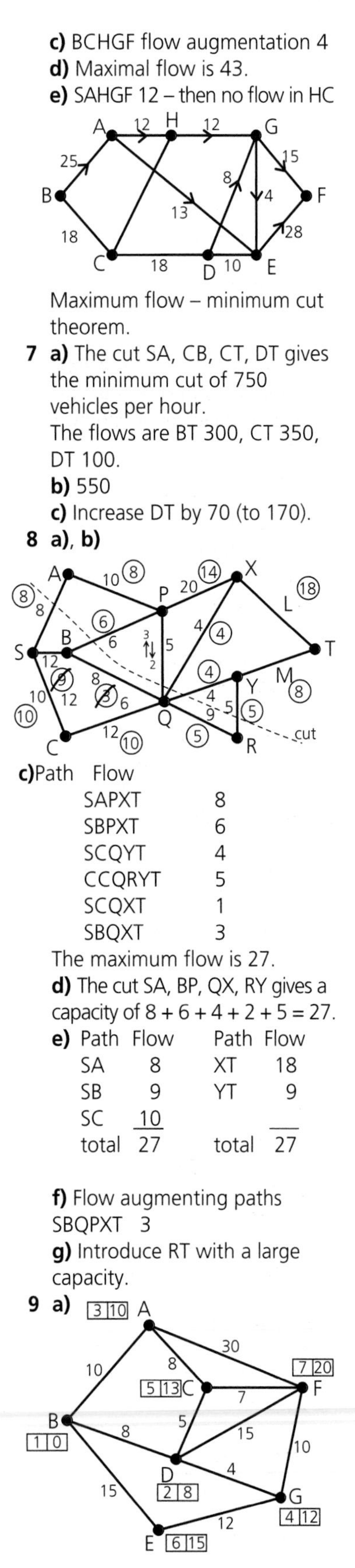

c) BCHGF flow augmentation 4

d) Maximal flow is 43.

e) SAHGF 12 – then no flow in HC

Maximum flow – minimum cut theorem.

7 a) The cut SA, CB, CT, DT gives the minimum cut of 750 vehicles per hour.
The flows are BT 300, CT 350, DT 100.

b) 550

c) Increase DT by 70 (to 170).

8 a), b)

c)

Path	Flow
SAPXT	8
SBPXT	6
SCQYT	4
CCQRYT	5
SCQXT	1
SBQXT	3

The maximum flow is 27.

d) The cut SA, BP, QX, RY gives a capacity of 8 + 6 + 4 + 2 + 5 = 27.

e)

Path	Flow	Path	Flow
SA	8	XT	18
SB	9	YT	9
SC	10		
total	27	total	27

f) Flow augmenting paths
SBQPXT 3

g) Introduce RT with a large capacity.

9 a)

b)

Tracing back gives route BDCF length 20.

Path	Flow	Saturated
BAF	2	AF
BACF	1	BA
BDCF	2	DC
BDF	2	DF
BEFG	2	BE, EG, GF
Total flow	9	

The cut BA, DC, DF, GF has capacity 3 + 2 + 2 + 2 = 9.
Value of flow = capacity of cut; therefore by 'maximum flow – minimum cut theorem' this flow is a maximum.

c) Dijkstra on network above.
BDCF flow 2

% flow = $\frac{2}{9} \times 100 = 22\%$

10

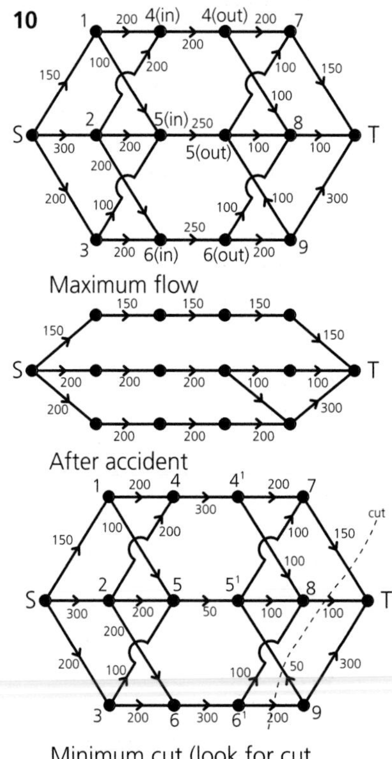

Maximum flow

After accident

Minimum cut (look for cut < 550)
7T, 8T, 5′9, 6′9

Capacity 150 + 100 + 0 + 200 = 450 < 550
Maximum flow ≤ capacity of any cut, therefore the demand cannot be satisfied.

DECISION MATHS

Glossary

activity network
see precedence network

activity on arc network
precedence network in which the arcs represent the activities, while the vertices represent events

activity on vertex network
precedence network in which the activities are represented by the vertices of a network, and edges show the order of precedence

adjacency matrix
matrix representing the vertices and edges of a graph

algorithm
systematic process for finding a solution to a problem

alternating path
path in a bipartite graph that joins vertices from one subset to vertices in the other, in such a way that alternate edges only are in the initial matching, and initial and final vertices are not incident with an edge in the matching

analytical solution
the solution of a recurrence relation given as a formula as opposed to a sequence of numbers

arc (or edge)
line connecting two vertices in a graph in a directed network, an edge that is directed

artificial backward capacity
record of flow being sent down an arc in a network, made with a reversed arrow along the arc

ascending bubble sort
see bubble sort

auxiliary equation
the equation obtained for the base m when $x_n = m^n$ is substituted into a homogeneous recurrence relation

back substitution stage
this stage of the Gaussian elimination method is the process of working back through the equations solving for the variables

bin
'space' forming the containers for bin-packing

bin-packing
sorting methods, *see page 14*

bipartite graph
graph which can be divided into two subsets in such a way that each edge of the graph joins a vertex from one subset to a vertex in the other

Boolean variable
can only take two values (0 or 1, true or false)

bubble sort
a sorting method, *see page 6*

cascade activity number
numbers allocated to activities in a project to ensure a logical sequence through the activities

cascade chart
diagram showing precedence relations, in the form of bars drawn against a time scale, that helps to make best use of all resources

combinations
unordered selections without repetition ^nC_r

complete bipartite graph
bipartite graph in which each vertex in one subset is joined to every vertex in the other by an edge

complete graph
a graph where each vertex is connected to every other vertex

connected graph
a graph with at least one route between any pair of vertices

constraints
limits on the values of variables in linear programming

cost function
relation linking the cost to the variables that are included in the cost

cost path
cheapest, (shortest) path through a cost network

critical activities
those activities that must be started on time to avoid delaying the whole project

critical path
minimum time needed to complete a project, following the longest path through the precedence network and including all the critical activities

cut
a line dividing a network into two parts, one containing the source and the other containing the sink

cycle
a path that completes a loop and returns to its starting point

degree of a vertex
the number of edges incident at the vertex

descending bubble sort
see bubble sort

deterministic model
model that assumes chance events do not occur

directed network
network in which there is a set direction of movement

discrete situation
a situation modelled by a discrete sequence of numbers

distance matrix
matrix representing a network of distances

dummy arc
arc introduced when an activity has to be preceded by more than one other activity; they are marked as dotted lines and have zero duration

edge (or arc)
line connecting two vertices in a graph

edge-disjoint
paths between two vertices are edge-disjoint if they have no edges in common

Elimination stage
this stage of the Gaussian elimination method is the process of subtracting or adding a multiple of one row from another

enumerations problems
counting problems and listing problems are types of enumeration problems

Eulerian cycle
a cycle that includes every edge of a graph exactly once

event
finish of one activity and the start of another, in a precedence network

exhaustive search
finding the shortest path by looking for every possible path

face
region of the plane bounded by edges of a planar graph

feasible flow
a flow through a network that satisfies given capacities

feasible region
region within a set of straight-line graphs showing linear programming constraints, within which the solution lies

first-fit algorithm
a sorting algorithm that places item in the first available bin

float
maximum time by which the start of an activity can be delayed without delaying the whole project

flow augmenting paths
paths that consist entirely of unsaturated arcs

full-bin algorithm
a sorting algorithm that fills bins one at a time

Gaussian elimination method
method of solving simultaneous equations by eliminating one variable and solving for the other – it can also be used for systems of more that two equations, using a similar principle

graph
diagram of vertices and edges representing how objects are related to each other

greedy algorithm
an algorithm that chooses the best option at each stage

Hamiltonian cycle
a cycle that passes through every vertex of a graph once, and only once, and returns to the starting point

heuristic algorithm
algorithm which produces a good solution to a problem using logical and intuitive steps

independent float
when the activities with float are independent of each other

infinite face
area outside the edges of a planar graph

interfering float
when the start time for one activity with a float affects the float for other activities

isomorphic
a graph having the same number of vertices as another graph, with the degrees of corresponding pairs of vertices being the same

iteration
the step-by-step process of solving problems

linear function
relation between two variables that can be expressed as an equation and drawn as a straight line

linear programming problem
problem made up of a linear objective function and constraints formed by linear inequalities

matching
a graph that links some of the vertices in one subset of a bipartite graph to some of the vertices in the other subset

matrix (matrices)
an array of numbers that can be used to convey information in a condensed form; also used in the solution of simultaneous equations

maximum capacity
maximum number of 'items' that can pass along an arc in a network

maximum matching
the optimum solution linking as many pairs of vertices as possible in a bipartite graph

minimum cut
the cut with the minimum capacity

minimum-connector problem
spanning tree of minimum length

multiplier
number used to multiply one (or more) equation, to enable the solution of simultaneous equations

nearest neighbour algorithm
see page 60

network
a graph in which every edge has a value called its weight

node
see vertex

non-linear
an equation or inequality in which the powers of the variables are not unity

objective function
main function in a linear programming problem that has to be minimised or maximised

optimal
best solution according to given conditions

pass
one iteration of a sorting algorithm

path
a route through a graph which does not go along any edge more than once or visit any vertex more than once.

Petersen graph
see page 52

pivot column
column with the smallest (i.e. 'most negative') value in the bottom row of the simplex tableau

pivot element
element in the simplex tableau which falls in both the pivot row and pivot column

pivot row
when the entries in the right hand column of the tableau are divided by the entries in the pivot column, the small value obtained gives the pivot row

planar graph
graph that can be drawn in the plane with no edges crossing

platonic solid
three-dimensional solid with plane faces

precedence network
diagram showing the precedence relations

precedence relations
the relationship showing the order in which activities must be done

predecessor
immediate preceding activity

quicksort
a sorting method, *see page 10*

resource histogram
diagram to show the number of people needed at any stage of a project

resource levelling
system of managing a project to

balance the amount of work to be done and the resources available

risk path
safest path through a network

saturated
carrying the maximum capacity

simplex method
process of using Gaussian elimination steps to solve linear programming problems

simplex tableau
equations of a linear programming problem written in tabular form

simulation
process that can be used to test what might happen in situations where an experiment with real subjects may be too long or too dangerous

sink
finishing point of a directed network

slack variables
variables introduced in linear programming problems which change the constraints in the form of given inequalities into equations; the solution of the problem occurs at one of the corners of the intersection of the planes

source
starting point of a directed network

spanning tree
a subgraph that includes all the vertices in the original graph and is also a tree

stochastic model
model that allows for an element of chance in the outcome

subdivision of a graph
formed by inserting vertices of degree 2 into a graph

subgraph
part of a graph which is itself a graph

supersink
artificial sink introduced into a network and fed from all the sinks

supersource
artificial source introduced into a network feeding all the sources

tour
route which visits every vertex in the network

tree
a connected graph that contains no cycles

trivial constraints
limits on values of variables that give little information as to how they are related, such as $x \geq 0, y \geq 0$

vertex
point on a graph representing a point where edges meet; also called a node

weight
value on the edge of a network

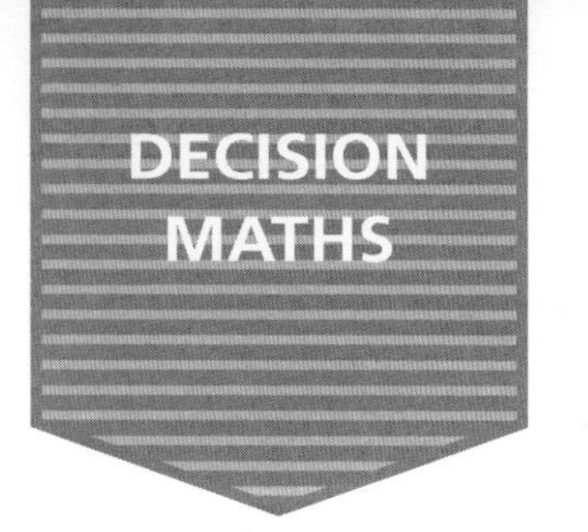

Index

This index covers the text of chapters 1 to 12 of the book, excluding the exercises.
A glossary of terms is present on pages 255–257. These references are not included in the index.